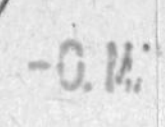

MODERN MATHEMATICS Made Simple

The Made Simple Series
has been created
especially for self-education
but can equally well
be used as
an aid to group study.
However complex the subject,
the reader is taken
step by step,
clearly and methodically,
through the course. Each volume
has been prepared by experts,
taking account of
modern educational requirements,
to ensure the most
effective way of acquiring knowledge.

In the same series

Accounting
Acting and Stagecraft
Additional Mathematics
Administration in Business
Advertising
Anthropology
Applied Economics
Applied Mathematics
Applied Mechanics
Art Appreciation
Art of Speaking
Art of Writing
Biology
Book-keeping
British Constitution
Business and Administrative Organisation
Business Economics
Business Statistics and Accounting
Calculus
Chemistry
Childcare
Commerce
Company Law
Computer Programming
Computers and Microprocessors
Cookery
Cost and Management Accounting
Data Processing
Dressmaking
Economic History
Economic and Social Geography
Economics
Effective Communication
Electricity
Electronic Computers
Electronics
English
English Literature
Export
Financial Management
French
Geology
German
Housing, Tenancy and Planning Law
Human Anatomy
Human Biology
Italian
Journalism
Latin
Law
Management
Marketing
Mathematics
Metalwork
Modern Biology
Modern Electronics
Modern European History
Modern Mathematics
Money and Banking
Music
New Mathematics
Office Practice
Organic Chemistry
Personnel Management
Philosophy
Photography
Physical Geography
Physics
Practical Typewriting
Psychiatry
Psychology
Public Relations
Rapid Reading
Retailing
Russian
Salesmanship
Secretarial Practice
Social Services
Sociology
Spanish
Statistics
Teeline Shorthand
Twentieth-Century British History
Typing
Woodwork

MODERN MATHEMATICS Made Simple

Patrick Murphy
MSc, FIMA

Made Simple Books
HEINEMANN : London

Printed and bound in Great Britain
by Richard Clay (The Chaucer Press) Ltd, Bungay, Suffolk
for the publishers, William Heinemann Ltd.,
10 Upper Grosvenor Street, London W1X 9PA

SBN 434 98544 9 casebound
SBN 434 98545 7 paperbound

Editorial: Robert Postema, F. G. Thomas
Cover Illustration: Julia Logsdail

Foreword

This book has been written for any student, parent or teacher who is interested in modern mathematics. No previous knowledge of the subject matter is assumed but the general standard and contents of the book are appropriate for anyone studying for O level GCE or seeking a solid foundation for further study in this field.

The topics of modern mathematics at this level have now become reasonably well established and indeed some, like sets, vectors and matrices, appear almost traditional in the eyes of many students. I have tried to probe this more familiar material with as much depth and variety as a text of this length will allow.

A discussion of elementary mathematical logic and switching circuits has been included in an attempt to encourage the reader to move further along the path of modern mathematics, not necessarily to introduce some aspect of a simple computer science course but more in the spirit of revealing a new interplay of mathematical ideas. On this point it is probable that some readers may find it more helpful and instructive to read Chapters 2 and 3 in reverse order since the practical outlet of the notation may render the ideas of elementary logic more meaningful. Similarly, I hope that the inclusion of a separate chapter on tessellations will enable some readers to employ artistic feelings for mathematical advantage by practising a roving eye in preparation for the more formal presentation of motion geometry in the following chapter.

Having emphasised the constant classification and definitive nature of mathematics throughout this book, I deliberately placed the chapter on rubber sheet geometry at the end so that the reader might fully appreciate that mystery in our subject whereby so many ideas arise from the most elementary starting point. The book is meant to be read in the order printed: certainly everyone should start with Chapter 1 but thereafter, for those wishing to capture the modern nuance in the shortest possible time, I suggest at least Chapters 6 and 12.

I wrote this book with great interest and enthusiasm in the hope that this would somehow reach out from the printed page to guide readers in their mathematical activity.

PATRICK MURPHY

Contents

1

SETS

The language of sets, together with the simplicity of combining sets according to elementary rules, gives us the opportunity to examine basic mathematical ideas and to illustrate some aspects of mathematical language which are common to all its branches. The study of sets enables us to attach mathematical overtones to everyday terms and experiences which are not always apparent in normal conversation. This is not to turn us into bores but more to begin to make us appreciate the definitive nature of our subject, for while much may be discovered by comparisons with the real world the fact remains that the activity of mathematics is a constant search for, and identification of, its results. A subsequent classification of those elements from which the results are derived is the starting point of almost any mathematical thinking.

Definition 1.1. A set is a collection of distinct objects. If it is always possible to decide whether or not any object whatsoever belongs to the collection then we say that the set is **well defined.**

The objects which belong to the set are called the **elements** of the set. Two sets are considered to be the same if they contain exactly the same objects, in which case we also say that the sets are **equal** to each other.

Notation

Usually a set will be represented by capital letters like A, K, P, X, and the elements of a set by what are called lower-case letters, such as b, m, n, t. A set may be defined by enclosing a list of its elements, in any order, between two brackets { and }. For example, the set of vowels in the alphabet may be called the set V and listed as

$$V = \{a, e, i, o, u\}$$

and we shall read this statement as '*V is the set whose elements are a, e, i, o, u*'. Some typical remarks about the set V are:

(i) 'u is an element of V' or 'u belongs to V'.
(ii) 'The elements of V are a, e, i, o, u'.
(iii) 'There are five elements in the set V'.
(iv) 'b does not belong to V'.

We take our notation further by substituting an abbreviated form for statements like (i) and (iv) above.

For the statement '*u belongs to V' we write*

$$u \in V$$

That is, the symbol $\in$ means '*belongs to*', or '*is an element of*', or '*is a member of*'.

Adopting the usual practice of striking out a symbol to represent its negation, we use the symbol $\notin$ to mean 'does not belong to', or 'is not an element of', so that for the statement 'b does not belong to V' we write $b \notin V$.

Further examples of this notation are given by suggesting that

$$A = \{1, 2, 3, a, b, c, \text{England}\}$$

and commenting as follows:

$$13 \notin \text{A}; a \in A; e \notin A; \text{France} \notin A.$$

Similarly, with

$$B = \{\text{All European capital cities}\}$$

we may comment:

$$\text{Paris} \in B; \text{Rome} \in B; \text{Sydney} \notin B; a \notin B$$

Observe that the sets A and B are well defined since we can always decide whether or not an object belongs to the set.

The Definition 1.1 above referred to distinct objects, i.e. no two objects of a set are the same, so that when, for example, we list the set of letters in the phrase 'little letters' we have the set $\{l, i, t, e, r, s\}$. This is the same as listing the typewriter keys which need to be struck in order to type out the phrase 'little letters'.

The objects in our sets may be anything we please, as the following examples suggest:

1. The numbers 2, 4, 6. We may write this as $A = \{2, 4, 6\}$.
2. {a pin, a marble, a shoe, a passport}.
3. {All the rivers in Brazil}.
4. {All wives who are older than their husbands}.
5. $B = \{\text{All even numbers between 7 and 1}\}$.

Examination of example 5 shows that $B = \{6, 4, 2\}$, and having written $A = \{2, 4, 6\}$ in example 1 this implies $A = B$, and that A and B are two names for the same set.

Notice in the above examples of five sets that each of the first two sets is defined by having their elements listed, whereas the last three sets are defined by stating a property which the elements have in common.

A property definition of a set is usually more convenient, especially when it avoids the need to list a large number of elements. We can still compare two sets when defined in this manner: for example, suppose that D and K are two sets given by the following statements,

$$D = \{\text{All odd numbers between 1.7 and 99 999.3}\}$$
$$K = \{\text{All odd numbers between 2.8 and 100 000}\}$$

Without listing all the members of the sets a mental inspection shows that $K = D$ and suggests that there may be more than one way to describe a set since K and D are the same set.

Exercise 1.1

State whether or not the following sets are well defined:

1. $A = \{\text{All odd numbers between 2 and 12}\}$.
2. $B = \{\text{The two best presidents of the USA}\}$.

3. $C = \{1, 2, 3, 4\}$.
4. $D = \{$The three worst films ever made$\}$.
5. $E = \{$All even numbers between 1 and 71$\}$.
6. $F = \{$The composers of the five best tunes$\}$.
7. $G = \{$All the rearrangements of the three letter word, *pat*$\}$.

List all the elements in each of the following sets:

8. $P = \{$All letters in the word *beginning*$\}$.
9. $Q = \{$All letters in the word *ginger*$\}$.
10. R is the set of all letters which are common to the two words *letters* and *elements*.
11. $T = \{$All letters which belong to the set P as well as the set $Q\}$.
12. $U = \{$All elements of P which do not belong to the set $Q\}$.
13. $V = \{$All elements of Q which do not belong to $P\}$.
14. $W = \{$All the digits in the largest number which is less than 135$\}$.

Give a property definition for each of the following sets.

15. {July, January, June}.
16. {10, 12, 14, 16}.
17. {3, 5, 7, 9, 11}.
18. {1, 4, 9, 16, 25}.
19. {5, 10, 15, 20, 25}.
20. {pet, etp, tpe, pte, tep, ept}.

Substitute the symbol $\in$ or $\notin$ for each '_' in the following sentences to make a correct statement for the sets in the above questions.

21. $3 \_ A$
22. $5 \_ C$
23. $69 \_ E$
24. $a \_ P$
25. $g \_ Q$
26. $r \_ R$
27. $e \_ T$
28. $b \_ U$
29. $g \_ V$
30. $7 \_ W$

Numbers

At first sight the sets

$P = \{$square, quadrilateral, circle, triangle$\}$,
$Q = \{$cat, dog, rabbit, mouse$\}$,
$R = \{a, e, b, f\}$,

do not appear to have anything in common. We can, however, take any two of the sets and, as suggested by either of the following diagrams, match one set to the other by linking each element of the first set to a different element of the second set:

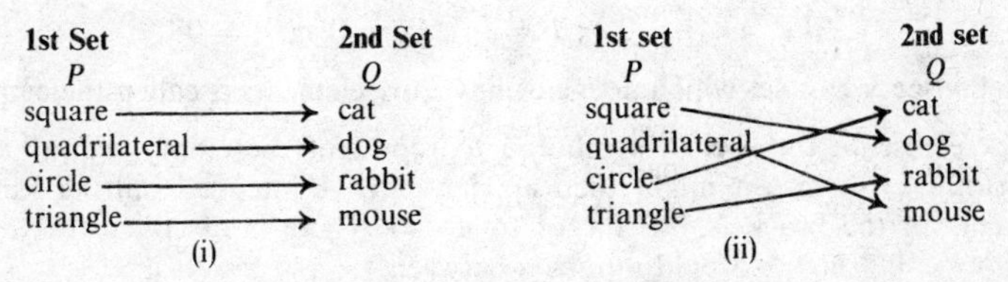

Either of these two procedures is described as *matching set P to set Q*, and we use the arrows to link the elements which we intend to correspond in the matching.

For example, in (i)

cat corresponds to square, so we say that cat and square are **corresponding elements** in this matching,

whereas in (ii)

cat corresponds to circle, so we say that cat and circle are **corresponding elements** in this matching.

Be sure to notice that if $A = B$ then set A can be matched to set B, but the converse result is not true: that is, just because set A can be matched to the set B it does not follow that $A = B$. (The two sets P and Q above illustrate this point.)

When a set A can be matched to a set B we say that A and B are **equivalent** sets.

We could have matched set Q to set R, or set R to set P in a similar manner to the above diagrams but it is more important to notice that any of the sets P, Q and R can be matched to the set of the first four natural numbers $\{1, 2, 3, 4\}$. Now, matching the elements of P to the numbers 1, 2, 3, 4 in that order is what we call counting the elements of the set P. Of course we conclude that there are four elements in each of the sets P, Q, R. The association of the word four or the numeral 4 with the set P is represented by writing $n(P) = 4$ and reading this as 'the number of elements in the set P is 4'. Clearly $n(Q) = 4$ and $n(R) = 4$ and so we see that every set which matches the set $\{1, 2, 3, 4\}$ is associated with the number four. We refer to 4 as the **cardinal number** of any set which, like P, can be matched with $\{1, 2, 3, 4\}$.

Similarly, every set which matches the set $\{1, 2, 3\}$ is associated with the number 3, thus $T = \{\text{sock, shoe, glove}\}$ is such a set and $n(T) = 3$. Finally, with $U = \{\text{bus, car}\}$ and $V = \{\text{flower}\}$ implying that $n(U) = 2$ and $n(V) = 1$, we ask how we might associate the number zero with a set.

The Empty Set

The method of defining a set by stating a property of its elements enables us to obtain a set which does not have any elements. Thus it is not possible to find any element which belongs to any of the following sets:

1. A = the set of all even numbers between 7 and 7.9,
2. B = the set of all odd numbers between 4 and 4.98,
3. C = the set of all words on this page which begin with the letter x,
4. D = the set of people known to have lived for 300 years,

so in each case we say that the set is empty and write

$$n(A) = 0, n(B) = 0, n(C) = 0, n(D) = 0$$

Definition 1.2. A set which does not have any elements is called the empty set.

We reserve the special symbol ϕ to represent the empty set rather than writing $\{\quad\}$ since it might then appear that we intended putting some elements in the brackets but forgot to do so. There is of course the obvious remark that $n(\phi) = 0$.

It is advisable to avoid saying that ϕ is the set with 'nothing in it' because this leads to the error of thinking that $\{0\}$ is the empty set.

Finite and Infinite Sets

A set is said to be finite if we can count the number of its elements and reach a last element in the counting process. Alternatively, we refer to a finite set as a set with a **definite number of elements.** Considering zero as a definite number enables us to refer to ϕ as a finite set.

Some further examples are:

1. A = {All the even numbers between 1 and 49}; n(A) = 24.
2. B = {All the multiples of 3 between 11 and 40}; n(B) = 10.
3. C = {All the pages in this book}; n(C) = ?
4. D = {All the pages in this book which hold a million letters}; n(D) = 0.
5. E = {All the living people on Earth at this moment}; n(E) is unknown but E is still finite.

An **infinite set** is a set which is not finite.

The notation for an infinite set which we mention here requires a certain amount of trust, as we shall now see. Thus, the infinite set N of the natural numbers is represented by

$$N = \{1, 2, 3, 4, 5, \ldots\}$$

where the run of three dots is intended to indicate the continuation of elements in the pattern of those already written. Here we trust that the next three elements are intended to be 6, 7, 8 and so on.

Similarly, the infinite set of whole numbers is represented by

$$W = \{0, 1, 2, 3, \ldots\}$$

where, as before, we trust that the next three elements are supposed to be 4, 5, 6 and so on.

Consider $S = \{1, 4, 9, 16, \ldots\}$. Here we trust that the elements of S are supposed to be the squares of the natural numbers so that a fuller statement about S might be $S = \{1, 4, 9, 16, 25, 36, 49, 64, \ldots\}$. Stating sets in this manner really presents us with a riddle rather than definition of the set but the exercises in numerical pattern spotting are still worthwhile.

Exercise 1.2

1. Suggest two sets, one of which is matched to $A = \{p, q, r\}$ and the other is matched to $B = \{k, l, m, n\}$ and display the matching in each case. Does either of your suggestions match the set $C = \{p, q, r, k, l\}$? If not give a reason why.
2. Write down all possible ways of matching set $A = \{p, q, r\}$ to set $D = \{1, 2, 3\}$.
3. Using the sets in the last question, give a matching from set D to set A.
4. In how many ways is it possible to match set X to set Y when both sets have (i) 2 elements, (ii) 4 elements.
5. The set {1, 2, 3, 4, 5} is matched to the set {2, 4, 6, 8, 10} as displayed in the following plan:

$$\begin{array}{ccccc} 1, & 2, & 3, & 4, & 5 \\ \downarrow & \downarrow & \downarrow & \downarrow & \downarrow \\ 2, & 4, & 6, & 8, & 10 \end{array}$$

   Suggest an arithmetic reason for matching the elements in this order.
6. Following the idea in the last question give a matching from the set {1, 2, 3, 4, 5} to the set {5, 10, 15, 20, 25} based on an arithmetic reason.
7. The set $X = \{1, 2, 3, 4\}$ is matched to a set Y based on the arithmetic reason that the elements of Y are 6 more than the corresponding elements of X. List the elements of Y.

State the number of elements in each of the following sets:

8. E = {All the even numbers between 1 and 33}.
9. F = {All the odd numbers between 6 and 72}.
10. G = {All the natural numbers between 3.4 and 4.3}.
11. H = {All the even numbers which are odd}.
12. J = {All the numbers which are equal to their squares}.

State the next three elements in each of the following sets:

13. $K = \{1, 4, 7, 10, \ldots\}$.
14. $L = \{22, 20, 18, \ldots\}$.
15. $M = \{65, 70, 75, \ldots\}$.
16. $Q = \{8, 16, 24, 32, \ldots\}$.
17. $R = \{0, 4, 16, 36, \ldots\}$.
18. $S = \{1, 2, 3, 5, 8, 13, \ldots\}$.
19. $T = \{1, 2, 3, 5, 7, 11, \ldots\}$.
20. $U = \{1/2, 2/3, 3/4, 4/5, \ldots\}$.

Subsets

Given the elements of a set we may always make a selection of these elements so as to form new sets. For example, consider starting with $A = \{l, m, n\}$ and making selections from the three elements l, m, and n to form new sets:

Using all three elements of A we get the set $\{l, m, n\}$, which is the set A itself.
Using two elements of A we get the sets $\{l, m\}$; $\{m, n\}$; $\{n, l\}$.
Using one element of A we get the sets $\{l\}$; $\{m\}$; $\{n\}$.
Using no elements of A we get the set $\{\ \} = \phi$, making a total of 8 sets in all, and since these sets are derived from set A we call them subsets of A.

Notice that the creation of subsets of A by selecting 3, 2, 1 and 0 of its elements at a time forces us to accept that

(i) A is a subset of A,
(ii) ϕ is a subset of A.

More formally we define subsets as follows:

Definition 1.3. A set B is a subset of set A if every element of B is also an element of A.

From the definition, since 'every element of A is also an element of A', it follows that A is a subset of A which means that *every set is a subset of itself.* We shall have a little more to say about this peculiarity later on.

The bigger the set (i.e. the more elements in the set), the more subsets it can yield. We shall see that the addition of one extra element to a set will double the number of subsets available.

With $n(A) = 3$ there are $8 = 2^3$ subsets in A, as we saw above. Now suppose we add a new element to A to make the set $B = \{k, l, m, n\}$ so that $n(B) = 4$. How many subsets has the set B?

All the subsets of A will remain as subsets of B and the only new subsets will all contain the new element k. These will be obtained by merely adding k to each of the subsets of A as shown in the following lists.

Subsets of A	**Subsets of B**
$\{l, m, n\}$	$\{k, l, m, n\}$
$\{l, m\}, \{m, n\}, \{n, l\}$	$\{k, l, m\}, \{k, m, n\}, \{k, n, l\}$
$\{l\}, \{m\}, \{n\}$	$\{k, l\}, \{k, m\}, \{k, n\}$
ϕ	$\{k\}$

We see therefore that set B has twice as many subsets as set A. Thus, with $n(B) = 4$ there are $8 \times 2 = 2^4$ subsets in set B. Keeping this doubling up process in mind, we see that we may extend the results as follows:

If $n(C) = 5$ then there are $32 = 2^5$ subsets in set C.
If $n(D) = 6$ then there are $64 = 2^6$ subsets in set D,

and hence, generally,

If $n(S) = N$ then there are 2^N subsets in S.

Fortunately we are usually only interested in specially chosen subsets for which we have already stated a property definition.

Suppose we have $A = \{2, 3, 4, 5, 6, 7, 8, 9, 10\}$ then we could define the following subsets:

$B = \{$All the even numbers in $A\} = \{2, 4, 6, 8, 10\}$.
$C = \{$All the odd numbers in $A\} = \{3, 5, 7, 9\}$.
$D = \{$All the prime numbers in $A\} = \{2, 3, 5, 7\}$.
$E = \{$All the numbers in A which are divisible by $4\} = \{4, 8\}$.
$F = \{$All the numbers in A which are multiples of $11\} = \phi$
$G = \{$All the numbers in A which are greater than $9.3\} = \{10\}$.

With $n(A) = 9$ the set A has $2^9 = 512$ different subsets so it is clear that the suggestions above are only a small fraction of those possible.

Notation

The notation for stating that B is a subset of A is

$$B \subseteq A$$

where the symbol $\subseteq$ is read 'is a subset of'; thus two quotes from the above list are $D \subseteq A$ and $\phi \subseteq A$.

As already noted, all the subsets of A except A itself are smaller sets than A and in this sense it is appropriate to refer to these as *proper subsets of A*. For such cases we have a slightly different notation, writing,

$$B \subset A$$

if we wish to emphasise that B is a proper subset of A. Similarly, we may write $D \subset A$ and $\phi \subset A$ where the symbol $\subset$ is read 'is a proper subset of'.

Thus, in general, the statement $X \subseteq Y$ is telling us only that X is a subset of Y, in which case either X is a proper subset of Y or X is equal to Y, whereas the statement $X \subset Y$ is telling us that X is definitely a proper subset of Y. (There is an obvious comparision with the use of the symbols $\leqslant$ and $<$. For example, the following statements are all true: $0 \leqslant 2$, $4 \leqslant 4$, $1 < 5$, $0 < 2$, $1 \leqslant 5$.)

Looking back to the beginning of this section with $n(A) = 3$, we now see that the statement $P \subseteq A$ allows 8 possibilities for P whereas the statement $P \subset A$ allows only 7 possibilities for P.

We again use the / sign to negate the meaning of a symbol so that the statement $X \nsubseteq Y$ means that *X is not a subset of Y* and $X \not\subset Y$ means that *X is not a proper subset of Y.*

Finally we must avoid confusing $\in$ with $\subseteq$. The symbol $\in$ relates an element to a set whereas the symbol $\subseteq$ relates a set to a set. This appears to be straightforward, but it depends upon knowing what has been declared as the elements of a set, as we shall see in the next examples.

Taking $A = \{p, q, 1, 3\}$ we see that $p \in A$ is correct and so is $a \notin A$ but $p \subset A$ and $q \subseteq A$ are both meaningless.

Similarly $\{p\} \subset A$ is correct and so is $\{p, 2\} \not\subset A$ but $\{p\} \in A$ and $\{p, 1\} \in A$ are both meaningless.

A more involved example is suggested by $B = \{a, \{k\}, 6, \{7, 8\}\}$—that is, B is a set with four elements a, $\{k\}$, 6 and $\{7, 8\}$—and here we emphasise that $\{k\}$ and not k is the element of B. Similarly, $\{7, 8\}$ and not 7 or 8 is the element of B. Some typical correct comments about B are:

$$a \in B, \{k\} \in B, \{k\} \not\subseteq B, \{7, 8\} \not\subseteq B, \{a, 6\} \subset B, \{a, \{k\}\} \subset B, 7 \notin B$$

Exercise 1.3

List all the possible subsets for each of the following sets:

1. $P = \{a\}$ 2. $Q = \{a, b\}$ 3. ϕ

Taking K = {All the natural numbers less than 20} list the elements in each of the following subsets of K:

4. A = {All the odd numbers in K}.
5. B = {All the multiples of 3 in K}.
6. C = {All the numbers in K which are divisible by 3}.
7. D = {All the numbers in K which are divisible by 5}.
8. E = {All the prime numbers in K}.

Consider the three sets $X = \{1, 2, 3\}$; $Y = \{2, 3, 4, 5\}$; $Z = \{2, 5\}$ and then substitute the symbol $\subset$ or $\not\subset$ for each '_' to get a true statement in each of the following:

9. $X \_ Y$
10. $Z \_ Y$
11. $X \_ Z$
12. $\phi \_ Z$
13. $\{1, 3\} \_ X$
14. $\{5\} \_ Z$
15. $\{1, 5\} \_ X$

With the set L given by $L = \{1, \{1, 2\}, \{1\}, 3, 4\}$ state which of the following remarks are true (T) or false (F):

16. $1 \in L$
17. $2 \in L$
18. $\{3\} \subset L$
19. $\{3\} \in L$
20. $\{3, 4\} \subset L$
21. $\{2\} \subset L$
22. $4 \in L$
23. $\{1, \{1\}\} \subset L$
24. $\{1, 3, 4\} \in L$
25. $\{1, 2\} \in L$
26. $\{1, 2\} \subset L$
27. $\{1\} \not\subset L$
28. $\{4, 3\} \not\subset L$
29. $5 \notin L$
30. $\{1, 2, 3\} \subset L$

Universal Sets

The sets which arise during the discussion of any problem may be regarded as subsets of one overall set which is fixed for that discussion. We call this set the **universal set** or **the universe of discourse** and we usually represent it by U. The choice of a universal set will depend on the problem being considered.

For example, any discussion about imported and home-produced cars in the matter of design, top speed, economy, insurance, etc., may have U = {cars} as the universal set. Usually it is not necessary to be more definite than this since the context of the problem will be sufficient to indicate that we mean U = {all cars ever built}. Any restriction of the discussion to speaking about 'cars built since 1938' needs to be stated but since {all cars built since 1938} $\subset$ {all cars ever built} it is still acceptable to take U = {cars}.

Similarly, any discussion which compares people on the subject of age, intelligence, height, weight, etc., may have U = {people} which means all people ever born and now dead or alive, so that {living people} $\subset$ {people}.

Finally, a discussion of various triangles like right-angled triangles, equilateral triangles, isosceles triangles, scalene triangles will have

$U = \{\text{triangles}\}$ as a universal set, meaning that every triangle which can be drawn is a member of U.

We shall now examine some ways of describing or translating symbolic expressions about subsets.

Example 1. The universal set $U = \{\text{people}\}$ has the following subsets:

$I = \{\text{intelligent people}\}$,
$M = \{\text{males}\}$,
$F = \{\text{females}\}$,
$C = \{\text{children}\}$,
$H = \{\text{hardworking people}\}$,
$S = \{\text{students}\}$.

Translate the relations suggested by the following symbolic expressions:

(i) $F \subset I$
(ii) $M \subset H$
(iii) $S \not\subset M$
(iv) $I \not\subset H$
(v) $a \in I$ and $a \in C$
(vi) $b \notin M$ and $b \in H$

Solution

(i) $F \subset I$ states that all elements of F are also elements of I—that is, all females are intelligent people.

(ii) $M \subset H$ states that all elements of M are also elements of H—that is, all males are hardworking.

(iii) $S \not\subset M$ states that not all of the elements of S belong to M—that is, not all students are males. There is a temptation (to be resisted) of being too free with the translation and making claims which are not suggested by the symbolic expression. Here we might be tempted to believe that S definitely contains males and females, which is an incorrect conclusion since $S \not\subset M$ is correct for two possible cases: (*a*) there are males and females in S, (*b*) there are no males in S.

(iv) $I \not\subset H$ states that not all intelligent people are hardworking. As in (iii) this allows two possible cases: (*a*) there are hardworking people in I and also some not hardworking people in I, (*b*) there are no hardworking people in I.

(v) $a \in I$ and $a \in C$ state that a is an intelligent person and a is a child. We can conveniently conclude that a is an intelligent child.

(vi) $b \notin M$ and $b \in H$ state that b is not a male and b is a hardworking person. We conclude that b is a hardworking female.

Complement

A diagram like Fig. 1.1 illustrates a relation between the universal set U and one of its subsets S, for it is appropriate to consider that all the elements of U are somewhere in the rectangle while those elements which belong to S are

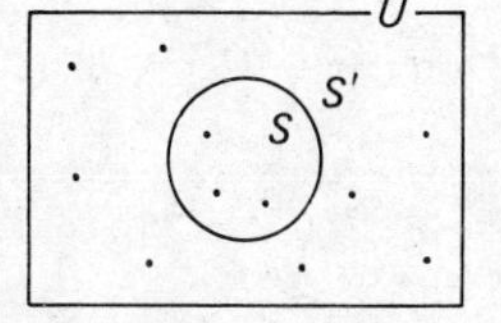

Fig. 1.1

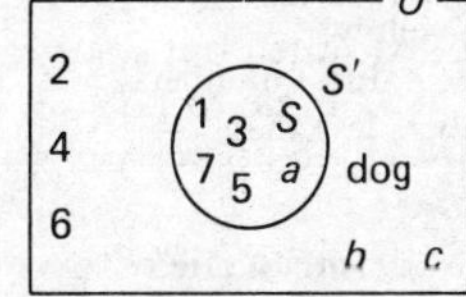

Fig. 1.2

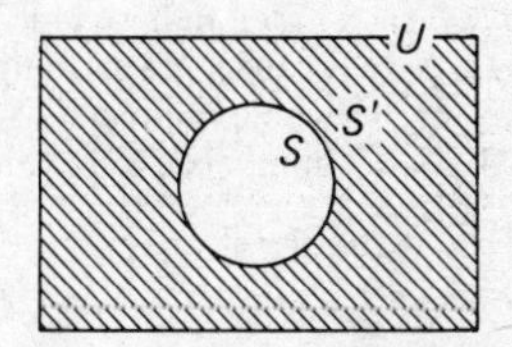

Fig. 1.3

somewhere in the circle. The two sets are identified by writing U and S on or near the perimeters and the presence of the elements may be suggested by the dots, although we do not always bother to insert these. If we know the elements then it might be convenient to indicate this knowledge by writing the elements on the figure, as done in Fig. 1.2, where $U = \{1, 2, 3, 4, 5, 6, 7, a, b, c, \text{dog}\}$ and the subset $S = \{1, 3, 5, 7, a\}$.

It now becomes apparent that in Fig. 1.3 the shaded region of the rectangle will contain all the elements of U which do not belong to S. This is another subset of U, sufficiently important to be given the special name of 'the complement of S'.

Definition 1.4. The complement of the set S is the set of all elements in U which do not belong to S. This set is written as S' and read as 'S dash'. Sometimes when necessary we shall also refer to the set S' as the complement of S *with respect to the set* U.

In Fig. 1.2 we saw that $S' = \{2, 4, 6, b, c, \text{dog}\}$. In the previous example on page 9 we had $U = \{\text{people}\}$, $M = \{\text{males}\}$, $F = \{\text{females}\}$. Since all people are either male or female we are able to deduce that

$$M' = F \text{ and } F' = M$$

and it is an interesting challenge to appreciate that $M'' = M$.

Exercise 1.4

Which of the following statements are true?

1. $\{\text{rabbits}\} \subset \{\text{animals}\}$
2. $\{\text{sparrows}\} \subset \{\text{birds}\}$
3. $\{\text{coalminers}\} \subset \{\text{people}\}$
4. $\{\text{traffic wardens}\} \subset \{\text{men}\}$
5. $\{\text{even numbers}\} \subset \{\text{numbers}\}$
6. $\{\text{fractions}\} \subset \{\text{numbers}\}$
7. $\{\text{positive odd numbers}\} \subset \{\text{natural numbers}\}$
8. $\{\text{squares}\} \subset \{\text{rectangles}\}$
9. $\{\text{rectangles}\} \subset \{\text{squares}\}$
10. $\{\text{equilateral triangles}\} \subset \{\text{isosceles triangles}\}$

Using the information in the example on page 9, translate the following suggestions:

11. $H \subset I$
12. $S \subset I$
13. $C \not\subset H$
14. $M \not\subset S$
15. $k \in H$ and $k \notin F$
16. $p \in S$ and $p \notin H$ and $p \notin I$
17. $F' \subset H$
18. $F' \not\subset I$
19. $I \not\subset S$
20. $S' \not\subset I'$

Find S' in each of the following cases:

21. $U = \{\text{people}\}$, $S = \{\text{adults}\}$
22. $U = \{\text{violet, indigo, blue, green, yellow, orange, red}\}$
    $S = \{\text{orange, green, indigo, yellow}\}$
23. $U = \{\text{whole numbers}\}$, $S = \{\text{natural numbers}\}$
24. $U = \{\text{numbers}\}$, $S = \{\text{numbers less than 7}\}$
25. $U = S$

Intersection

Here we are concerned with elements of a universal set U which have more than one characteristic property so that they are qualified for membership of more than one set. The description of the element usually refers to the sets to which it belongs, thus describing an element of $U = \{\text{people}\}$ as a hardworking female tells us that this element not only belongs to the set of hardworking people but also belongs to the set of females. Likewise, describing an element of $U = \{\text{numbers}\}$ as divisible by 6 tells us that this number not only belongs to the set of numbers divisible by 2 but also to the set of numbers divisible by 3, assuming of course that we see 6 as 2×3.

Such elements with dual membership, form what is called the **intersection** of the two sets concerned.

Definition 1.5. The intersection of set A and set B is the set of all those elements which belong to both set A and set B at the same time.

The intersection is written $A \cap B$ and is read as 'A intersection B'. We may also represent this intersection by writing $B \cap A$—that is, taking $A \cap B = B \cap A$. We shall note that in particular $A \cap \phi = \phi$.

Example 2.

$$\{1, 2, 3, 4, 5\} \cap \{2, 4, 6, 8\} = \{2, 4\}.$$
$$\{\text{c, e, k, l, r, s}\} \cap \{\text{k, r, s, y, z}\} = \{\text{k, r, s}\}$$
$$\{\text{car, van, cycle, lorry}\} \cap \{\text{man, woman, child}\} = \phi$$

Example 3.

$$U = \{0, 1, 2, 3, 4, 5, 6, 7\}; \quad S = \{0, 2, 4, 6\}; \quad S' = \{1, 3, 5, 7\}$$
$$\therefore \quad S \cap S' = \phi$$

This last result is not only true for this particular set S: it is true in general, i.e. it is always true for any set.

It is helpful to practise the use of reasons to argue this type of result. Recall the definition of the complement of any set S on page 10. Elements of U only belong to S' if they do not belong to S, so that no element can belong to both sets S and S', consequently $S \cap S' = \phi$.

A simple illustration of the relation between sets was introduced on page 9 for the complement of a set. Such illustrations like Figs. 1.1, 1.2, 1.3 are called **Venn diagrams** and although they are not a substitute for arguments based on the definitions the visual presentation of the relations in this manner does offer an easy opportunity for an intuitive appreciation of the results. Let us consider the suggestions in Figs. 1.4–1.5 for representing possible results for $A \cap B$. We have illustrated three possibilities:

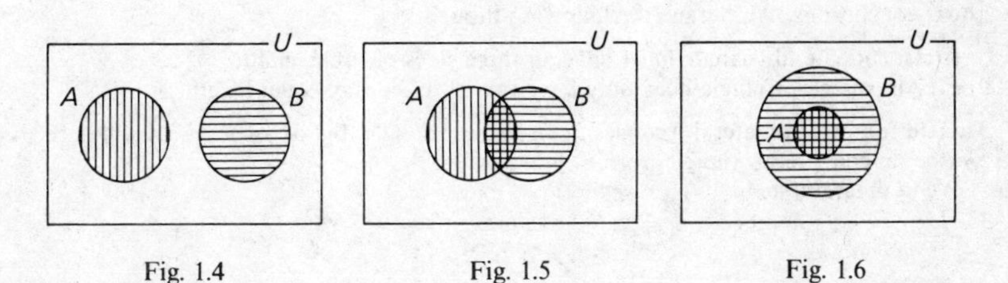

Fig. 1.4 Fig. 1.5 Fig. 1.6

Fig. 1.4: Here sets A and B do not intersect and so $A \cap B = \phi$ is represented by this diagram. In this case no element of A belongs to B.

Fig. 1.5: Here sets A and B do intersect and the intersection is given by the region with the double lines and so $A \cap B \neq \phi$ is represented by this diagram. In this case some elements of set A belong to set B, and some elements of set B belong to set A.

Fig. 1.6: Here we have $A \subset B$ and with the intersection showing double lines the diagram suggests that $A \cap B = A$. In this case *all* of the elements of set A belong to set B and *some* of the elements of set B do not belong to set A. An interchange of B with A in this figure would illustrate the suggestion $A \cap B = B$.

Example 4. Use a Venn diagram to represent the relation between the sets

$$U = \{1, a, 2, b, 3, c, 4, d\}; \quad A = \{1, 2, c, d\}; \quad B = \{a, b, 1, 2\}$$

We start by drawing two intersecting circles within a rectangle and then fill in the elements by inspection; by so doing we obtain Fig. 1.7.

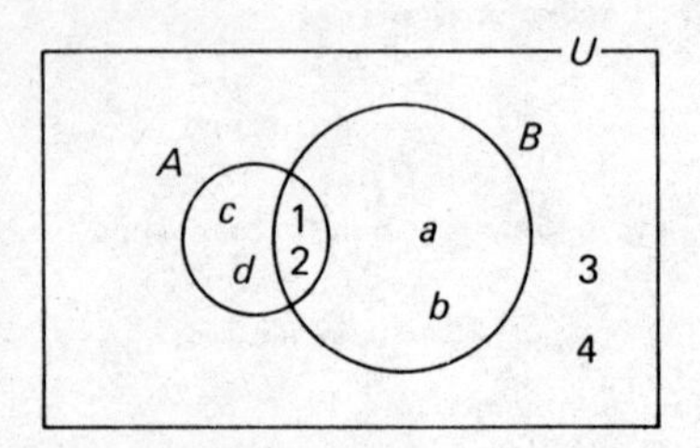

Fig. 1.7

In completing the Venn diagram we have shown that $A \cap B = \{1, 2\}$.

Notice that the diagram merely provides us with regions into which we write the elements, it does not solve anything. However, it does enable us to identify and direct our thoughts in answering the question as to which elements belong to $A \cap B$ and the relation between all three sets is made very clear. The diagram is also quite helpful at a higher level of abstraction in the next example.

Example 5. U = {plane triangles}; E = {equilateral triangles}; I = {isosceles triangles}; R = {right-angled triangles}. Illustrate the relation between these sets with Venn diagrams.

Solution. A Venn diagram will merely summarise our conclusions: it will not produce those conclusions. We pursue the following thoughts:

(i) An equilateral triangle must have all three sides of equal length.
(ii) An isosceles triangle need only have two of its sides of equal length.

Conclusion: All equilateral triangles are isosceles triangles but an isosceles triangle need not be an equilateral triangle.

Venn diagram:

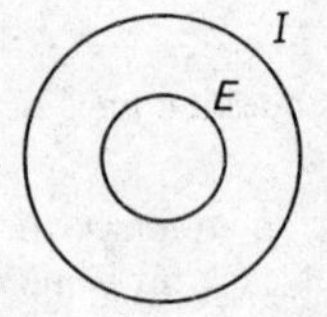

(iii) An equilateral triangle has each of its angles 60° but a right-angled triangle must have an angle of 90°.
Conclusion: No equilateral triangle can be a right-angled triangle,
Venn diagram:

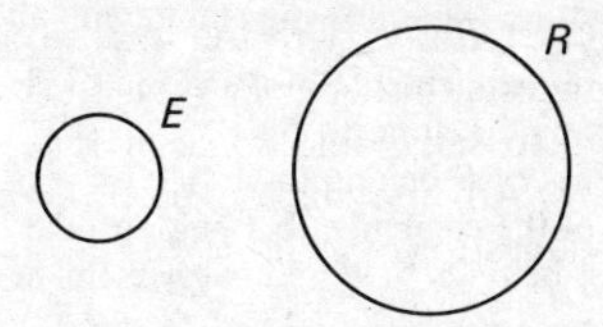

(iv) We know that it is possible to have a triangle with angles of 45°, 45°, 90°.
Conclusion: Some isosceles triangles can be right-angled triangles but not all right-angled triangles are isosceles triangles—for example, a triangle with angles of 30°, 60°, 90° is not an isosceles triangle.
Venn diagram:

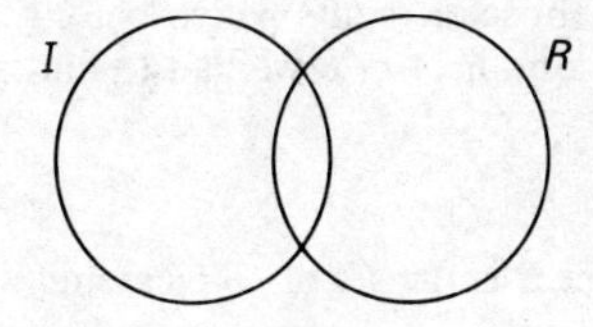

The complete Venn diagram is therefore, Fig. 1.8.

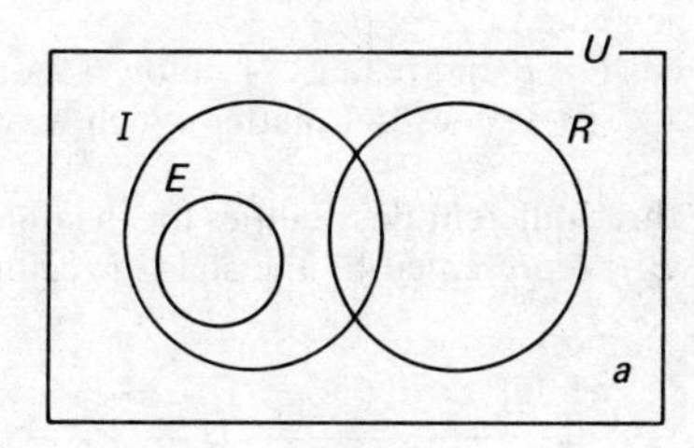

Fig. 1.8

Notice that it is possible to have a triangle as element a in the figure which does not belong to I or R—for example, a triangle with sides 2, 3, 4.

Exercise 1.5

$U = \{\text{natural numbers from 1 to 15 inclusive}\}$
$A = \{1, 2, 3, 4, 5, 6, 7, 8, 9, 10\}$, $B = \{1, 2, 3, 5, 7, 11\}$, $C = \{2, 4, 6, 8, 10\}$, $D = \{4, 8\}$.
Use the above sets to answer the following questions:

1. $A \cap B$
2. $B \cap C$
3. $C \cap D$
4. $A \cap D$
5. $B \cap A$
6. $B \cap B$
7. $D' \cap D'$
8. $A' \cap B$
9. Let $E = A \cap B$ and then find $E \cap C$
10. Using the idea in the previous question find $A \cap D \cap C$
11. $D \cap \phi$
12. Is $A \cap B$ a proper subset of A?
13. Is $A \cap B$ a proper subset of B?
14. Give a property definition for the sets (i) B, (ii) C, (iii) $B \cap C$.

15. If U = {birds}, A = {parrots}, B = {green birds}, T = {talking birds} give a description of the elements in the following sets:

    (i) $A \cap B$, (ii) $A \cap T$, (iii) $A \cap B \cap T$, (iv) $B \cap T'$, (v) $A' \cap B \cap T'$.

16. Give a diagram to illustrate the following statements about two sets A and B:

    (i) Some, but not all of the elements of B belong to A.
    (ii) All of the elements of A belong to B.
    (iii) None of the elements of B belong to A.
    (iv) Some, but not all of the elements of A belong to B.
    (v) Every element of A belongs to B and every element of B belongs to A.

Union

The union of two sets A and B is formed on a less selective basis than their intersection for here we merely unite the two lists of elements and delete any repeated elements. For example, with $A = \{1, 5, a, b\}$ and $B = \{1, 5, 7, a, d\}$ the united list is given by $\{1, 5, 7, a, d, b\}$ after deleting the repeated elements—that is, we now have all those elements which belong to A *or* B.

Recall that in order to obtain $A \cap B$ we had to find all those elements which belonged to A *and* B.

Definition 1.6

The **union** of set A and set B is the set of all elements which belong to set A or to set B.

We note that this set includes all those elements which belong to both *set A and set B*, i.e. those elements in $A \cap B$.

This union is written $A \cup B$ and read as 'A union B'. A further point to note is that $A \cup B = B \cup A$, i.e. it does not matter which way round we write the result.

Figs. 1.9–1.11 give three different possibilities for the union of two sets A and B. In each case $A \cup B$ is represented by the shaded region.

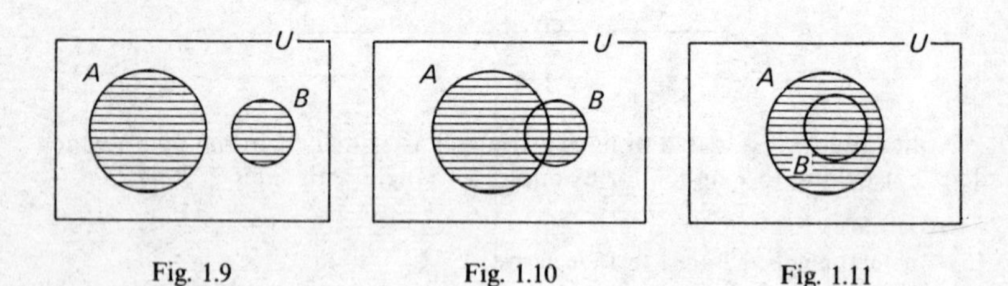

Fig. 1.9 Fig. 1.10 Fig. 1.11

Fig. 1.9: Here the two sets are *disjoint*—that is, there is no overlap of the two regions representing the sets A and B so that $A \cap B = \phi$.

Fig. 1.10: Here $A \cup B$ has elements which belong to both set A and set B at the same time. This diagram is suggesting that $A \cap B \neq \phi$.

Fig. 1.11: Here $B \subset A$ and the shaded region is the same as for A, so that this diagram is suggesting that $A \cup B = A$. An interchange of A and B on this diagram would have resulted in the suggestion that $A \cup B =$ B.

Written examples to which the above Venn diagrams correspond may be suggested as follows:

Example 6. For Fig. 1.9: $U = \{\text{people}\}$; $A = \{\text{male children}\}$, $B = \{\text{female children}\}$. Here we see that sets A and B are disjoint, i.e. $A \cap B = \phi$, since no child is both male and female. We may write $A \cup B = \{\text{children}\}$.

Example 7. For Fig. 1.10: $U = \{\text{stamps}\}$; $A = \{\text{Australian stamps}\}$, $B = \{\text{stamps issued before 1960}\}$. Here we have $A \cap B = \{\text{Australian stamps issued before 1960}\} \neq \phi$ and $A \cup B = \{\text{stamps which are either Australian or issued before 1960 or both}\}$. Notice that $A \cap B$ is contained in this set.

Example 8. For Fig. 1.11: $U = \{\text{House furniture}\}$; $A = \{\text{chairs}\}$, $B = \{\text{bent wood chairs}\}$. Here we have $B \subset A$ because every element in B is already in A so that $A \cup B = A$.

The following are six standard results associated with the union of two sets A and B:

(i) $A \cup A' = U = A \cup A'$
(ii) $A \cup \phi = A = \phi \cup A$
(iii) $U \cup \phi = U = \phi \cup U$
(iv) $\phi \cup \phi = \phi$
(v) $A \cup U = U$
(vi) $U \cup U = U$

Inclusive and exclusive 'or'

The description of $A \cup B$ is often given as 'A or B' but we must take note of two possible meanings for 'or':

(i) In mathematics we use 'or' in the sense of 'and/or' and we refer to this as as the inclusive 'or'. Thus when $A \cup B$ is read as '*A or B*' we are using 'or' in the inclusive sense, meaning that an element of $A \cup B$ may come from any of the three sets A, B, $A \cap B$.

(ii) The exclusive 'or' is used to refer to the elements in the two sets A, B which do not belong to $A \cap B$.

Each of these two uses of the word 'or' is illustrated by Fig. 1.12 and 1.13.

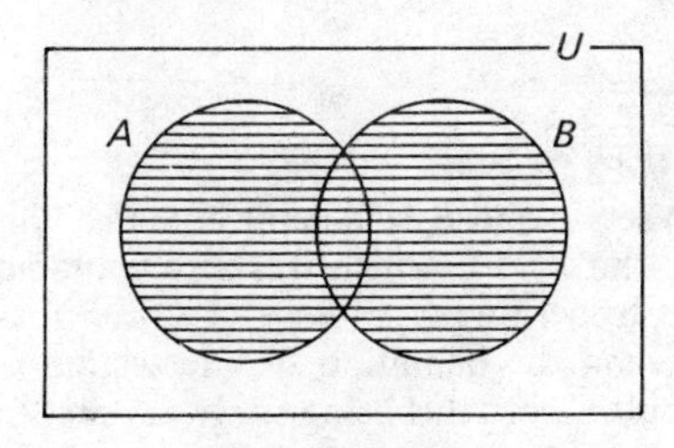

$A \cup B$, inclusive or

Fig. 1.12

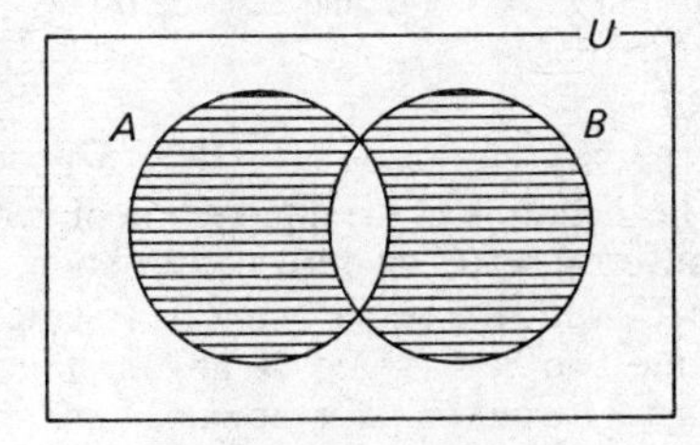

$A \Delta B$, exclusive or

Fig. 1.13

This difference in the meaning of the two uses of 'or' justifies a different notation to indicate which one we require so we write $A \cup B$ for the inclusive and $A \Delta B$ for the exclusive meaning.

In fact, we go even further in our description of $A \Delta B$ by calling $A \Delta B$ the symmetric difference of the two sets A, B. We shall consider the symmetric difference of two sets in more detail on page 21.

Definition 1.7. The **symmetric difference** of two sets A and B is the set of elements which belong to either set A or set B but not to $A \cap B$.

Notice that $A \Delta B = B \Delta A$, and furthermore that if A and B are disjoint sets (i.e. $A \cap B = \phi$) then $A \Delta B = A \cup B$.

Exercise 1.6

$U = \{1, 2, 3, 4, 5, 6, 7, 8, 9\}$; $A = \{3, 5, 7, 9\}$; $B = \{2, 4, 6, 8\}$; $C = \{3, 6, 9\}$; $D = \{5\}$; $E = \{4, 8\}$.

Use the above sets to list the elements of the following sets:

1. A'
2. B'
3. $B \cup A$. Write $F = (B \cup A)$ and find the elements in F'.
4. $D \cup B$. Write $G = (D \cup B)$ and find the elements in G'.
5. $D \cup E$
6. $A' \cup A$
7. $A' \cup B$
8. $C \cup \phi$
9. $A \cup B$, $A \cap B$, $A \Delta B$.
10. $B \cup C$, $B \Delta C$.
11. $B' \cap A$, $A' \cap B$. Write $H = (B' \cap A)$ and $K = (A' \cap B)$ and find the elements in $H \cup K$.
12. $B \Delta A$. Compare your result with $H \cup K$ in question 11.

With U = {fruit}; R = {ripe fruit}; G = {green fruit}; A = {apples}; C = {cherries}; L = {lemons} describe the following sets:

13. $A \cap G$
14. $L \cup C$
15. $A \cup L$, $A \Delta L$
16. $L \cup R$
17. $L \Delta R$
18. $C \cup R'$
19. Put $K = A \cap R$ and describe $K \cup G$.
20. Put $H = A \cap R'$ and describe $H \cup G$.

Basic Operations on Sets

The definition of the intersection of two sets A and B each being contained in a universal set U enabled us to obtain a third set C which was also contained in U, with the result being represented by writing $A \cap B = C$. Given these same two sets A and B, anyone following the definition of intersection on page 11 would obtain the same final result C, and this being so we say that the result C is **unique**.

Now any method of combining two sets so as to produce a unique set for the answer is called a **mathematical operation** on the sets. This means that we shall describe $\cap$ as an *operation defined on the universal set U*. We can describe $\cap$ in more detail by noting that its definition brings together only two sets at a time, in which case it is appropriate to call $\cap$ a **binary operation**. Furthermore, since the unique answers like C are always contained inside U it is appropriate to call $\cap$ a **closed** binary operation on U.

For a more familiar example of a binary operation consider the operation of multiplication $\times$ defined on $U = \{\text{numbers}\}$. Some simple examples which come to mind are:

(i) $3 \times 4 = 12 = 4 \times 3$
(ii) $2.1 \times 6.73 = 14.133 = 6.73 \times 2.1$
(iii) $\frac{8}{9} \times \frac{2}{3} = \frac{16}{27} = \frac{2}{3} \times \frac{8}{9}$

We generalise these results with a statement like $a \times b = c = b \times a$, noting that there is only one answer c and that this belongs to U. This now describes $\times$ as a closed binary operation on $U = \{\text{numbers}\}$. (The same remarks apply to the operation $+$ on U.)

Since it does not matter which way round we do the multiplication (i.e. 3×4 or 4×3), we say that the operation $\times$ is **commutative.**

The question now arises of how to deal with a multiplication such as $2 \times 3 \times 4$, for multiplication is a binary operation combining only two numbers at a time, but in $2 \times 3 \times 4$ we have three numbers, so where do we start and how do we proceed?

We start by agreeing that $(2 \times 3) \times 4$ means work out the bracket which associates 2 with 3 first, and produces the first step of $(2 \times 3) \times 4 = 6 \times 4$, and a final step of $6 \times 4 = 24$. Thus, the procedure is to multiply two numbers at a time using brackets to indicate which two numbers are involved at each stage.

Another way of finding $2 \times 3 \times 4$ is to start by associating 3 with 4 to produce a first step given by $2 \times (3 \times 4) = 2 \times 12$ and a final step of $2 \times 12 = 24$.

Since $(2 \times 3) \times 4 = 2 \times (3 \times 4) = 24$, we write $2 \times 3 \times 4 = 24$ knowing that 24 is the only result which can be obtained. As we are able to obtain the same result no matter which numbers are associated in the steps of the multiplication, we say that the operation $\times$ is **associative**, or obeys the associative law. (Again, the same remarks apply to the operation of addition, $+$, on U.)

Returning to the operation $\cap$ on sets, consider the question of evaluating $A \cap B \cap C$. As in the case of multiplication, since $\cap$ is a binary operation only, we ask where do we start and how do we proceed? As before we overcome the problem by the use of brackets, starting with $(A \cap B) \cap C$.

Putting $A \cap B = E$ we see that all the elements in E are in both sets A and B at the same time. Therefore

$$(A \cap B) \cap C = E \cap C$$

Putting $E \cap C = D$, we see that all the elements in D are in set E and set C at the same time. This means that all the elements in D are in set A and set B and set C at the same time.

We obtain the same result when considering $A \cap (B \cap C) = D$, so we have found that

$$(A \cap B) \cap C = A \cap (B \cap C) = D$$

and thereby deduced that the operation $\cap$ is associative or obeys the associative law.

Since it makes no difference where we start we may write

$$A \cap B \cap C = D \text{ (without brackets)}$$

So far we have deduced that $\cap$ is a closed binary operation which obeys both the commutative law and the associative law. Similarly, we can show that $\cup$ is a closed binary operation which obeys both the commutative law and the associative law.

Example 9. We can illustrate the associative law using the sets $A = \{p, q, r, s\}$, $B = \{q, r, s, t\}$, $C = \{r, s, t, u\}$.

$$A \cap B = \{q, r, s\} \qquad B \cap C = \{r, s, t\}$$
$$\therefore \quad (A \cap B) \cap C = \{q, r, s\} \cap C \text{ and } A \cap (B \cap C) = A \cap \{r, s, t\}$$
$$= \{r, s\} \qquad\qquad = \{r, s\}$$
$$\therefore \quad (A \cap B) \cap C = A \cap (B \cap C) \text{ and we may write}$$
$$A \cap B \cap C = \{r, s\}$$

For the operation $\cup$ we have

$$A \cup B = \{p, q, r, s, t\} \qquad B \cup C = \{q, r, s, t, u\}$$
$$(A \cup B) \cup C = \{p, q, r, s, t\} \cup C \text{ and } A \cup (B \cup C) = A \cup \{q, r, s, t, u\}$$
$$= \{p, q, r, s, t, u\} \qquad\qquad = \{p, q, r, s, t, u\}$$
$$\therefore \quad (A \cup B) \cup C = A \cup (B \cup C)$$

and we may write

$$A \cup B \cup C = \{p, q, r, s, t, u\}.$$

Example 10. With U = {people}, A = {black-haired people}, B = {people who wear spectacles}, C = {blue-eyed people}, we shall illustrate $A \cap B \cap C$ and $A \cup B \cup C$ with Venn diagrams.

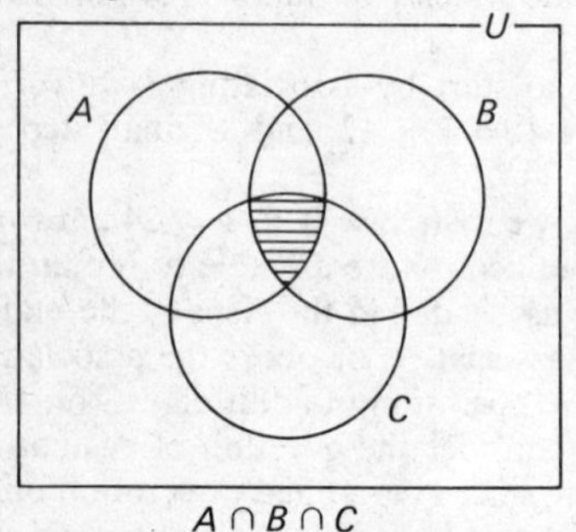

Fig. 1.14

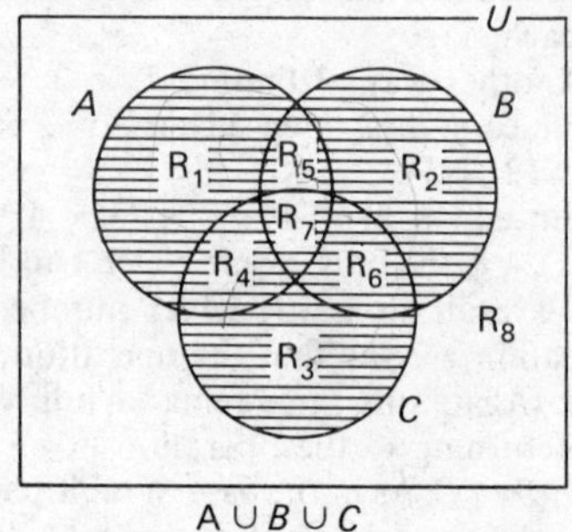

Fig. 1.15

Fig. 1.14: Here the elements of $A \cap B \cap C$ would lie in the shaded region which is common to all three sets A, B and C. We would describe $A \cap B \cap C$ as the set of people who are black-haired, blue-eyed and wear spectacles. It is possible to be a little more emphatic by describing $A \cap B \cap C$ as the set of people who not only have black hair and blue eyes but also wear spectacles.

Fig. 1.15: Here $A \cup B \cup C$ is represented by the shaded region and may be described as the set of people who are black-haired or blue-eyed or wear spectacles where we are using the inclusive 'or'. Note that an element in the region labelled R_5 indicates a person who has black hair and wears spectacles but does not have blue eyes.

Example 11. With $U = \{-3, -2, -1, 0, 1, 2, 3\}$, $A = \{-2, 1, 3\}$, $B = \{-3, -1, 1, 2, 3\}$, $C = \{-3, -1, 0, 1, 2\}$, we wish to find the complement of an intersection and the complement of a union of sets as follows:

We have $\quad A' = \{-3, -1, 0, 2\}$, $B' = \{-2, 0\}$

Thus $\quad A \cap B = \{1, 3\}$ and $(A \cap B)' = \{-3, -2, -1, 0, 2\}$

and $\quad A' \cup B' = \{-3, -1, 0, 2\} \cup \{-2, 0\} = \{-3, -2, -1, 0, 2\}$

We have shown for these two sets that $(A \cap B)' = A' \cup B'$. (i)

Again $\quad A \cup B = \{-3, -2, -1, 1, 2, 3\}$ so that $(A \cup B)' = \{0\}$
and $\quad A' \cap B' = \{-3, -1, 0, 2\} \cap \{-2, 0\} = \{0\}$
$\therefore \quad (A \cup B)' = A' \cap B'$ (ii)

The two results (i) and (ii) are called De Morgan's laws. We have not proved them, just merely shown that they are satisfied in this example. Now consider $(A \cap B \cap C)'$ *and* $(A \cup B \cup C)'$.

$$A \cap B \cap C = (A \cap B) \cap C = \{1, 3\} \cap C = \{1\}$$
$$\therefore \quad (A \cap B \cap C)' = \{-3, -2, -1, 0, 2, 3\}$$
$$A \cup B \cup C = (A \cup B) \cup C = \{-3, -2, -1, 1, 2, 3\} \cup C = U$$
$$\therefore \quad (A \cup B \cup C)' = U' = \phi$$

The Distributive Law

This law involves a relation between two operations and we use a more familiar example with multiplication and addition to identify the meaning of this law.

Example 12. The statement $5 \times (8 + 3)$ means that the bracket should be evaluated first and multiplication of 5 carried out second, so as to get

$$5 \times (8 + 3) = 5 \times 11 = 55$$

But we find that

$$(5 \times 8) + (5 \times 3) = 40 + 15 = 55$$

Hence

$$5 \times (8 + 3) = (5 \times 8) + (5 \times 3) = (8 + 3) \times 5$$

On the right-hand side we say that the operation $\times$ has been distributed over the operation $+$ in the bracket.

The comparable result in sets to

$$5 \times (8 + 3) = (5 \times 8) + (5 \times 3) \text{ is}$$
$$A \cap (B \cup C) = (A \cap B) \cup (A \cap C)$$
$$= (B \cup C) \cap A \text{ for any three sets } A, B \text{ and } C.$$

This result may be illustrated with Venn diagrams (see Exercise 1.7).

We say that $\cap$ satisfies the distributive law with respect to $\cup$. We may also prove that $\cup$ satisfies the distributive law with respect to $\cap$, since

$$A \cup (B \cap C) = (A \cup B) \cap (A \cup C)$$
$$= (B \cap C) \cup A \text{ for any three sets } A, B \text{ and } C.$$

Example 13. Here we show that $+$ is not distributive over $\times$ by considering

$$5 + (8 \times 3) = 5 + 24 = 29$$

and noting that

$$(5 + 8) \times (5 + 3) = 13 \times 8 = 104$$

Thus, the comparable result in numbers to

$$A \cup (B \cap C) = (A \cup B) \cap (A \cup C), \text{ which is true}$$

is

$$5 + (8 \times 3) = (5 + 8) \times (5 + 3), \text{ which is false}$$

It follows that the relation between $\cap$ and $\cup$ is more selective than the relation between $\times$ and $+$.

Exercise 1.7

$U = \{0, 1, 2, 3, 4, 5, 6, 7, 8, 9\}$, $A = \{0, 2, 4, 6, 8\}$, $B = \{1, 2, 3, 4\}$, $C = \{3, 4, 5, 8, 9\}$, $D = \{0, 6, 7, 8\}$

Using the above sets list the elements in the following sets. Recall that the bracketed parts of a statement must be worked out first.

1. A'
2. B'
3. C'
4. D'
5. $A \cap B \cap C$
6. $B \cap C \cap D$
7. $B \cup C \cup D$
8. $A \cup B \cup D$
9. $(A \cap B) \cup C$
10. $A \cap (B \cup C)$
11. $(A \cup B \cup D)'$

Show that each of the following results are true, using the same sets above:

12. $A \cup (B \cap C) = (A \cup B) \cap (A \cup C)$
13. $A \cap (B \cup C) = (A \cap B) \cup (A \cap C)$
14. $(A \cup B)' = A' \cap B'$
15. $(A \cap B)' = A' \cup B'$
16. $(A \cup C \cup D)' = A' \cap C' \cap D'$
17. $(B \cap C \cap D)' = B' \cup C' \cup D'$

Using Fig. 1.15 (page 18), state which regions represent the following sets:

18. $A \cap B$
19. $B \cap C$
20. $A \cap B \cap C'$
21. $A' \cap B' \cap C$
22. $(A \cup B \cup C)'$

Equations

Elementary 'find x' problems can often be solved mentally. A typical problem would be 'I am thinking of a number and when I add 11 to it the answer is 19. What is the number?' Most people can answer this question without using algebraic suggestions like 'Let x be the number' and representing the given statement by the equation

$$x + 11 = 19$$

before finding $x = 8$ by guesswork or otherwise.

Equations like $2 \times 5 = x$ or $x = 21 + 10$ could also be solved mentally because in each case knowing two of the quantities enables us to find the third. Speaking in deeper mathematical terms, we say that we are solving equations in $U = \{\text{numbers}\}$ with the operations $+$ and $\times$.

Now the purpose of this section is to examine similar situations in sets with the operations of $\cap$ and $\cup$—that is, given two of the sets in equations like $A \cap B = C$ and $A \cup B = C$ we ask: 'Will it be possible to find the third set?'

We have already seen that a knowledge of A, B enables us to find a unique set C such that $A \cap B = C$, but suppose we know A, C and try to find B?

Example 14. With $U = \{\text{letters of the alphabet}\}$, $A = \{a, b, c, d, e\}$ and $C = \{b, c, d\}$ can we find B such that $A \cap B = C$? (This is similar to finding b when $7b = 21$, say.)

We are trying to find B so that $\{a, b, c, d, e\} \cap B = \{b, c, d\}$.

By inspection (i.e. intelligent guesswork) we see that $B = \{b, c, d\}$ is a possible solution. On further inspection we begin to realise that other solutions are $\{b, c, d, f\}$, $\{b, c, d, g\}$, $\{b, c, d, f, g\}$ and so on. We can now see that the elements of B must be b, c and d together with any number of elements of the alphabet except a or e. There are $2^{21} = 2\,097\,152$ solutions to this problem.

Example 15. With $U = \{\text{dogs}\}$, $B = \{\text{Boxers}\}$, $C = \{\text{Corgies}\}$, can we find A such that $A \cap B = C$?

We are trying to find A such that $A \cap \{\text{Boxers}\} = \{\text{Corgies}\}$. Since no Boxer is a

Corgi it follows that it is not possible to suggest a set A to satisfy this equation. In other words, this equation does not have a solution.

Example 16. With $U = \{\text{numbers}\}$, $A = \{2, 4, 6, 8\}$, $C = \phi$ can we find B so that $A \cap B = C$?

We are trying to find B so that $\{2, 4, 6, 8\} \cap B = \phi$.

For an empty intersection it is clear that B can be any subset of A'. There are an infinite number of such sets since B may be any set of numbers which does not contain any of the elements of the set A.

We conclude that the operation of $\cap$ on sets is very different from the operations of $+$ and $\times$ on numbers since these elementary equations with sets do not allow similar elementary solutions to those we obtain for the comparable algebraic equations.

As in the case of $\cap$ we have seen that knowing A, B enables us to find a unique set C such that $A \cup B = C$, but again suppose we know A, C and try to find B in the following examples.

Example 17. With $U = \{\text{alphabet}\}$, $A = \{a, b, c\}$, $C = \{a, b, c, d, e\}$ can we find B such that $A \cup B = C$?

By inspection we see that we must have at least the elements d and e of C together with any of the eight selections from a, b and c.

Thus $\{d, e\}$, $\{a, d, e\}$, $\{b, d, e\}$, $\{a, b, d, e\}$ are four of the eight possible solutions for B.

Example 18. With $U = \{\text{numbers}\}$, $B = \{1, 5, 9\}$, $C = \{2, 6\}$ can we find A in $A \cup B = C$?

Here we require A so that $A \cup \{1, 5, 9\} = \{2, 6\}$.

Since $A \cup \{1, 5, 9\}$ will always contain the elements 1, 5 and 9 no matter what we suggest for A it follows that this equation does not have a solution.

We conclude that the operation of $\cup$ on sets is also very different from the operation of $+$ and $\times$ on numbers.

We shall attempt to identify some of those differences in Chapter 6.

Less fuss has been made of the other operation Δ on sets but it has a surprise in store for us when we examine similar equations to those above—that is, finding A when knowing B and C in $A \,\Delta\, B = C$. (Recall that $A \,\Delta\, B$ is the set of elements in A or B but not in $A \cap B$, i.e. $A \,\Delta\, B$ is $A \cup B$ with the elements in $A \cap B$ eliminated.)

Example 19. With $U = \{\text{alphabet}\}$, $A = \{a, b, c, d, e\}$, $C = \{b, c, d, f, g\}$ we wish to find B in $A \,\Delta\, B = C$ if it exists.

Here we are trying to find B so that $\{a, b, c, d, e,\} \,\Delta\, B = \{b, c, d, f, g\}$.

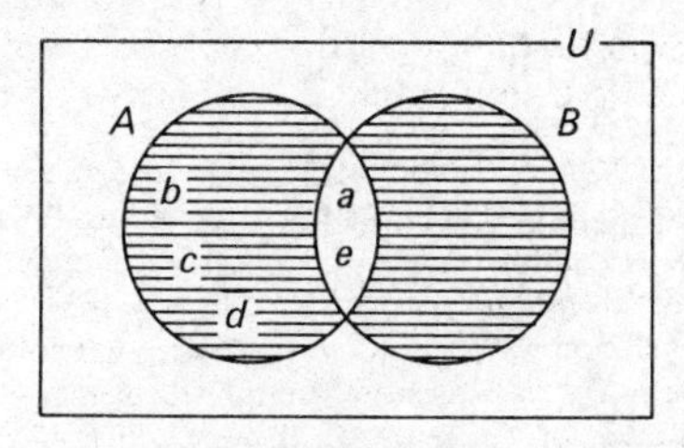

Fig. 1.16

The shaded region of Fig. 1.16 represents $A \,\Delta\, B = \{b, c, d, f, g\}$ so that these, and only these, elements may be placed in the shaded region. It is this point which determines the arrangement of the elements of set A which is shown on the figure. It now follows that we can insert f and g into the remainder of the region representing B. Hence, we have one and only one solution for B, namely $B = \{a, e, f, g\}$.

Example 20. With $U = \{\text{numbers}\}$, $A = \{2, 4, 6, 8\}$, $C = \phi$ solve the equation $A \Delta B = C$.

Here we are trying to find B so that $\{2, 4, 6, 8\} \Delta B = \phi$. A little devoted inspection suggests that $B = \{2, 4, 6, 8\}$ is the one and only solution to this equation.

Example 21. Using the same sets as for Example 15, $U = \{\text{dogs}\}$, $B = \{\text{Boxers}\}$, $C = \{\text{Corgies}\}$, suppose we now ask 'is there a set A such that $A \Delta B = C$?'.

Here we are trying to find a solution to $A \Delta \{\text{Boxers}\} = \{\text{Corgies}\}$. Since there are no Boxers in $C = \{\text{Corgies}\}$ we must eliminate Boxers by having them in $A \cap B$. Hence $A = \{\text{Corgies, Boxers}\}$ is our one and only solution.

It is possible to prove that the equation $A \Delta B = C$ has one and only one solution for A whenever we are given B and C.

Example 22. With $U = \{1, 2, 3, 4, 5\}$, $B = \{1, 2, 3, 4\}$, $C = \{1, 2, 3, 4\}$ find a solution to the equation $A \Delta B = C$.

We are trying to solve $A \Delta \{1, 2, 3, 4\} = \{1, 2, 3, 4\}$. If any member of B belonged to A then that element would not have belonged to the result $A \Delta B$. Therefore $A = \phi$.

It can be seen that although the operations $\cap$, $\cup$ and Δ might appear to be combining sets in a similar manner only the operation Δ will produce unique solutions to equations in sets of the type that has been discussed in the above examples. It was necessary to draw attention to this variety in mathematical operations in order to appreciate the well-behaved operations of $+$ and $\times$ on the set of numbers in everyday arithmetic.

Finding the one and only one set C when we know A and B in each of the equations $A \cap B = C$, $A \cup B = C$, $A \Delta B = C$ is sometimes referred to as 'doing' the operation. Finding the one and only one set A in the same equations when we know B and C is referred to as 'undoing' the operation. What we have suggested from the above examples is that it is always possible to 'undo' the operation of Δ but this is certainly not possible for the operations $\cap$ and $\cup$.

Exercise 1.8

For questions 1 to 5 take $U = \{0, 1, 2, 3, 4, 5, 6, 7\}$ and find set B to satisfy each of the following six equations:

(i) $A \cap B = C$, (ii) $A \cup B = C$, (iii) $A \Delta B = C$, (iv) $C \cap B = A$, (v) $C \cup B = A$, (vi) $C \Delta B = A$.

1. $A = \{3, 5, 6\}$, $C = \{0, 2, 4\}$
2. $A = \{0, 1, 2, 3, 4\}$, $C = \{0, 1, 2\}$
3. $A = \{1, 2, 7\}$, $C = \{1, 2, 3, 4\}$
4. $A = \{1, 2, 3\}$, $C = \{1, 2, 3, 4\}$
5. $A = \phi$, $C = \{1, 2\}$

Find the set X in each of the questions 6 to 10:

6. $X \cap A = \phi$
7. $X \cup \phi = \phi$
8. $X \Delta X' = U$
9. $X \Delta A = \phi$
10. $X \Delta A = A$

Solving Problems with Venn Diagrams

Venn diagrams are a useful aid in directing our thoughts when examining data which can be classified in terms of sets and their intersections. We shall see that the diagrams also enable us easily to check the numerical consistency of the data. In some ways we really invent the problems so as to find a use for Venn diagrams and in this respect the problems have an artificial quality but the method of using them has a mathematical purpose which is real enough since the alternative is so very tedious and obscure.

Up to now the Venn diagrams have only been used to show the presence of the elements in the sets but here we use the diagrams to indicate the **number** of elements in the respective regions—that is, we now put 4 in the region corresponding to $A \cap B$ to indicate that there are 4 elements in this region and, not as before, to indicate that 4 is an element of this region. When used in this manner we shall refer to the diagram as a n(set) Venn diagram.

Example 23. In a class of 30 children, 15 can play tennis and 19 can swim, but 6 can neither play tennis nor swim. How many children can both swim and play tennis?

Solution. The question indicates that we shall need to discuss the sets $S = \{\text{swimmers}\}$ and $T = \{\text{tennis players}\}$, the letters S and T being chosen for easy identification. Since we shall also need to discuss non-swimmers, S', and non-tennis players, T', it follows that we need to discuss all 30 children in the class as our universal set U.

The Venn diagram of Fig. 1.17 is therefore drawn with S and T overlapping within the region U so as to admit that there may be some members of U who can both swim and play tennis.

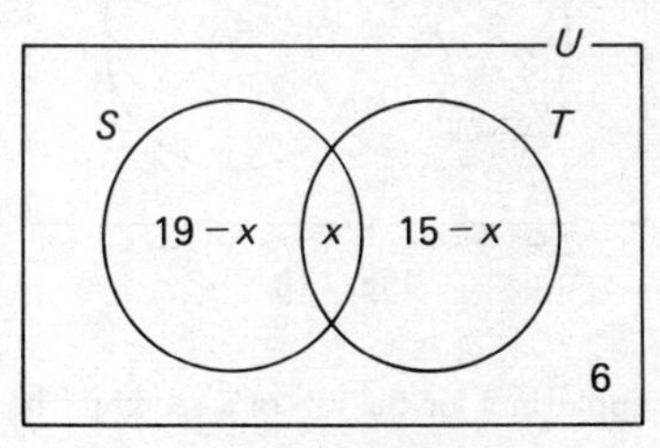

Fig. 1.17

Having designed the Venn diagram we now insert some algebraic details. Let $x = n(S \cap T)$ be inserted on the n(set) Venn diagram. From this we are now able to insert $19 - x$ to represent the number who can swim but not play tennis. Similarly, we have $15 - x$ to represent the number who can play tennis but cannot swim. Finally, we insert 6 to indicate the number who cannot do either sport. This completely describes how we obtain the entries in Fig. 1.17.

Since the total number of elements in U is 30 we now have the algebraic equation:

$$\begin{aligned} 19 - x + x + 15 - x + 6 &= 30 \\ 40 - x &= 30 \\ x &= 10 \end{aligned}$$

i.e. 10 children can swim and play tennis. *Answer*

Now some readers may be able to see immediately that there are only 24 children who can play tennis or swim so that the result

$$19 + 15 = 34$$

indicates that 10 children have been counted twice and this in turn implies that there are 10 children who can both swim and play tennis. In the language of sets we have,

$$n(S \cup T) = n(S) + n(T) - n(S \cap T)$$

and the above remarks amount to

$$24 = 19 + 15 - x \text{ and hence } n(S \cap T) = 10$$ *Answer*

Example 24. Someone has suggested that in a class of 20 children, 6 are good at mathematics and 4 are good at physics while 2 are no good at either subject. Find the number of children who are good at both.

Solution. We shall make the class of 20 children our universal set U. Again we use appropriate lettering by suggesting that

$$M = \{\text{children good at Mathematics}\}, P = \{\text{children good at Physics}\}$$

In drawing the Venn diagram of Fig. 1.18 we must overlap the sets M and P in order to admit that there might be some children who really are good at both subjects.

Since the total number of elements in U is 20 we now have the algebraic equation:

$$6 - x + x + 4 - x + 2 = 20$$

i.e.

$$12 - x = 20$$

which shows that we cannot find a positive value for x to satisfy this equation. We therefore conclude that the numerical content of the suggestion in the question is inconsistent.

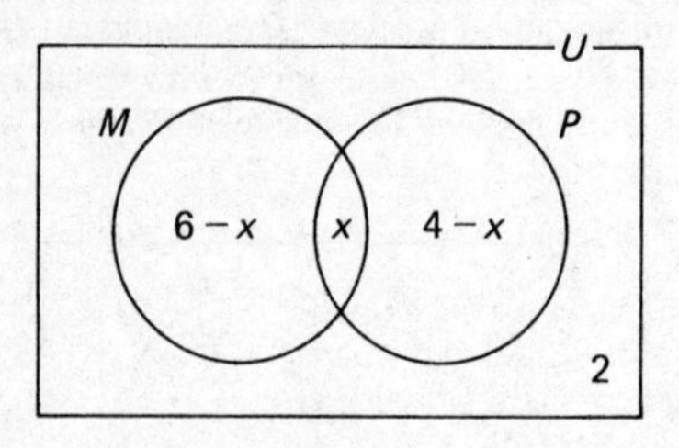

Fig. 1.18

Example 25. Each of 50 applicants for the job of a secretary has been asked to answer only two questions:

(i) 'Can you type at 100 wpm?' (35 replied yes.)
(ii) 'If the answer to (i) is no, can you take shorthand at 100 wpm?' (11 replied yes.)

Show that if the only applicants suitable must be able to type and take shorthand at 100 wpm then these two questions were a waste of time.

Solution. Let T = {applicants who can type at 100 wpm} and S = {applicants who can take shorthand at 100 wpm}, so that with U = {applicants} the Venn diagram of Fig. 1.19 can be used to represent the problem.

We have again suggested that we let $x = n(S \cap T)$ and since $n(T) = 35$ it follows that $35 - x$ of the applicants can type but not take shorthand at the required speed.

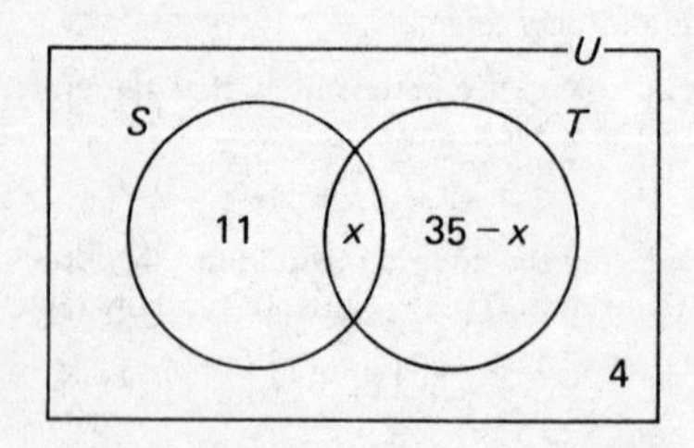

Fig. 1.19

The equation given by the entries on the diagram is

$$11 + x + 35 - x + 4 = 50$$

which reduces to

$$50 = 50$$

which is true for all values of x. In other words $x = 0, 1, 2, 3, \ldots, 35$ will satisfy the requirements of the Venn diagram but does not solve the problem. The conclusion is that x is indeterminate due to insufficient data.

Obviously it would have been more sensible to ask the question, 'Can you type and take shorthand at 100 wpm?' in the first place. *Answer*

The technique of mixing algebra with n(set) Venn diagrams is practically the same for problems involving three sets. For more than three sets the diagrams become much more difficult to master.

Example 26. A random sample of 100 adults was asked if they regularly bought a morning paper each weekday, an evening paper each weekday or a Sunday paper and the results were tabulated as follows below. The numerical statements represent the result in set notation, where M = {buyers of a weekday morning paper}, E = {buyers of a weekday evening paper}, S = {buyers of a Sunday paper}:

(i) 60 bought a morning paper— $n(M) = 60$
(ii) 23 bought an evening paper— $n(E) = 23$
(iii) 62 bought a Sunday paper— $n(S) = 62$
(iv) 8 bought a morning and an evening paper— $n(M \cap E) = 8$
(v) 45 bought a morning and a Sunday paper— $n(M \cap S) = 45$
(vi) 5 bought an evening and a Sunday paper— $n(E \cap S) = 5$
(vii) 2 bought all three papers— $n(M \cap E \cap S) = 2$

Find the number of adults who buy (a) one and only one paper, (b) no paper at all.

Solution. Having asked so many questions the reader might be correct in thinking that two more would not have done any harm. We are, however, making eight questions or eight pieces of information do the work of ten. Questionnaires tend to get very tiresome to fill in so it is a good idea to practise asking the minimum number of questions to gain the maximum amount of information.

We must start with the most general Venn diagram for three sets, like Fig. 1.15 (page 18) in which each set is drawn to overlap the other two, in order to admit the possibility that each of the eight regions may contain some elements.

The first piece of information to deal with is item (vii) because this involves all three sets so that in conjunction with items (iv), (v), (vi) we shall be able to find the number of elements in four of the eight regions at once. Thus we see that we need 3 elements in the adjacent region to make up the required number of elements for $n(E \cap S)$. Similarly, we need 43 + 2 to make up the number of elements in $n(M \cap S)$. In this way we gradually build up the information content of the diagram.

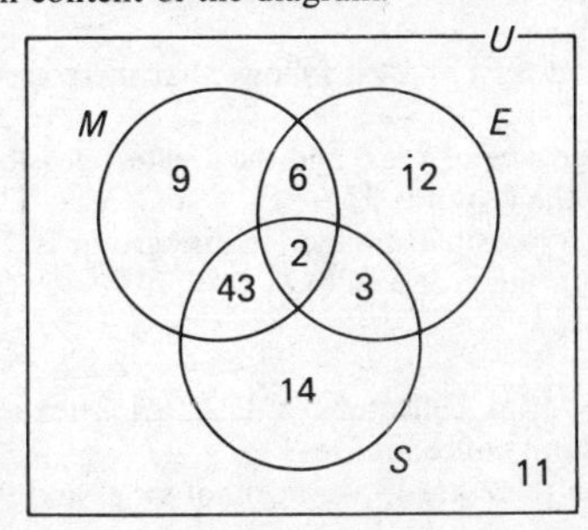

Fig. 1.20

For example, we now have 43 + 2 + 3 elements for S and we know that $n(S) = 62$. Therefore with $62 - 43 - 2 - 3 = 14$ we now complete the insertions for all the regions of the set S. The results of 9 and 12 are similarly obtained on Fig. 1.20.

We now observe by counting from the diagram that

$$n(M \cup E \cup S) = 9 + 6 + 12 + 43 + 2 + 3 + 14 = 89$$

and since there were 100 adults in the survey it follows that $100 - 89 = 11$ of them do not buy any of the papers.

Our answers are therefore:

(*a*) $9 + 12 + 14 = 35$ adults regularly buy one and only one paper.
(*b*) 11 adults do not buy a paper. *Answer*

Example 27. (*a*) A group of army recruits were examined for bad eyesight, poor hearing and flat feet. From the whole group, 37 passed for eyesight, 33 passed for hearing and 24 passed for feet. Given that only 5 recruits failed all three tests find the least and greatest possible number of recruits in the group.

(*b*) Find the least and greatest possible numbers in the group when the following facts are added to those already known in part (*a*). $n(E \cap H) = 18$, $n(E \cap F) = 11$ where $E = \{\text{recruits passing eyesight test}\}$, $F = \{\text{recruits passing feet test}\}$, $H = \{\text{recruits passing hearing test}\}$.

Solution. (*a*) The least possible number in the group is found by suggesting that the largest set of passes contains all the other passes, that is F and H are both subsets of E. Hence the least possible number in the group is $37 + 5 = 42$.

The greatest possible number in the group is found by suggesting that no recruit passes more than one test and so the greatest possible number in the group is $37 + 33 + 24 + 5 = 99$.

(*b*) Consider the Venn diagram of Fig. 1.21 to represent the problem with the regions named as for Fig. 1.15. Allow x elements in the region common to F and H outside region R_7, and in R_6.

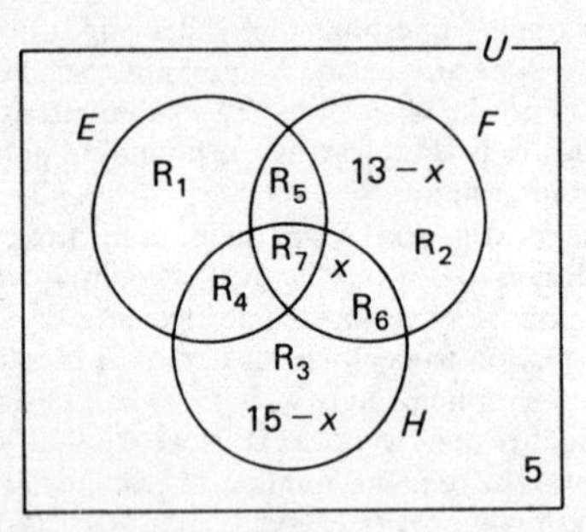

Fig. 1.21

With $n(E \cap H) = 18$ and $n(H) = 33$ it follows that there are $15 - x$ elements in the region R_3.

With $n(E \cap F) = 11$ and $n(F) = 24$ it follows that there are $13 - x$ elements in the region R_2.

Hence the least possible value of x is 0 and the greatest possible value of x is 13. Since the number of recruits in the group is $37 + 15 - x + x + 13 - x + 5 = 70 - x$ it now follows that the least possible number in the group is $70 - 13 = 57$ and the greatest possible number in the group is $70 - 0 = 70$. *Answer*

Exercise 1.9

1. Thirty people regularly drink coffee or tea. If 23 drink tea and 11 drink coffee, how many regularly drink both coffee and tea?
2. In a class of 21 children there are 13 who cannot swim and 9 who can play tennis. If

11 of the children cannot swim or play tennis, how many children can swim but not play tennis?

3. In the list of 60 examination candidates, 48 have marks of and above 50% and 27 have marks of and below 60%. How many candidates have obtained marks between 60% and 50% inclusive?
4. In a set of 100 people, 63 have blue eyes and 75 have fair hair. What is the greatest and least possible number of fair-haired people with blue eyes to satisfy this information?
5. In a party of 25 guests only 1 guest cannot eat cheese. There are only 10 males in the party. How many males can eat cheese? (There are two solutions.)
6. Three newspapers A, B and C are published in a particular town. A survey of the residents in the town indicated the following: paper A is read by 20%, paper B is read by 16%, paper C is read by 18%; 7% read papers A and B, 3% read papers B and C, 5% read papers C and A; only 1% read A, B and C.

   Find the percentage of the residents who read (a) none of the three papers, (b) one and only one of the three papers.
7. After an inspection of 100 soldiers it was found that everyone of them had to have a haircut, or a visit to the dentist or new boots. The information is tabulated as follows in the obvious notation:

   $$n(H) = 73,\ n(D) = 28,\ n(B) = 21,\ n(H \cap D) = 12,\ n(D \cap B) = 9,\ n(B \cap H) = 8.$$

   Find (i) $n(H \cap D \cap B)$ and (ii) $n(H' \cap D \cap B)$.
8. Complaints about a canteen fall into three types concerning the cooking, the menu and the service. A summary of the 56 complaints received is as follows:

   $$n(C) = 36,\ n(M) = 42,\ n(S) = 11,\ n(C \cap M) = 25,\ n(M \cap S) = 7,\ n(S \cap C) = 6.$$

   Find (i) $n(C \cap M \cap S)$ and (ii) $n(C' \cap M' \cap S)$.

2

ELEMENTARY MATHEMATICAL LOGIC

In general we use two different styles of reasoning in mathematics—namely, inductive reasoning and deductive reasoning.

If we reach a conclusion based on our own observations or experiences then we are using **inductive reasoning.** For example, having noticed that each time we add two odd numbers the result is an even number we conclude, inductively, that this may always be true, but of course we have not really proved this result for all possible pairs of odd numbers. **Deductive reasoning** starts with assumptions and then reaches a conclusion based on those assumptions. In short, we generally discover a mathematical result by inductive reasoning but we prove it by deductive reasoning.

Inductive Reasoning

Inductive reasoning is the creative part of mathematical activity and it is usually the first step towards finding out what might be worth proving. We have already referred to the example of the addition of two odd numbers always resulting in an even number, so let us pursue this in a little more detail.

We experiment by listing a few results:

$$1 + 3 = 4$$
$$3 + 5 = 8$$
$$5 + 7 = 12$$
$$7 + 9 = 16$$
$$9 + 11 = 20 \quad \text{and so on.}$$

Now, thinking inductively about the above list of results we might believe that the result of adding two odd numbers was always a multiple of 4. Are we entitled to suggest that this result is always true? Not on the basis of these five observations but then how many observations must be made before concluding inductively that this result must be true: 100, 1000?—it will have to be a matter of experience which eventually decides. Interestingly, the thousandth example in the above pattern of examples would be

$$1999 + 2001 = 4000$$

so are we now able to say that two odd numbers have a sum which is a multiple of 4? The answer is 'no' because we have not varied the choice of numbers sufficiently. Indeed, as soon as we test

$$7 + 11 = 18$$

we withdraw the original conclusion, and yet we could have tried over 1000 examples before experimenting with $7 + 11 = 18$.

The uncertainty lies in not knowing whether we have explored sufficient possibilities to justify our conclusions. This is the weakness of inductive reasoning; it usually lacks generality, i.e. the assurance of having dealt with every eventuality.

As another example consider the Goldbach conjecture that 'every even number greater than 4 may be expressed as the sum of two odd prime numbers'. We start by experimenting with a few examples.

$$\begin{aligned} 6 &= 3 + 3 \\ 8 &= 3 + 5 \\ 10 &= 3 + 7 \\ 12 &= 5 + 7 \\ 14 &= 3 + 11 \\ 16 &= 3 + 13 \\ 18 &= 5 + 13 \quad \text{and so on.} \end{aligned}$$

No one has ever found an even number greater than 4 which cannot be expressed as the sum of two odd prime numbers but there again no one has actually proved the result. Our inductive reasoning says that the result is true but as our previous example warns us, if we have not exhausted all possible examples then how can we be sure?

We have already met this line of inductive reasoning when suggesting the sequence of terms in the infinite sets on page 5. Recall that on the evidence of a few terms we speculated on those which followed.

Exercise 2.1

1. Write down the next line and the tenth line in each of the following lists. If possible describe what is being represented.

(i) $1 + 3$
$1 + 3 + 5$
$1 + 3 + 5 + 7$

(ii) $1 + 1 \times 3$
$1 + 2 \times 4$
$1 + 3 \times 5$

(iii) $5^2 - 4^2$
$13^2 - 12^2$
$25^2 - 24^2$
$41^2 - 40^2$

(iv) $5^2 - 3^2$
$10^2 - 8^2$
$17^2 - 15^2$
$26^2 - 24^2$

(v) $1 \rightarrow 2$
$2 \rightarrow 4$
$3 \rightarrow 8$

2. Find the value of $\frac{n}{2}(n + 1)$ when $n = 1, 2, 3, 4$ and relate the results to the following four lines:

(i) 1
(ii) $1 + 2$
(iii) $1 + 2 + 3$
(iv) $1 + 2 + 3 + 4$

Find the value of the 200th line: $1 + 2 + 3 + \ldots + 199 + 200$.

3. Examine the following four lines, compare them with the results in Question 2 and attempt to find the formula which gives the sum of the numbers in each line:

(i) 1
(ii) $1^3 + 2^3$
(iii) $1^3 + 2^3 + 3^3$
(iv) $1^3 + 2^3 + 3^3 + 4^3$

Find the value of the 20th line; $1 + 2^3 + 3^3 + \ldots + 20^3$.

4. Examine the value of $\frac{n}{6}(n + 1)(2n + 1)$ for $n = 1, 2, 3, 4$ and speculate on what this formula might represent.

5. Discover whether or not the expression $n^2 - n + 41$ always represents a prime number for n equal to any whole number (i.e. put $n = 0, 1, 2, 3, \ldots$ in $n^2 - n + 41$ and see if the result is a prime number).
6. Investigate the value of $6n - 1$ and $6n + 1$ for n equal to any whole number. Can you make a conclusion?

Deductive Reasoning

We have noted that with inductive reasoning we may reach a conclusion which is very plausible but still unproved because it has been impossible to test that conclusion for all possible cases. Even when our conclusion is true for a considerable number of examples this will not constitute a proof unless these examples have exhausted all possible cases. For instance, there are a million examples to support the claim that 1 000 001 is the largest of the positive integers since there are a million integers in between 0 and 1 000 001 which are smaller than 1 000 001, but we should not mislead ourselves and claim that 1 000 001 is the biggest integer for we know a greater integer, namely 1 000 002. Thus, even though we had a million examples to support the claim this was still not enough to prove the claim. To illustrate the point of being able to exhaust all possibilities consider the following example.

Example 1. Prove that the number 359 873 cannot be a perfect square.

Solution. Any positive integer must end in one of the digits

$$0, 1, 2, 3, 4, 5, 6, 7, 8, 9 \qquad \text{(i)}$$

When such a number is squared, the result will end in one of the digits

$$0, 1, 4, 9, 6, 5, 6, 9, 4, 1 \qquad \text{(ii)}$$

Hence, no squared number ends in 3 and so 359 873 cannot be a perfect square. Because we exhausted all possibilities in reaching the conclusion, the result has been proved.

Answer

The above is an example of deductive reasoning—that is, the reaching of a conclusion based on known or assumed facts. The known facts were that *all* integers end in a digit from 0 to 9 inclusive. From these facts we deduced that *all* squared integers must end in a digit from $\{0, 1, 4, 5, 6, 9\}$. Hence, we could have come to the stronger conclusion that positive integers which end in 2, 3, 7, 8 cannot be perfect squares. (Do note that this does *not* say that a number which ends in 4 (say) *is* a perfect square—e.g. 14, 24, 44 are not perfect squares.)

Certain statements can be proved by counter-example. For example, the statement that 'all prime numbers are odd numbers' can be proved to be false by merely offering one counter-example—namely, that 2 is a prime number. We have shown that 'not all prime numbers are odd numbers' by this one example.

But what is a proof? Well certainly it must be a convincing argument but unfortunately some people are more easily convinced than others so we merely shift to the need to describe 'convincing'. The only scheme we are able to approve in the manner of deductive proof is that we must start by assuming some facts which are relevant to the problem and on the basis of these facts we reach a conclusion by argued steps which are consistent with known mathematical ideas. In other words, we assume a body of knowledge, which we call **A** for short, and deduce a conclusion which we call **B** for short. The whole argument is described in shorthand by the statement

'If A then B'

Sometimes the assumed knowledge is not explicitly stated—that is, certain aspects of background knowledge are assumed without acknowledgement and although the deductive arguments are sound and convincing the proof may well be regarded as informal. When all the assumed knowledge is explicitly stated then the proof is described as formal.

Example 2. Give an informal proof that two odd numbers always have an even sum.

Solution. Any even number can be represented by a rectangle of width 2 but any length such as Fig. 2.1.

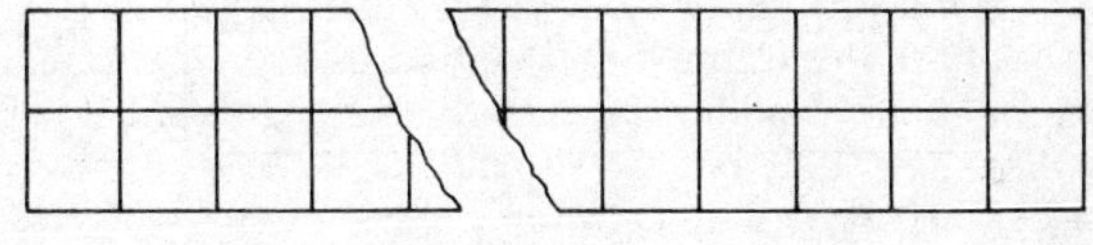

Fig. 2.1

Any odd number can be represented by a rectangle of width 2 plus an odd square such as Figs. 2.2 and 2.3.

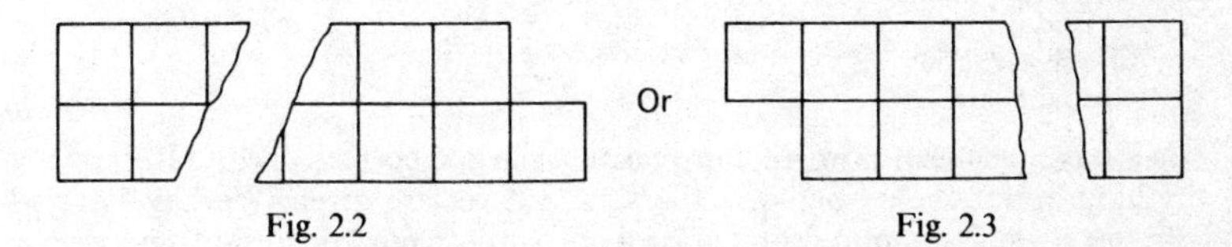

Fig. 2.2 Fig. 2.3

The sum of two odd numbers is therefore given by Fig. 2.4, which makes a rectangle of width 2, thus representing an even number. Hence two odd numbers have an even sum.
Answer

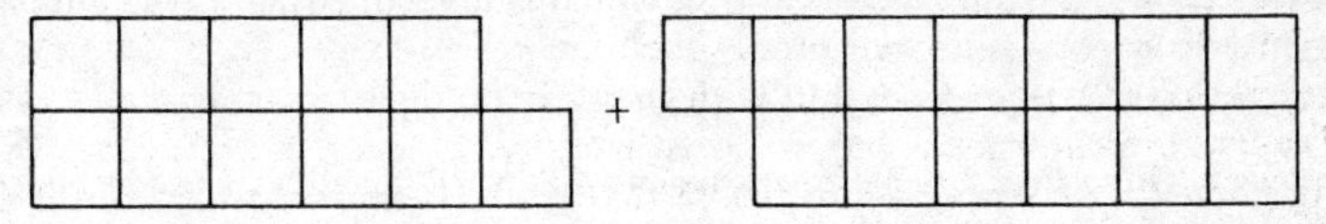

Fig. 2.4

Example 3. Give a formal proof that two odd numbers always have an even sum.

Solution. Let n and k be any two whole numbers so that $2n + 1$ and $2k + 1$ are any two odd numbers. Suppose that the sum of these two odd numbers is S.

$$\begin{aligned} \therefore \quad S &= 2n + 1 + 2k + 1 \\ &= 2n + 2k + 2 \\ &= 2(n + k + 1) \end{aligned}$$

Hence S is divisible by 2 for all values of n and k and so it follows that S is an even number.
Answer

One of the many advantages about formal deductive proofs is that we can reach correct conclusions in a subject using elements of which we know nothing, always provided that we are familiar with the mathematical steps which lead from the assumptions to the conclusions. Consider this situation in the following example.

Example 4. In the land of the Shandufal, the relations displayed in a Venn diagram are the same as in this book. Several of the spar diagrams used in the Shan culture are interrelated. Thus no quid is a trid but every pard and every zoid is a quid although some zoids are not pards. Given that every elch is a trid deduce formally that no zoid is an elch.

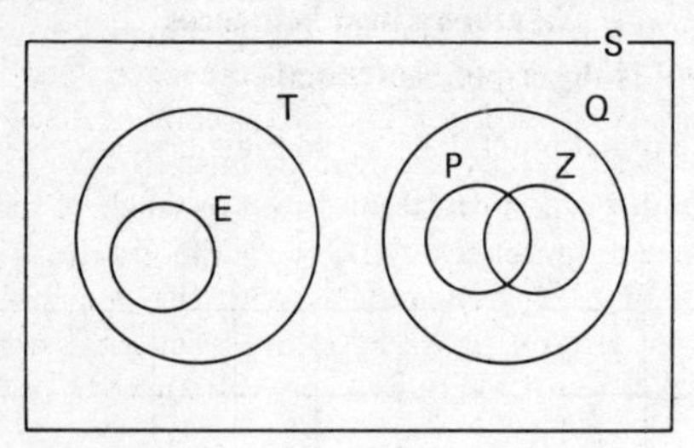

Fig. 2.5

Solution. The Venn diagram which registers the given information is in Fig. 2.5, and we can use this to guide our thoughts to the conclusion. Thus:

All zoids are quids
No quid is a trid
$\therefore$ No zoid is a trid
All elchs are trids
$\therefore$ No zoid is an elch

Answer

We have just seen how we can reason or argue correctly without knowing anything about the elements in the sets that we are given. We have started with the given assumptions and argued correctly to our conclusion, and we say that the argument is valid whether or not the assumptions or the conclusions make any sense. Thus, the validity (i.e. the correctness) of an argument has nothing to do with the truth value of the assumption. For example, it is possible to argue from an assumption which is false to either a true or a false conclusion using a valid argument.

Let us look at this idea again with the next example.

Example 5. Given that $a = b$ we may assume that equal quantities x can be added to both sides and the statement reads

$$a + x = b + x$$

Prove that if $3 = 8$ then $10 = 15$.

Solution. If $3 = 8$, then $3 + 7 = 8 + 7$

$$\text{i.e.} \quad 10 = 15$$

We have a false assumption leading to a false conclusion via a valid argument.

Answer

Exercise 2.2

1. Give an informal and a formal proof that if each of two positive numbers is a multiple of 3 then so is their sum.
2. Prove that in the game of noughts and crosses with both players playing to win, if the first player starts by occupying the centre square then he cannot lose.

3. Using the information of example 4 above prove that: (i) no elch is a quid. (ii) no elch is a pard.
4. Two integers m and n do not have a common factor. Prove that it is not possible to have $2m^2 = n^2$. (Use the type of approach of example 1 above.)

Mathematical Sentences

We describe ideas with the aid of sentences. In everyday events and language we use sentences like 'It is raining', 'The bus is late' and sometimes we might say these sentences whether they are true or false. Some sentences like 'How old are you?' are neither true nor false.

In mathematics we use sentences like 'Five is the greatest number in the set', 'Thirty is four more than twenty' and, as with the previous examples, these may be true or false. Again it is possible to give a sentence which is neither true nor false—for example, 'How small is x?' or 'A number plus five is equal to seven'.

We also use symbols to achieve brevity and in the hope of gaining clarity, so that 'Thirty is four more than twenty' would be written as

$$30 - 4 = 20$$

and 'A number plus five is equal to seven' would be written as

$$x + 5 = 7$$

each of these abbreviations being a mathematical sentence. In order to classify sentences we now define a statement.

Definition 2.1. A statement is a sentence which can be classified as true or false but not both. (Notice that the exclusive 'or' is being used here—see page 15.)

This assumption that statements are either true or false, but not both, is referred to as the 'law of the excluded middle' and a good description of our intended use of statements is that we are going to work with a two-state logic.

The following sentences are statements and referred to as either T = True or F = False.

(i) This is on page 33 (T)
(ii) London is in England (T)
(iii) $3 + 7 = 11$ (F)
(iv) $9 \times 4 = 36$ (T)

The following sentences are not statements.

(i) 'What is the time?'
(ii) 'Who is coming for a walk?'
(iii) $x - 2 = 7$
(iv) $2x = x + 3$

We shall avoid meaningless statements like '4.6 = a circle'. In sentence (iii) $x - 2 = 7$ above, the letter x does not represent any specified number. Substitution of 8 for x creates a false statement whereas substitution of $x = 9$ creates a true statement so that with x unspecified the sentence $x - 2 = 7$ is neither true nor false and therefore cannot be a statement, but unlike (i) or (ii) it is a sentence which can become a statement after a suitable substitution.

Recall that when we say 'Solve the equation $x - 2 = 7$' we mean find all the values of x which make a sentence $x - 2 = 7$ true.

Again, the sentence $x^2 = 9$ is not a statement but when we say 'Solve the equation $x^2 = 9$' we find $x = +3$ and -3 so as to make the sentence true. A letter used in this way to represent a variety of numbers is called a **variable.**

A comparable example in everyday language is 'She is the captain of the team' since the sentence cannot be classified as true or false until we have said who 'she' is.

Returning to our sentence $x - 2 = 7$, we observe that the symbols 2 and 7 represent particular numbers which will remain unchanged no matter what value is suggested for the variable x; consequently it is appropriate to call these symbols **constants.** In algebra we usually make letters like a, b, c at the beginning of the alphabet represent constants and letters like x, y, z at the end of the alphabet represent variables.

Replacement and Solution Sets

Asked to solve the equation $x^3 - 7x^2 + 14x - 8 = 0$, you would need some skill in order to start the search for all the values of the variable x which make the sentence true. We can make the task easier and more appropriate for this book by restricting the numbers with which we may replace x. The question would therefore be reworded as suggested in Example 6, below, in which the values of x would be restricted to being members of what we call a **replacement set.**

Any replacement which makes a sentence true is called a **solution** and the set of all solutions that we get from the replacement set is called its **solution set.** Obviously, if the replacement set does not include a solution of the equation then the solution set will be ϕ.

Example 6. Use the replacement set $\{0, \frac{1}{2}, 1, 1\frac{1}{2}\}$ to solve the equation

$$2x^2 + 3x - 2 = 0$$

Solution. As instructed by the question, we replace x by 0, $\frac{1}{2}$, 1 and $1\frac{1}{2}$ in turn.

(i) $x = 0$; $2x^2 + 3x - 2 \neq 0$, so $x = 0$ is not a solution.
(ii) $x = \frac{1}{2}$; $2x^2 + 3x - 2 = \frac{2}{4} + 1\frac{1}{2} - 2 = 0$, so $x = \frac{1}{2}$ is a solution.
(iii) $x = 1$; $2x^2 + 3x - 2 \neq 0$, so $x = 1$ is not a solution.
(iv) $x = 1\frac{1}{2}$; $2x^2 + 3x - 2 = \frac{9}{2} + \frac{9}{2} - 2 \neq 0$, so $x = 1\frac{1}{2}$ is not a solution.

We have exhausted the replacement set and we have found the solution set to be $\{\frac{1}{2}\}$.

We may note that $x = -2$ is also a solution but this is not acceptable under the conditions of the example because -2 is not an element of the replacement set. Thus, solution sets are dependent upon the given replacement set. *Answer*

Example 7. Use the replacement set $\{0, 1, 2, 3, 4\}$ to solve the equation

$$x^3 - 7x^2 + 14x - 8 = 0$$

Solution. We do just what the question says—namely, replace x by 0, 1, 2, 3 and 4 in turn.

(i) $x = 0$; $x^3 - 7x^2 + 14x - 8 = -8 \neq 0$, so $x = 0$ is not a solution.
(ii) $x = 1$, $x^3 - 7x^2 + 14x - 8 = 1 - 7 + 14 - 8 = 0$, so $x = 1$ is a solution.
(iii) $x = 2$; $x^3 - 7x^2 + 14x - 8 = 8 - 28 + 28 - 8 = 0$, so $x = 2$ is a solution.
(iv) $x = 3$; $x^3 - 7x^2 + 14x - 8 = 27 - 63 + 42 - 8 \neq 0$, so $x = 3$ is not a solution.
(v) $x = 4$; $x^3 - 7x^2 + 14x - 8 = 64 - 112 + 56 - 8 = 0$, so $x = 4$ is a solution.

We have used all the elements of the given replacement set and we have found the solution set to be $\{1, 2, 4\}$. *Answer*

Exercise 2.3

Say which of the following are statements and say whether they are either true (T) or false (F):

1. $4 + 3 = 7$
2. $4 + 4 = 9$
3. $x + 4 = 9$
4. $7 < 13$
5. $7 \nless 13$
6. $x \nless 13$
7. How old are you?
8. How much is x?
9. 9933 is a perfect square.
10. 1047 is a not perfect square.

Using the replacement set $\{-4, -3, -2, -1, 0, 1, 2, 3, 4\}$ obtain the solution sets for the following equations:

11. $x^2 = 16$
12. $x + 9 = 8$
13. $x^2 + x + 1 = 0$
14. $x^2 = 2x + 8$
15. $x^3 = x^2$

Compound Statements

In the previous section we discussed what may be called simple statements. In order to produce a deductive argument for a proof we shall need a combination of simple statements.

In deductive reasoning it is usual to join statements with words like 'and', 'or', 'if', 'then' called **connectives**. Each time this is done we produce a compound statement. In everyday language compound statements could be:

'It is sunny and the pond is warm.'
'If I do not do overtime tonight then I will go to the cinema or to the club.'
'If all the sides are equal in length then each angle must be sixty degrees.'
'If $9 + 3 = 17$ then I have made a mistake somewhere.'

In deductive reasoning there are well-established ways of forming compound statements and, furthermore, for mathematical purposes, the method of deciding whether or not the resulting compound statement is true or false is also well established.

Conjunctions

When two or more statements are joined by the word 'and' a compound statement is formed which is called the **conjunction** of the statements. The various shades of meaning which everyday language gives to some combinations of words will sometimes enable a conjunction to be suggested without using the word 'and'. It is possible to have a conjunction like the following:

(i) 'All footballers are ball players but some ball players are not footballers.'

In this example (i) we see that the conjunction of two statements can be suggested without the presence of 'and'. Here we have used 'but', which in turn could have been replaced by 'whereas'.

(ii) 'In the long-distance endurance test Rolls-Royces, Rovers and Jaguars came first, second and third respectively.'

In example (ii) the conjunction of three statements is disguised by the use of commas and the word 'respectively'. The meaning intended is briefly: 'Rolls-Royces came first, *and* Rovers came second, *and* Jaguars came third'.

(iii) 'Bill is my big brother.', i.e. 'Bill is big and Bill is my brother' is intended (we think!).

We now need to suggest how a conjunction becomes true or false according to its component statements being true or false.

Naturally, it would be helpful if the definitions complied with the unambiguous use of that everyday language which most of us use to convey our ideas, but it must be understood that the definitions of mathematical logic are more concerned with removing ambiguities from the language of mathematics. Fortunately, as far as the word 'and' is concerned there will be very little conflict although the conjunctions (i), (ii) and (iii) above indicate that some skill in linguistic interpretation may be required.

To speak of a conjunction in general it is appropriate to refer to one statement as x and another statement as y in describing a conjunction as being any statement of the type

'x and y'

thereafter insisting that the truth or falsity of the conjunction be established by the following definition:

Definition 2.2. The conjunction 'x and y' is true if both x and y are true; otherwise the conjunction is false.

We shall simplify the writing of the conjunction by using the symbol $\wedge$ in order to replace 'x and y' by $x \wedge y$. We note that $x \wedge y = y \wedge x$.

Since x, y only have two states, either true (T) or false (F), we can conveniently list all four possible combinations in what will be called a truth table, i.e. a table which shows what makes the conjunction either true or false. The last column in Table 2.1 is called the set of truth values of $x \wedge y$.

x	y	$x \wedge y$
F	F	F
F	T	F
T	F	F
T	T	T

Table 2.1

The use of 'and' as defined here corresponds very closely to ordinary usage of the word 'and'. For example, we can match the row entries of the table with conjunctions like:

(i) '$3 + 6 = 7$ (F) and $8 \times 9 = 4$ (F)' is false (F)
(ii) '$3 + 6 = 7$ (F) and $3 \times 4 = 12$ (T)' is false (F)
(iii) '$3 \times 4 = 12$ (T) and $3 + 6 = 7$ (F)' is false (F)
(iv) '$3 \times 4 = 12$ (T) and $4 \times 9 = 36$ (T)' is true (T)

The truth table for $\wedge$ is so convenient that we sometimes use it to define the conjunction rather than the written description given by Definition 2.2.

Disjunctions

When two or more statements are joined by the word 'or' we form a compound sentence called a **disjunction** and, as in the case of sets (page 15), we use the inclusive 'or' which leads to the next definition.

Definition 2.3. The disjunction 'x or y' is false if both x and y are false;

otherwise the disjunction is true. We shall simplify the writing of the disjunction by using the symbol $\vee$ in order to replace 'x or y' by '$x \vee y$'. We note that $x \vee y = y \vee x$, and that the inclusive 'or' is used in this definition.

In subsequent writing the exclusive 'or' will be indicated by the use of the two words 'either ... or'.

The truth table for a disjunction $x \vee y$ is given in Table 2.2. This may be used to define $x \vee y$ instead of the written definition 2.3. The last column in Table 2.2. is called the set of truth values of $x \vee y$.

x	y	$x \vee y$
F	F	F
F	T	T
T	F	T
T	T	T

Table 2.2

When comparing the two truth tables (2.1, 2.2) above it is noticeable that for $x \wedge y$, (F) is dominant whereas for $x \vee y$, (T) is dominant. Clearly any long deductive argument will involve the combination of more than two statements. At the moment we have only used $\wedge$ and $\vee$ in a binary capacity exactly like the introduction of $\cap$ and $\cup$ in the previous chapter (pages 10, 14) and as we saw there it will be necessary to introduce brackets in order to deal with more than two statements at a time.

Example 8. Show that $(x \wedge y) \wedge z = x \wedge (y \wedge z)$.

Solution. When we use the $=$ sign in this context we mean that the truth values for $(x \wedge y) \wedge z$ are the same as for $x \wedge (y \wedge z)$. As seen in the definition, $x \wedge y$ is true if both x and y are true; otherwise $x \wedge y$ is false.

Treating $x \wedge y$ as the one bracketed statement $(x \wedge y)$ we can say that:

$(x \wedge y) \wedge z$ is true if both $(x \wedge y)$ and z are true, otherwise $(x \wedge y) \wedge z$ is false.

$\therefore$ $(x \wedge y) \wedge z$ is true if all three x, y and z are true, otherwise $(x \wedge y) \wedge z$ is false.

With similar reasoning we deduce that $x \wedge (y \wedge z)$ is true if all three x, y and z are true; otherwise $x \wedge (y \wedge z)$ is false.

Hence $(x \wedge y) \wedge z = x \wedge (y \wedge z)$ and we may write $x \wedge y \wedge z$ without ambiguity since it makes no difference where we put the brackets. This means that the operation $\wedge$ obeys the associative law (page 17).

We can see this result in a more familiar context by considering the following illustrations:

For a number to be divisible by 6 it must be divisible by both 2 and 3.
Let w represent the statement '2002 is divisible by 2' (T).
Let x represent the statement '2002 is divisible by 3' (F).

Here we see that w is true, x is false and so $w \wedge x$ is false, i.e. 2002 is not divisible by 6.

Continuing with this illustration:

Let y represent the statement '2002 is divisible by 7' (T).
Let z represent the statement '2002 is divisible by 11' (T).

Here we see that $y \wedge z$ is true, and we can go on to note that $w \wedge (y \wedge z)$ is true. In other words, 2002 is divisible by 2 and 7 and 11, i.e. 2002 is divisible by 154.

Again, $p \wedge x$ is false no matter what statement is represented by p. For example, suppose p represents the statement '2002 is greater than 2000' or '2002 is a negative integer'. Such is the consequence of definition 2.2.

As we increase the number of operations and statements so we are able to generate an increasing variety of more involved compound statements. We now call the compound statement a **proposition** and refer to its individual substatements as the **variables of the proposition**.

The truth value of a proposition depends upon the truth values of its variables so that all we have to do is construct a truth table for the proposition which exhausts all the different combinations of the truth values of its variables.

A proposition involving three variables will require $2^3 = 8$ rows for its truth table and likewise 16, 32, 64, 128 rows to register the different combinations for 4, 5, 6 or 7 variables respectively. Although foolproof, a truth table is not without tedium.

When a proposition is always true for every combination of values of its variables it is called a **tautology**.

When the proposition is likewise always false it is called a **contradiction**.

In the next example we consider a proposition which contains both the connectives $\wedge$ and $\vee$.

Example 9. Obtain the set of truth values for the proposition

$$(x \wedge y) \vee (z \wedge y)$$

Solution. For three variables we need $2^3 = 8$ rows in the truth table 2.3. Notice how the variable values are listed. For x in the first column we write 4 F's followed by 4 T's; for y in the second column we write 2 F's followed by 2 T's; for z in the third column we write 1 F followed by 1 T.

x	y	z	$x \wedge y$	$z \wedge y$	$(x \wedge y) \vee (z \wedge y)$	$x \vee z$	$(x \vee z) \wedge y$
F	F	F	F	F	F		
F	F	T	F	F	F		
F	T	F	F	F	F		
F	T	T	F	T	T		
T	F	F	F	F	F		
T	F	T	F	F	F		
T	T	F	T	F	T		
T	T	T	T	T	T		

Table 2.3

The table is compiled column by column working from left to right until the set of truth values of the proposition $(x \wedge y) \vee (z \wedge y)$ is completed in column six. (Columns seven and eight are left for the reader—see Question 11, Exercise 2.4.)

Exercise 2.4

In each of the first eight questions form the conjunction and the disjunction of the given statements and give the truth values of each result:

1. x: $9 < 13$
   y: $13 < 24$

2. x: 77 is divisible by 7
   y: 77 is divisible by 11
3. x: A circle is not a square
   y: A circle is a square
4. x: $4 < 0$
   y: $4 = 0$
   z: $0 < 4$
5. x: $a < 0$
   y: $a = 0$
   z: $0 < a$
6. x: $a < 5$
   y: $a = 5$
   z: $5 < a$
7. x: $a = 1$ is a solution to $a^2 = 1$
   y: $a = -1$ is a solution to $a^2 = 1$
8. x: $a = 2$ is a solution to $(a - 2)(a - 3)(a - 4) = 0$
   y: $a = 3$ is a solution to $(a - 2)(a - 3)(a - 4) = 0$
   z: $a = 4$ is a solution to $(a - 2)(a - 3)(a - 4) = 0$
9. Find the set of values of a which make $x \wedge y$ true:

   x: a is a solution of $(p - 1)(p - 2)(p - 3) = 0$
   y: $2 \leqslant a$
10. Find the set of values of k which makes $x \wedge y$ true:

    x: $k \leqslant 10$
    y: $5 \leqslant k$
11. Return to Table 2.3 and complete the seventh and eighth columns. Compare the results in columns six and eight and say what you deduce.
12. Complete a truth table similar to Table 2.3 to show that:

    (i) $x \vee (y \vee z) = (x \vee y) \vee z$
    (ii) $x \vee (y \wedge z) = (x \vee y) \wedge (x \vee z)$

Negation

A negation converts a true statement into a false statement and a false statement into a true statement. More formally we have the following definition.

Definition 2.4. A statement which always has a different truth value from x is called a negation of x.

The negation of x is represented by writing $\bar{x}$, which is read as either 'not x' or 'x bar'. A truth table for $\bar{x}$ is given in Table 2.4

x	$\bar{x}$	$\bar{\bar{x}}$
F	T	
T	F	

Table 2.4

Example 10. With x representing $3 + 4 = 9$ we can represent $\bar{x}$ by either of the following statements:

(i) $3 + 4 \neq 9$
(ii) It is false that $3 + 4 = 9$

Example 11. In the following three statements:

(i) The town of Craj is in Smov,
(ii) Craj is not in Smov,
(iii) It is false that Craj is in Smov

we note that (ii) and (iii) are each the negation of (i), so that if (i) is true then (ii) and (iii) are each false. Likewise if (i) is false then (ii) and (iii) are each true.

In considering the negation of a proposition we shall have to use brackets in order to gain the correct meaning. Thus $\bar{x} \vee \bar{y}$ is different from $\overline{x \vee y}$ since $\overline{x \vee y}$ means find $(x \vee y)$ first and then negate the result. Similarly, $\bar{x} \wedge \bar{y}$ is different from $\overline{x \wedge y}$ since $\overline{x \wedge y}$ means find $(x \wedge y)$ first and then negate the result.

We notice in particular that because $\bar{x}$ always has a different truth value from x it follows that

$x \wedge \bar{x}$ is always false

and $x \vee \bar{x}$ is always true

A double negation leaves the original statement unchanged, i.e. $\bar{\bar{x}} = x$ as can be confirmed by returning to Table 2.4 and filling in the column for $\bar{\bar{x}}$.

Example 12. Construct a truth table for

$$\overline{x \wedge y},\ \overline{x \vee y},\ \bar{x} \wedge \bar{y},\ \bar{x} \vee \bar{y}$$

and confirm DeMorgan's laws that

$$\text{(i) } \overline{x \wedge y} = \bar{x} \vee \bar{y}$$
$$\text{(ii) } \overline{x \vee y} = \bar{x} \wedge \bar{y}$$

Solution. Since there are only four different combinations for the truth values of x, y, the four-row truth table 2.5 will give all the required information.

x	y	$\bar{x}$	$\bar{y}$	$x \wedge y$	$\overline{x \wedge y}$	$x \vee y$	$\overline{x \vee y}$	$\bar{x} \wedge \bar{y}$	$\bar{x} \vee \bar{y}$
F	F	T	T	F	T	F	T	T	T
F	T	T	F	F	T	T	F	F	T
T	F	F	T	F	T	T	F	F	T
T	T	F	F	T	F	T	F	F	F

Table 2.5

Working from left to right each column is completed in turn using the relevant definitions which have been discussed so far. Comparing the columns for $\overline{x \wedge y}$ and $\bar{x} \vee \bar{y}$ we find that each have the same set of truth values, and we indicate this by writing

$$\overline{x \wedge y} = \bar{x} \vee \bar{y}$$

Similarly, by comparing the column for $\overline{x \vee y}$ and $\bar{x} \wedge \bar{y}$, we conclude that

$$\overline{x \vee y} = \bar{x} \wedge \bar{y}$$

Since we have deduced these results by examining all possible cases it follows that we have proved DeMorgan's laws with respect to any two statements x and y. *Answer*

Example 13. Find the truth value of the proposition $(x \vee y) \wedge (\overline{x \wedge y})$ and consider the possibility of a substitute proposition using the given column seven of Table 2.6.

Solution. Completion of the table now follows the usual column by column procedure.

x	y	$x \vee y$	$x \wedge y$	$\overline{x \wedge y}$	$(x \vee y) \wedge \overline{(x \wedge y)}$	$x \equiv y$
F	F	F	F	T	F	T
F	T	T	F	T	T	F
T	F	T	F	T	T	F
T	T	T	T	F	F	T

Table 2.6

The set of truth values for the proposition is shown in column six. Now suppose we define a new connective $\equiv$ relating x to y with truth values given in column seven without mentioning how this relation is to be read. By comparing columns six and seven we see that

$$(x \vee y) \wedge \overline{(x \wedge y)} = \overline{x \equiv y} \qquad \textit{Answer}$$

Exercise 2.5

1. Use Table 2.5 to obtain the truth values of each of the following:
   (i) $(x \wedge y) \wedge (x \vee y)$
   (ii) $(x \wedge y) \vee \overline{(x \wedge y)}$
   (iii) $\overline{(x \wedge y)} \wedge (x \vee y)$
   (iv) $\bar{x} \wedge (x \vee y)$
   (v) $\bar{x} \vee (\bar{x} \wedge \bar{y})$

2. Find the truth values of the following:
   (i) $\overline{x \wedge y}$
   (ii) $\overline{\overline{x \wedge y}}$
   (iii) $\overline{x \wedge y} \vee (x \vee y)$
   (iv) $(x \vee y) \wedge \overline{(x \vee y)}$

3. A proposition p has truth values T, F, F, T. What are the truth values for $\bar{p}$? Identify $\bar{p}$ from the tables in this section.

Conditional Statements

Even though it can be proved that we may express any proposition we please entirely in terms of $\wedge$ together with negation, a consideration of Example 13 above suggests that it may be useful to define a few more logical connections in order to obtain simpler substitute propositions. Any new connective can be defined by listing its truth values just as we did in Table 2.6 for $\equiv$ without any previous discussion and certainly without any reference to everyday language. With a similar lack of restriction consider the formal definition of the conditional.

Definition 2.5. With x and y representing statements, the conditional statement, written $x \rightarrow y$, is true except for the one case when x is true and y is false.

The truth table 2.7 also defines the conditional $x \rightarrow y$.

x	y	$x \rightarrow y$	$y \rightarrow x$	$(x \rightarrow y) \wedge (y \rightarrow x)$
F	F	T		
F	T	T		
T	F	F		
T	T	T		

Table 2.7

As a productive exercise the reader should now complete columns four and five of this table and then compare with the final column of Table 2.6. Any conclusion? (See Table 2.8.)

Hopefully we have been able to compare two results involving two new connectives $\equiv$ and $\rightarrow$ without having been told what to call them. This merely emphasises our proper obedience to the definition given via the truth table.

We may go on to unravel the truth values for any proposition like

$$\{(x \rightarrow y) \rightarrow z\} \wedge \{(\overline{y \rightarrow z}) \equiv x\}$$

by simply obeying the brackets and referring to the truth tables. There would be no difficulty in reading these propositions in Tables 2.6 and 2.7.

Now one indisputable feature of a valid proof is that if it starts with a true assumption it must reach a true conclusion, so it would appear that $x \rightarrow y$ is a suitable symbol for registering a valid step in a mathematical proof because the truth table says that x true leading to y false makes $x \rightarrow y$ false. We shall of course need to support this suggestion by examining the case when x is false.

The common language connectives 'if' . . . 'then' come closest to an adequate expression of what we require in order to translate $x \rightarrow y$ into everyday terms—so much so that the definition of the conditional is usually couched in these terms.

Definition 2.6. With x and y representing statements; the compound statement 'if x then y' is called a conditional statement, being false only when x is true and y is false; otherwise the conditional is true. We write '$x \rightarrow y$' to represent 'if x then y'.

The proposition 'if x then y' is often called an **implication**, symbolised by $x \rightarrow y$ and alternatively described as 'x implies y'. Examples of common usage which we are prepared to accept as an adequate description of our logical intentions are as follows:

(i) 'If the bus does not come in the next five minutes then I will be late for work.'

(ii) 'If there is a queue then I will not go to the match.'

(iii) 'If the numbers are even then so is their sum.'

(iv) 'If the sides of a triangle are 3, 4 and 5 then the triangle is right-angled.'

(v) 'If $x = 1$ is a solution of $4x = a$ then $a = 4$.'

Now consider the definition of $x \rightarrow y$ in more detail. We have seen that in a deductive proof we must proceed from an assumption to a conclusion by valid mathematical steps, whether the assumptions are true or false. When the assumptions are true, a valid argument must produce a conclusion which is true.

On page 32 we saw that when the assumption is false a valid argument may produce a conclusion which is false.

Example 14. Suppose that x represents the false statement $-3 = 3$ and is taken as our assumption.

A valid step allows us to square both sides of an equation and retain the equality, so we can outline an argument as follows:

	x: $-3 = 3$	assumption (F)
$\therefore$	$(-3)^2 = 3^2$	valid step
	y: $9 = 9$	conclusion (T)

The conditional 'if x then y' is true, i.e. it is possible to reach a true conclusion by a valid argument based on a false assumption.

Example 15. Using the same false assumption as in Example 14 we may employ the valid step of multiplying both sides of an equation by the same quantity without changing the relation between them:

x: $-3 = 3$	assumption (F)
$-3 \times 2 = 3 \times 2$	valid step
y: $-6 = 6$	conclusion (F)

These Examples 14 and 15 merely confirm the plausibility of the truth table entries, which state that $x \to y$ is true if x is false and y is true or false.

Example 16. Consider the following argument:

x: $9 = 9$	assumption (T)
$\therefore \quad (+3)^2 = (-3)^2$	(T)

Taking square roots of both sides

y: $+3 = -3$	conclusion (F)

The very close relation between $x \to y$ and the idea of valid mathematical proof is revealed in this example. It tells us that since we have started with a true assumption and reached a false conclusion there must be something wrong somewhere. For a valid step we should be 'taking *positive* square roots of 9' on each side of the assumption x to yield y: $+3 = +3$ (T).

Thus, in a valid proof, we know that a true assumption must yield a true conclusion. In other words, we try to make certain that the line

x	y	$x \to y$
T	F	F

will never occur.

Example 17. Consider the following statements: x: $ABCD$ is a square (T)
y: angle $ABC = 90°$
z: Diagonals $AC = BD$

$\therefore (x \to y) \wedge (x \to z)$ is true whenever the first statement is given as true. This is saying that when $ABCD$ is a square, ABC will be a right angle *and* when $ABCD$ is a square, the diagonals will be of equal length. Alternatively, we could say $x \to (y \wedge z)$, i.e. when $ABCD$ is a square it follows that angle $ABC = 90°$ and $AC = BD$.

Usually, a proof is made up of several small conditional steps each leading from one true statement to the next true statement so that with the original assumption being true we have a sequence of steps like

$$u \to v,\ v \to w,\ w \to x,\ x \to y$$

which make the overall proof

$$u \to y$$

Since a mathematical argument must be valid, at each step in the argument the third row in Table 2.7 must not arise. In other words, all steps must only have statements x, y such that $x \to y$ is true, whatever the truth value of x. The resulting restriction is called a **logical implication**, which we indicate by $\Rightarrow$ instead of $\to$. Thus $x \Rightarrow y$ has the truth table of $x \to y$ with the third row omitted.

Examination of the truth table for $x \to y$ shows that whenever the conditional is true

the truth of x guarantees the truth of y

or, to put it in its more usual form, we say that

x is a sufficient condition for y

meaning that the truth of x is sufficient to ensure the truth of y and that the truth of y is a necessary condition for the truth of x.

An example of this wording is given in the following illustrations:

For a rectangle to be a square it is sufficient that all its sides be of equal length.

For a triangle to be an isosceles triangle it is sufficient that two of its angles be the same size.

For a quadrilateral to be a square it is necessary that all of its angles be right-angles. (Not sufficient because a rectangle may not be a square.)

For a number to be divisible by 14 it is sufficient that it be divisible by 2 and 7, i.e. (divisibility by 14) $\leftrightarrow$ (divisibility by 2 and 7).

Exercise 2.6

Say whether the following conditionals are true or false:

1. If $6 \times 4 = 24$ then $6 \times 8 = 48$.
2. If $6 \times 4 = 25$ then $6 \times 8 = 50$.
3. If $x = 5$ then $4x = 20$.
4. If 4 is a prime number then $4 + 3 = 7$.
5. If London is in France then Paris is in England.
6. If New York is in the USA then Paris is in Australia.
7. If sets A, B are disjoint (T) then $A \cap B = A \cup B$.
8. If $x \neq 0$ and $x \not< 0$ then $x > 0$.
9. For a quadrilateral to be a square it is sufficient that all of the angles be right-angles.
10. A necessary condition for a quadrilateral to be a square is that all its sides be of equal length.

Biconditional Sentences

As may be expected by the name, a biconditional $x \leftrightarrow y$ is an implication which is read two ways—that is, not only does x imply y but y implies x. The notation of $x \leftrightarrow y$ for a biconditional is therefore very appropriate.

Definition 2.7. The conjunction $(x \to y) \wedge (y \to x)$ is called the biconditional of x, y and is written $x \leftrightarrow y$.

$x \leftrightarrow y$ is true when x and y have the same truth value.
$x \leftrightarrow y$ is false when x and y have different truth values.

The biconditional is also read as 'x if and only if y' and this is sometimes abbreviated to 'x iff y'.

The truth table for $x \leftrightarrow y$ is given in column three or column seven of Table 2.8.

The biconditional is also referred to as the **equivalence** of x and y and the alternative notation $x \equiv y$ is sometimes used, as in the fourth column of Table 2.8. Thus $x \leftrightarrow y$ and $x \equiv y$ have the same set of truth values.

x	y	$x \leftrightarrow y$	$x \equiv y$	$x \rightarrow y$	$y \rightarrow x$	$(x \rightarrow y) \wedge (y \rightarrow x)$
F	F	T	T	T	T	T
F	T	F	F	T	F	F
T	F	F	F	F	T	F
T	T	T	T	T	T	T

Table 2.8

The relevance of a biconditional to mathematical proof is considerably important. It is concerned with theorems and converse theorems, a theorem being a proposition which is always true for a stated universal set. To outline the relation between theorems and converse theorems consider the next example.

Example 18. Taking x: the quadrilateral is a square.
y: the quadrilateral is a rectangle.

(Remember that all squares are rectangles but not all rectangles are squares.)

We have (i) Theorem: If a quadrilateral is a square then it is a rectangle, i.e. $x \rightarrow y$.
(ii) Converse theorem: If a quadrilateral is a rectangle then it is a square, i.e. $y \rightarrow x$.

Example 19. A biconditional statement of a theorem carries the responsibility of doing two things. For example, when we say 'For a and b, any two positive numbers $a^2 < b^2$ if and only if $a < b$', we mean

(i) If $a^2 < b^2$ then $a < b$,

as well as

(ii) If $a < b$ then $a^2 < b^2$.

Therefore, to prove a biconditional will require not only the proof of a theorem but also a proof of the converse theorem.

Now, with any theorem $x \rightarrow y$ we associate three main ideas which we list as follows:

(i) Theorem: $x \rightarrow y$
(ii) Converse: $y \rightarrow x$
(iii) Inverse: $\bar{x} \rightarrow \bar{y}$
(iv) Contrapositive: $\bar{y} \rightarrow \bar{x}$

We shall now explore some of the relations between these ideas.

Example 20. We continue with the substance of Example 18 in order to consider:

Inverse: (iii) 'If a quadrilateral is not a square then it is not a rectangle', i.e. $\bar{x} \rightarrow \bar{y}$ (when the quadrilateral is not a square it could still be a rectangle).

Contrapositive: (iv) 'If a quadrilateral is not a rectangle then it is not a square', i.e. $\bar{y} \rightarrow \bar{x}$.

As prospective mathematical thinkers we should now be asking ourselves 'Is a theorem the same as its contrapositive theorem?' and 'Is the converse theorem the same as the inverse theorem?' An examination of the following truth table 2.9 shows that the theorem and its contrapositive theorem have the same set of truth values and the converse theorem has the same set of truth values as the inverse theorem.

x	y	Theorem $x \to y$	Converse $y \to x$	Inverse $\bar{x} \to \bar{y}$	Contra-positive $\bar{y} \to \bar{x}$
F	F	T	T	T	T
F	T	T	F	F	T
T	F	F	T	T	F
T	T	T	T	T	T

Table 2.9

It follows therefore that it may be possible to take advantage of this result if the contrapositive form of a theorem is easier to prove than the theorem itself, and we should consider the next example with this in mind.

Example 21. Given that n is a positive integer prove that if n^2 is odd then n is odd.

Solution. Just to identify the foregoing remarks we suggest

x: n^2 is odd (T)
y: n is odd (T/F to be decided)

We are asked to prove the theorem given as $x \Rightarrow y$.

Consider instead the contrapositive theorem $\bar{y} \Rightarrow \bar{x}$, i.e. 'If n is not odd then n^2 is not odd'—which, of course, means the same as 'If n is even then n^2 is even'.

Proof: Because n is even put $n = 2k$, where k is a positive integer

$$\therefore \quad n^2 = 4k^2$$

which is divisible by 2.

$$\therefore \quad n^2 \text{ is even.}$$

We have proved that $\bar{y} \Rightarrow \bar{x}$ is true.

$$\therefore \quad x \Rightarrow y \text{ is true}$$

and since x is true it follows that y must be true.

i.e. Whenever n^2 is odd, so is n *Answer*

Exercise 2.7

State the converse, inverse and contrapositive for each of the following five theorems $x \to y$: with x (T) in each case.

1. x: it is raining
   y: the pavement is wet
2. x: the number is divisible by 6
   y: the number is divisible by 3
3. x: the number is odd
   y: twice the number is even
4. x: the rectangle has two equal sides
   y: the rectangle is a square
5. x: the triangle has two equal sides
   y: the triangle has two equal angles
6. Given that n is a positive integer prove that if n^3 is odd then n is odd.

We conclude this chapter with a set of miscellaneous worked examples.

Example 22. Consider the following proposition:

If up is down the east is west, and if east is west, then ice is hot (i)
Therefore, if up is down, then ice is hot (ii)

Construct a truth table and determine if the conclusion is true.

Solution. Let the sentences be notated as follows:

x: up is down
y: east is west
z: ice is hot

If up is down then east is west: $x \rightarrow y$
If east is west then ice is hot: $y \rightarrow z$
If up is down then ice is hot: $x \rightarrow z$

Line (i) in the question now reads: $(x \rightarrow y) \wedge (y \rightarrow z)$. The use of 'therefore' before line (ii) means that the proposition becomes (i) → (ii). The proposition finally reads

$$\{(x \rightarrow y) \wedge (y \rightarrow z)\} \rightarrow (x \rightarrow z)$$

x	y	z	$x \rightarrow y$	$y \rightarrow z$	(i) $(x \rightarrow y) \wedge (y \rightarrow z)$	(ii) $x \rightarrow z$	(i) → (ii)
F	F	F	T	T	T	T	T
F	F	T	T	T	T	T	T
F	T	F	T	F	F	T	T
F	T	T	T	T	T	T	T
T	F	F	F	T	F	F	T
T	F	T	F	T	F	T	T
T	T	F	T	F	F	F	T
T	T	T	T	T	T	T	T

Table 2.10

The tedium of constructing Table 2.10 may hide the obvious result from view because we have merely proved what is called a **transitive** law—namely, that from the relations $x \rightarrow y$, $y \rightarrow z$ we may deduce

$$x \rightarrow z$$

We can put this another way by saying that $\{(x \rightarrow y) \wedge (y \rightarrow z)\} \rightarrow (x \rightarrow z)$ is a tautology.

Answer

Example 23. Show that the proposition $(\bar{x} \vee y) \wedge (x \wedge \bar{y})$ is a contradiction.

Solution. With only two statements x, y to consider it is sometimes possible to obtain the required result by a simple inspection like the following:

We know that the proposition is true if and only if both $\bar{x} \vee y$ and $x \wedge \bar{y}$ are true.

For $x \wedge \bar{y}$ to be true we require both x (true) and y (false) and this compulsory choice will always make $\bar{x} \vee y$ false.

Hence, $(\bar{x} \vee y) \wedge (x \wedge \bar{y})$ is always false.

Answer

Example 24. Prove that $\overline{x \vee y \vee z} = \bar{x} \wedge \bar{y} \wedge \bar{z}$.

Solution. This result is an extension of DeMorgan's laws and is easily proved by constructing a truth table. We can also talk the proof out by saying

$$x \vee y \vee z$$

is only false when all three x, y, z are false,

$$\therefore \quad \overline{x \vee y \vee z}$$

is only true when all three x, y, z are false, etc. (To be completed as Question 8, Exercise 2.8.)

It is instructive to see a formal proof along the following lines.

Proof: Put $w = x \vee y$ so that $\bar{w} = \bar{x} \wedge \bar{y}$ (DeMorgan's law for two propositions—page 40).

$$\therefore \quad \overline{x \vee y \vee z} = \overline{w \vee z}$$
$$= \bar{w} \wedge \bar{z}$$
$$= (\bar{x} \wedge \bar{y}) \wedge \bar{z}$$
$$\therefore \quad \overline{x \vee y \vee z} = \bar{x} \wedge \bar{y} \wedge \bar{z}$$

Answer

Example 25. Prove that the proposition $\{(x \to y) \wedge \bar{x}\} \to \bar{y}$ is not always true.

Solution. We construct a truth table for the proposition in the usual manner, working out each column in turn from left to right.

x	y	$x \to y$	$\bar{x}$	$(x \to y) \wedge \bar{x} = a$	$\bar{y}$	$a \to \bar{y}$
F	F	T	T	T	T	T
F	T	T	T	T	F	F
T	F	F	F	F	T	T
T	T	T	F	F	F	T

Table 2.11

The truth table shows that the given proposition is not a tautology. *Answer*

Now since the proposition $a \to \bar{y}$ in Example 25 is not always true it follows that it cannot be used as a valid mathematical proof.

It is quite interesting to try to identify the fault, which is indicated in the second line of Table 2.11, by considering the geometric terms of Example 26.

Example 26. Suppose statements x, y are

x: The quadrilateral is a square

y: The angles of the quadrilateral are 90°

Give an example to illustrate how the proposition $\{(x \to y) \wedge \bar{x}\} \to \bar{y}$ does not represent a valid proof.

Solution. From Table 2.11 we see that $a = \bar{x}$ and this observation enables us to recast the proposition as $\bar{x} \to \bar{y}$.

A description of the statements in the suggested proof is therefore

$\bar{x}$: The quadrilateral is not a square

$\bar{y}$: All the angles of the quadrilateral are not 90°

Now consider being asked to prove the following:

Given that the quadrilateral *ABCD* is not a square, prove that all of its angles are not 90° . . . (i).

We can illustrate the second line of Table 2.11 by simply suggesting that *ABCD* is a rectangle.

This suggestion makes $\bar{x}$(T), $\bar{y}$(F) and produces the third line of Table 2.7 and so disqualifies (i). *Answer*

In the case of Example 26 it was reasonably straightforward to reveal the fault by considering a counterexample, but sometimes this type of invalid proof is not so easy to detect.

As we have tried to note throughout this chapter, the most important point is that, starting from a true assumption a valid argument must reach a true conclusion.

Exercise 2.8

Construct tables for each of the following seven propositions:

1. $\bar{x} \vee y$
2. $x \rightarrow y$
3. $(x \rightarrow y) \vee (y \rightarrow x)$
4. $(\bar{x} \vee y) \rightarrow (x \rightarrow y)$
5. $(x \rightarrow y) \wedge \overline{(y \rightarrow z)}$
6. $\overline{(x \equiv y)} \wedge (\bar{x} \rightarrow y)$
7. $((x \rightarrow y) \rightarrow z) \rightarrow x$

8. Prove that $\overline{x \wedge y \wedge z} = \bar{x} \vee \bar{y} \vee \bar{z}$
9. Consider the following proposition:

   If the price goes up then the people will not buy, and if the people will not buy then the price will come down. Therefore, if the price goes down then the people will buy. Obtain truth values for the proposition.

10. Discuss the validity of the following argument:

    The kitchen is the most dangerous room in the home, therefore it would be safer to cook in the bedroom.

3

SWITCHING CIRCUITS

A simple switch is represented by the type of diagram in Fig. 3.1, in which the switch is used to interrupt current flowing between P and Q. The two-state on–off form of switch enables it to register the true–false nature of a statement. This allows us to refer to the statement and the switch by the same letter symbol.

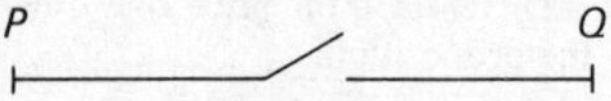

P Q

Fig. 3.1

For the common language of switching circuits we use variables like x, y and z to represent both the statement and its corresponding switch. A false statement is indicated by an open switch (off) and the representative variable is given the value 0. A true statement is indicated by a closed switch (on) and the representative variable is given the value 1.

In the switching diagrams we shall not show open or closed switches but merely write the representative variable for the switch as shown in Fig. 3.2.

Since a negation $\bar{x}$ is a separate statement from x we shall use a separate switch to represent its truth values, as displayed in Fig. 3.3, and the switch will be closed (on) when $x = 0$ and open (off) when $x = 1$.

x

Fig. 3.2

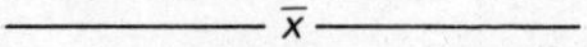

$\bar{x}$

Fig. 3.3

Connections in Series

Switches are said to be connected in series when anchored on the same stretch of wire as shown in Fig. 3.4. Reading from left to right we see this connection as x before y. Clearly the current will flow if and only if both switches are closed, i.e. $x = 1$ and $y = 1$.

We may use 0,1 to indicate not only the off or on state of a switch but also the non-flow or flow state of the circuit. This enables us to speak of a transmission or flow function of a circuit whose value will be given by a truth table in the same manner as for the propositions outlined in Chapter 2. Thus, for the circuit of Fig. 3.4 we can construct Table 3.1 giving the values of the flow function f in the third column.

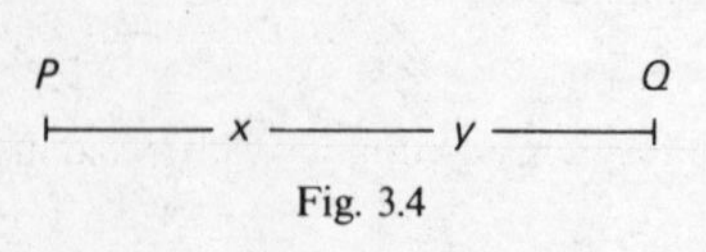

Fig. 3.4

x	y	f
0	0	0
0	1	0
1	0	0
1	1	1

Table 3.1

The entries of this table are exactly the same, writing F = 0, T = 1, as for Table 2.1 and so it follows that a connection of two switches in series will model the conjunction 'and', given by $x \wedge y$. In the notation of switching circuits we use . instead of $\wedge$ so that we state the value of the flow function of the circuit in Fig. 3.4 as

$$f = x.y$$

Clearly f will remain unchanged if we interchange the switches x and y, so we may write

$$f = x.y = y.x$$

If we now substitute $\bar{x}$ for y then one of the switches will always be open since $x = 0$ means $\bar{x} = 1$ while $x = 1$ means $\bar{x} = 0$.

$$\therefore \quad f = x.\bar{x} = 0$$

If x was kept permanently closed then we would have $x = 1$ and

$$f = x.y = 1.y = y$$

Likewise, if x was kept permanently open then we would have $x = 0$ and

$$f = x.y = 0.y = 0$$

Finally, it is clear that three switches w, x, y wired in series will give a flow function $f = w.x.y$ whose truth value may be found in the same manner as for $w \wedge x \wedge y$ on page 37—that is, by bracketing and finding that

$$(w.x).y = w.(x.y) = w.x.y$$

Connection in Parallel

This method of connection is represented by Fig. 3.5, where the switches x and y are seen to be anchored in different branches of wire so that current will flow from P to Q in either or both of the branches of the circuit according to whether x or y is closed, i.e. if $x = 1$ or $y = 1$. (Inclusive 'or'—see page 15.) The flow function f for this circuit is given in the third column of Table 3.2. By

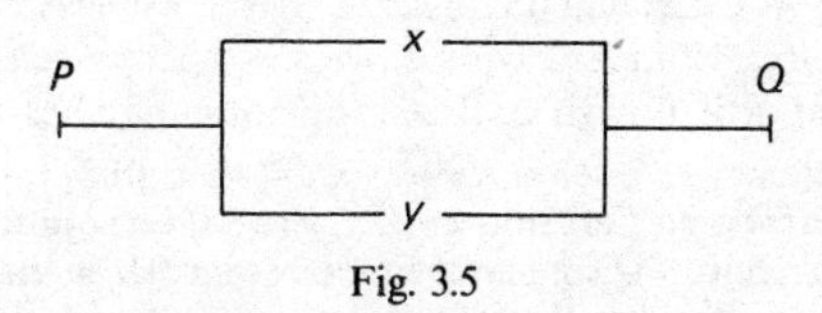

Fig. 3.5

x	y	f
0	0	0
0	1	1
1	0	1
1	1	1

Table 3.2

comparing this table with Table 2.2 we deduce that a connection of two switches in parallel will model the disjunction 'or' given in $x \vee y$.

In the notation of switching circuits we use + instead of $\vee$ so that we state the value of the flow function of the circuit in Fig. 3.5 as

$$f = x + y$$

It is clear that there will be no change in the flow function if we interchange the switches x and y, so we may write

$$f = x + y = y + x$$

If we now substitute $\bar{x}$ for y then one of the switches x, $\bar{x}$ will always be closed. Therefore

$$f = x + \bar{x} = 1$$

In other words, the current will always flow so we may as well wire P directly to Q and save the two switches in Fig. 3.5.

If x was kept permanently closed then we would have $x = 1$ and

$$f = x + y = 1 + y = 1$$

in which case there was no point in having the switch y. Indeed, we may as well save switch x also and connect P directly to Q.

If x was kept permanently open then we would have $x = 0$ and

$$f = x + y = 0 + y = y$$

Finally, it is clear that three switches w, x, y wired in parallel will give a flow function $f = w + x + y$ whose truth value may be found in the same manner as for $w \vee x \vee y$ on page 39—that is, by bracketing and seeing that

$$(w + x) + y = w + (x + y) = w + x + y$$

The general switching circuit will be made up of individual sections connected in series and parallel in much the same way as we created propositions in Chapter 2. Naturally, it will be possible for two different circuit arrangements to have the same flow function and this is appropriately recorded by writing $f_1 = f_2$ and saying that the two circuits are equivalent.

The method of combining the basic series and parallel connections is illustrated by the following examples.

Example 1. Obtain the flow function of the circuit in Fig. 3.6 and if possible find a simpler equivalent circuit.

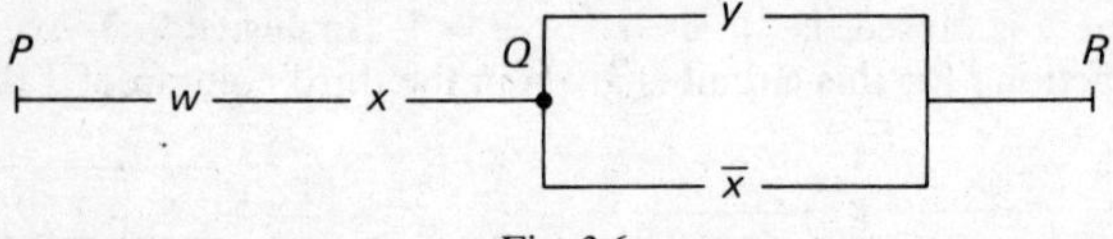

Fig. 3.6

Solution. The flow function f of a circuit may be given either by stating all its truth values in a truth table like Table 3.3 or by expressing f in terms of w, x, y and their negations.

The circuit in Fig. 3.6 is really two circuits: PQ connected in series with QR so that if f_1 is the flow function of PQ and f_2 is the flow function of QR then $f = f_1 . f_2$.

w	x	y	$\bar{x}$	f
0	0	0	1	0
0	0	1	1	0
0	1	0	0	0
0	1	1	0	0
1	0	0	1	0
1	0	1	1	0
1	1	0	0	0
1	1	1	0	1

Table 3.3

Since $f_1 = w.x$ and $f_2 = y + \bar{x}$ it follows that

$$f = f_1.f_2 = w.x.(y + \bar{x})$$

Alternatively, the truth table 3.3 above gives f by merely tracing our way through the circuit, realising that $f_1 = 1$ if and only if both w and x are closed. This will mean that $\bar{x} = 0$, so finally $f = 1$ if and only if $w = x = y = 1$.

We now try to see if it is possible to obtain f using fewer switches. Since we must have $x = 1$ to get from P to Q it follows that we must have $y = 1$ to get from Q to R. In other words, we have

$$f = w.x.y$$

and we can save the switch $\bar{x}$.

Alternatively, we may use the distributive law (pages 19, 54) to get

$$f = w.x.(y + \bar{x}) = w.x.y + w.x.\bar{x} = w.x.y \qquad \textit{Answer}$$

Example 2. Obtain the flow function of the circuit in Fig. 3.7 and simplify the circuit if possible.

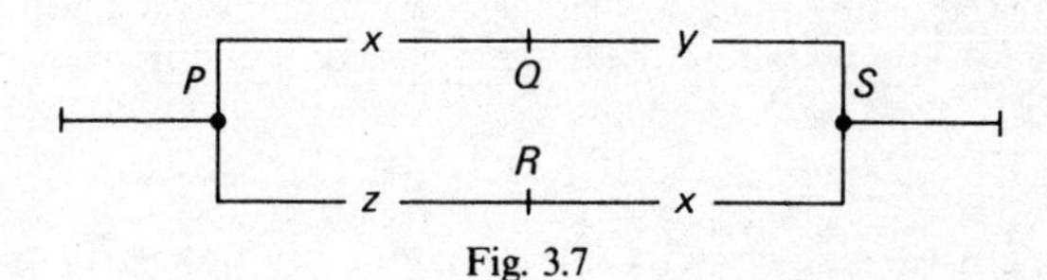

Fig. 3.7

Solution. Here we have two series circuits PQS and PRS wired in parallel. The flow function for PQS is given by $f_1 = x.y$. The flow function for PRS is given by $f_2 = z.x = x.z$.

The flow function for the whole circuit PS is therefore given by

$$f = f_1 + f_2$$
$$= x.y + x.z$$

By inspection we can see that if $x = 0$ then $f = 0$ and this suggests that we need only use one x switch in the simpler circuit of Fig. 3.8 to obtain the same flow function. In this simpler circuit we have circuit PT in series with circuit TS

$$f = x.(y + z) \text{ is the flow function}$$

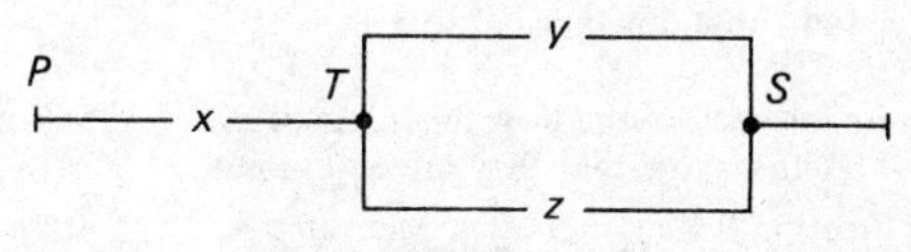

Fig. 3.8

Comparing the two forms we see that

$$x.(y + z) = x.y + x.z$$

In other words, we may distribute . over + to obtain the result on the right-hand side (see page 19 with . for $\cap$ and + for $\cup$).

Example 3. Suggest two circuits to show that

$$x + y.z = (x + y).(x + z)$$

Solution. Taking the flow function $f = (x + y).(x + z)$, we see that this will be obtained as shown in Fig. 3.9 from:

(i) switches x,y in parallel; circuit PQ,
(ii) switches x,z in parallel; circuit RS,
(iii) PQ and RS in series.

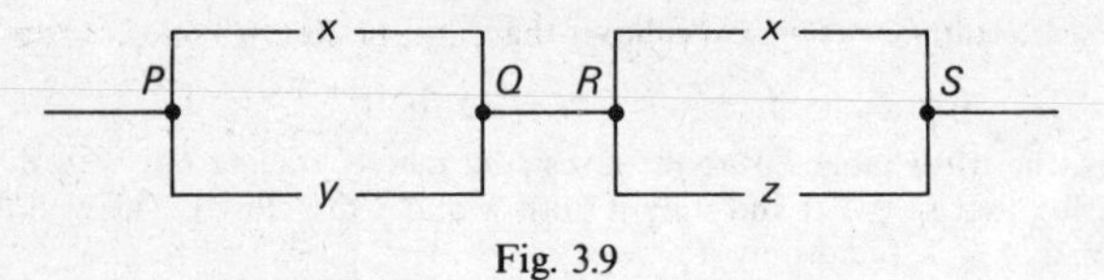

Fig. 3.9

Examining the possible current flow in this circuit we see that

$f = 1$ if and only if $(x = 1)$ or $(y = 1$ and $z = 1)$

i.e. $f = 1$ if and only if $(x = 1)$ or $(y.z = 1)$

Hence we may suggest $f = x + y.z$ with the corresponding circuit given by Fig. 3.10.

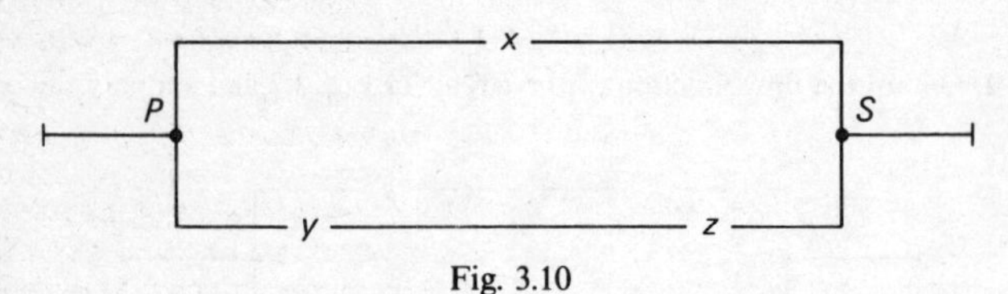

Fig. 3.10

$\therefore \quad x + y.z = (x + y).(x + z)$ (see below) *Answer*

Exercise 3.1

1. Explain why $x + x = x$ and $x.x = x$
2. Find $\bar{x} + \bar{x}$ and $\bar{x}.\bar{x}$

Simplify each of the following functions in Questions 3–11 and draw the corresponding circuits.

3. $x.(x + y)$
4. $(x + y).x$
5. $y.(x + y)$
6. $z.(x + y + \bar{z})$
7. $(\bar{x} + x).y$
8. $\bar{x}.x + y$
9. $(\bar{x} + y).(x + y)$
10. $(\bar{x} + y).(x + \bar{y})$ and show this is equal to $x \equiv y$
11. $(\bar{x} + x).(y + \bar{y})$
12. Draw the circuits for each of the flow functions (i) $x.z + y.z + y.\bar{x}$ and (ii) $(x + y).(\bar{x} + z)$ and show that they are equivalent.

The Distributive Law

The simplification of the flow function is easier to carry out by algebraic means than by the trial and error modification of the circuit.

The most useful rule for simplifying a function is the distributive law which has been discussed in page 19 and Examples 2 and 3 above. The result $x.(y + z) = x.y + x.z$ may be extended to

$$\text{(i)}\ x.(w + y + z) = (x.w) + (x.y) + (x.z)$$
$$\text{(ii)}\ (x + y).(w + z) = (x + y).w + (x + y).z = x.w + y.w + x.z + y.z$$

i.e. the operation . is distributed over the operation +.

In Example 3 above we proved the result $x + (y.z) = (x + y).(x + z)$. Here the process is described as distributing + over . in the statement on the left-hand side.

The result $x + (y.z) = (x + y).(x + z)$ may be extended to

$$\text{(iii)}\ x + (w.y.z) = (x + w).(x + y).(x + z)$$
$$\text{(iv)}\ (x.y) + (w.z) = ((x.y) + w).((x.y) + z)$$
$$= (x + w).(y + w).(x + z).(y + z)$$

i.e. the operation + is distributed over the operation . .

Notice that in (i), (ii), (iii) and (iv) if we interchange . with + (i) becomes (iii) and (ii) becomes (iv). This process of obtaining one result from another by merely interchanging the operations is called **dualising**. We call (iii) the **dual** of (i) or (i) the **dual** of (iii). Similarly, (ii) is the **dual** of (iv).

If we keep in mind one or two standard results such as

$$x + \bar{x} = 1 \qquad x.\bar{x} = 0$$
$$1 + z = 1 \qquad 1.z = z$$
$$x.y + y = y \qquad x.y + x = x$$

we can employ the distributive law to considerable advantage in simplifying flow functions.

Example 4. Use the distributive law to simplify

$$f = x.(y + z) + x.\bar{y}$$

Draw the circuit for f and discuss the simplification by current flow.

Solution

$$f = x.(y + z) + x.\bar{y}$$
$$= x.y + x.z + x.\bar{y} \qquad \text{(distributive law)}$$
$$= x.(y + z + \bar{y}) \qquad \text{(distributive law)}$$
$$= x.(1 + z) \qquad (y + \bar{y} = 1)$$
$$= x.1 \qquad (1 + z = 1)$$
$$\therefore\ f = x$$

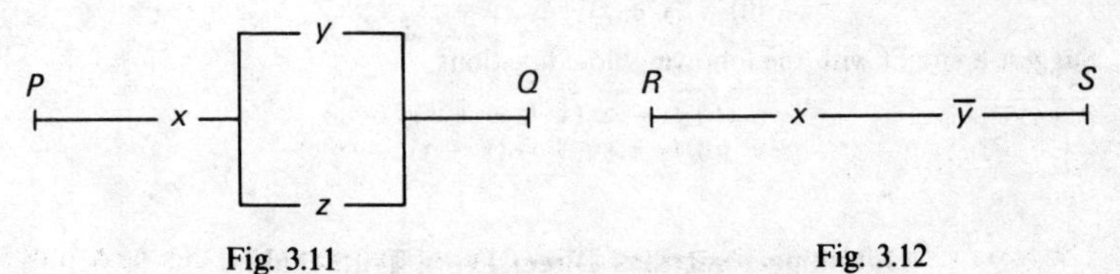

Fig. 3.11 Fig. 3.12

In Fig. 3.11 circuit PQ has the flow function $f_1 = x.(y + z)$.
In Fig. 3.12 circuit RS has the flow function $f_2 = x.\bar{y}$.

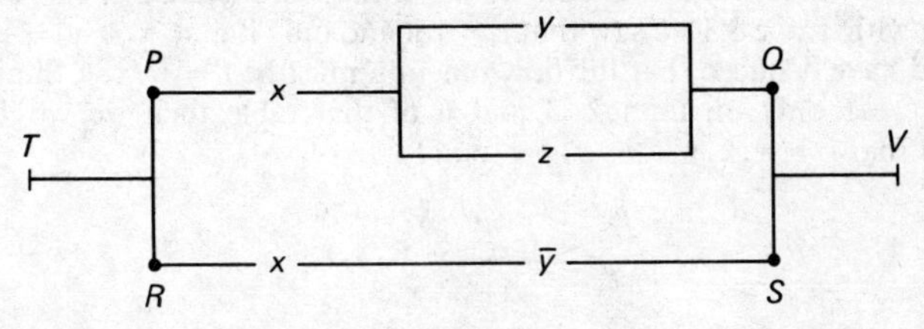

Fig. 3.13

$f = f_1 + f_2$ means that we must assemble these two circuits in parallel and this will yield the complete circuit TV shown in Fig. 3.13. Examining the circuit for current flow from T to V, we see that we must get past x, but once past x we are bound to reach V via y or $\bar{y}$. The switch y is not required, hence everything in the circuit depends on x—that is

$$f = 1 \text{ if and only if } x = 1$$
$$f = 0 \text{ if and only if } x = 0$$
$$\therefore \quad f = x \text{ as already obtained above}$$

Answer

Example 5. Draw a circuit with flow function

$$f = y.(\bar{x} + \bar{y} + x) + x.\bar{y}$$

Solution. Here we shall simplify the function first using the distributive law and then obtain a circuit for the final form of f.

$$\begin{aligned} f &= y.(\bar{x} + \bar{y} + x) + x.\bar{y} & \\ \therefore \quad f &= y.(\bar{y} + 1) + x.\bar{y} & (x + \bar{x} = 1) \\ &= y.1 + x.\bar{y} & (\bar{y} + 1 = 1) \\ \therefore \quad f &= y + x.\bar{y} & (y.1 = y) \end{aligned}$$

This is an appropriate form for using the distributive law

$$\begin{aligned} \therefore \quad f = y + x.\bar{y} &= (y + x).(y + \bar{y}) & \\ &= (y + x).1 & (y + \bar{y} = 1) \\ f &= y + x & \end{aligned}$$

to be obtained by wiring switches x and y in parallel (see page 51). *Answer*

Exercise 3.2

1. Expand (i) $(y + z).x$, (ii) $(y + z).y$ and (iii) $(y + z).z$.
2. Draw a circuit to represent the flow function $f = y.(y + z)$ and show that $f = y$.
3. Simplify the following flow functions using the absorption law displayed in question 2:

   (i) $(x + y).(x + z)$
   (ii) $(x + z).(z + y)$

4. Use the distributive law to simplify the following:

   (i) $x.(\bar{y} + z) + x.(y + \bar{z}) + x$
   (ii) $\bar{x}.(y + \bar{z}) + x.(\bar{y} + \bar{z})$

5. Suggest a circuit with the following flow functions:

   (i) $f_1 = x.(\bar{x} + y) + \bar{x}.\bar{y}$
   (ii) $f_2 = (x + \bar{y}).(\bar{x} + y)$

Obtaining Functions Direct From Truth Tables

In most of the examples so far we have started with f expressed in an algebraic form which has then been simplified and a final circuit obtained. Sometimes we can work direct from the truth values if these are available. For example, working with Table 3.3 we saw that $f = 1$ if and only if $w = x = y = 1$ and so we could have deduced that the flow function must be $f = w.x.y$. Similarly, if instead $f = 1$ only on lines 2, 3 and 6 of that table then we would have deduced that $f = \bar{w}.\bar{x}.y + \bar{w}.x.\bar{y} + w.\bar{x}.y$

$$\begin{aligned} \therefore \quad f &= (\bar{w} + w).\bar{x}.y + \bar{w}.\bar{y} & \\ &= (1).\bar{x}.y + \bar{w}.x.\bar{y} & (\bar{w} + w = 1) \\ &= \bar{x}.y + \bar{w}.x.\bar{y} & \end{aligned}$$

We conclude that it is possible similarly to obtain the flow function for any set of truth values we have in mind and thereafter by the usual simplification methods obtain a final circuit.

Example 6. Obtain a circuit for controlling a light by two switches.

Solution. This is the common control problem for the hall light of a house which can be switched on or off from upstairs or downstairs. A truth table for our requirements is given in Table 3.4. With both switches open we assume that the light is off and the value

x	y	f
0	0	0
0	1	1
1	0	1
1	1	0

Table 3.4

of the circuit will therefore be 0. Any change from this state will turn the light on so that $x = 0$, $y = 1$ gives $f = 1$ and $x = 1$, $y = 0$ must also give $f = 1$. Having switched the light on by these two arrangements it follows that one further change must switch the light off and Table 3.4 is now complete.

From the table we see that the flow function for the circuit is given by

$$f = \bar{x}.y + x.\bar{y}$$

because $f = 1$ if and only if $\bar{x}.y = 1$ or $x.\bar{y} = 1$. The circuit with this flow function is given in Fig. 3.14. The symbols on the diagram are the usual ones for representing a lamp and the mains supply. Of course, Fig. 3.14 does not give the real wire connections which an electrician might use. For real wire connections different types of switches are available and in this elementary case the real wiring would be given in Fig. 3.15. (Here, the moving part of the switch connects either AB or AC.) *Answer*

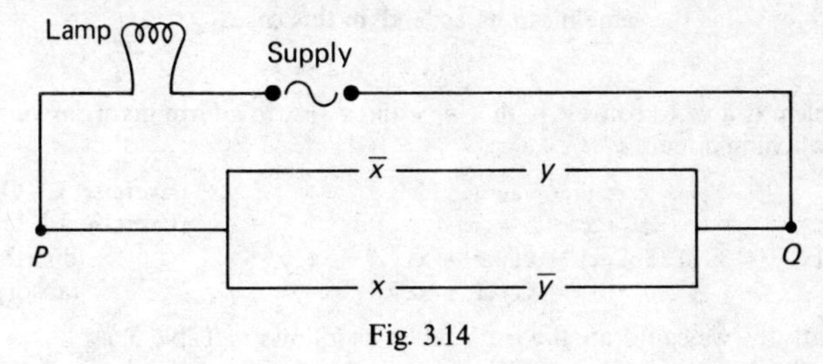

Fig. 3.14

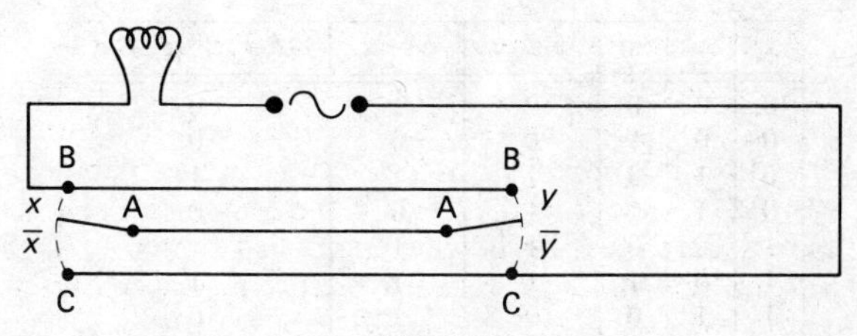

Fig. 3.15

Example 7. Draw the truth table for three variables x, y and z. Select any three lines and record these as $f = 1$ and the other five lines as $f = 0$. Construct a circuit for the flow function which has been chosen.

Solution. Suppose the three selected lines for $f = 1$ are given by Table 3.5.

Hence $$f = \bar{x}.y.z + x.\bar{y}.\bar{z} + x.y.\bar{z}$$

x	y	z	f
0	1	1	1
1	0	0	1
1	1	0	1

Table 3.5

$$\therefore \quad f = \bar{x}.y.z + x.\bar{z}.(\bar{y} + y)$$
$$= \bar{x}.y.z + x.\bar{z}.1 \qquad (\bar{y} + y = 1)$$
$$= \bar{x}.y.z + x.\bar{z}$$

The corresponding circuit is given in Fig. 3.16.

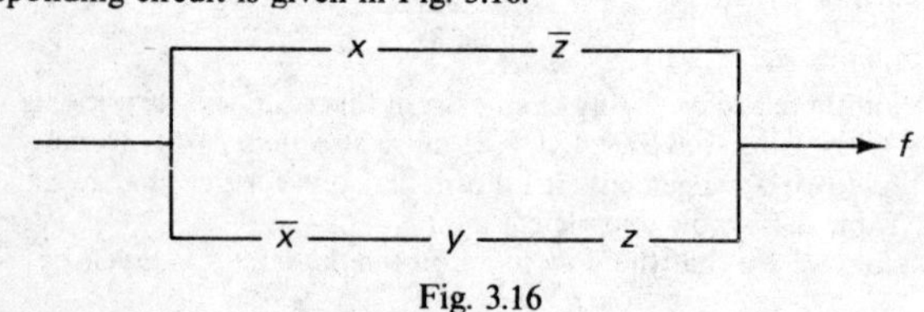

Fig. 3.16

Example 8. Obtain a switching circuit with the flow function given by the suggestion that x is not equivalent to y but x is implied by z.

Solution. We express the given suggestion in symbolic form:

(i) x is not equivalent to y; $x \not\equiv y$
(ii) x is implied by z: $z \to x$

The word 'but' has the same meaning as 'and' in this case.

$$\therefore \quad f = (x \not\equiv y).(z \to x)$$

Our problem is now to convert both $x \not\equiv y$ and $z \to x$ to a form involving only . and + for the switching circuit.

$$x \not\equiv y = \bar{x}.y + x.\bar{y} \qquad \text{(Exercise 3.3, Question 2)}$$
$$z \to x = \bar{z} + x \qquad \text{(Exercise 3.3, Question 2)}$$
$$\therefore f = (\bar{x}.y + x.\bar{y}).(\bar{z} + x) = \bar{x}.y.\bar{z} + x.\bar{y}.\bar{z} + x.\bar{y} \qquad \text{(distributive law)}$$
$$= \bar{x}.y.\bar{z} + x.\bar{y} \qquad \text{(absorption law)}$$

Alternatively, we can draw the truth tables as follows in Table 3.6.

x	y	z	$x \not\equiv y$	$z \to x$	$(x \not\equiv y).(z \to x)$
0	0	0	0	1	0
0	0	1	0	0	0
0	1	0	1	1	1
0	1	1	1	0	0
1	0	0	1	1	1
1	0	1	1	1	1
1	1	0	0	1	0
1	1	1	0	1	0

Table 3.6

$$\therefore \quad f = \bar{x}.y.\bar{z} + x.\bar{y}.\bar{z} + x.\bar{y}.z$$
$$= \bar{x}.y.\bar{z} + x.\bar{y}.(\bar{z} + z)$$
$$\therefore \quad f = \bar{x}.y.\bar{z} + x.\bar{y}.(1) \qquad (z + \bar{z} = 1)$$

The circuit is given in Fig. 3.17.

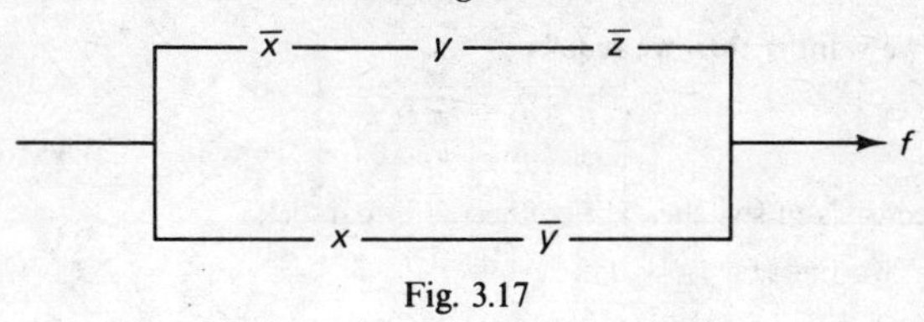

Fig. 3.17

Exercise 3.3

1. Using Table 3.6, express each of the following flow functions in terms of x, y, z and their negations:

   (i) $f_1 = 1$ on lines 1, 2, 7 only
   (ii) $f_2 = 1$ on lines 8, 7, 5 only
   (iii) $f_3 = 0$ on lines 1, 2, 3, 4, 6 only
   (iv) $f_4 = 1$ on lines 1, 3, 7 only

2. Draw a circuit for each of the following flow functions:

   (i) $f_1 = x \not\equiv y$
   (ii) $f_2 = x \rightarrow y$

3. Show that the switching circuit with flow function $f_1 = y.\bar{z}$ is not equivalent to the circuit of Fig. 3.17.
4. Find the flow function of the circuit which is obtained by wiring the circuits of Figs. 3.16 and 3.17 in (i) parallel, and (ii) series. Draw the final simplified circuit in each case.

Negation

In the previous section we saw how easy it was to select the appropriate lines of a truth table and produce the corresponding flow function.

With three variables there are eight possible combinations for the truth values. If six of these combinations yield $f = 1$ then the simplification process is involved with six different conjunctions like $x.y.z$ say. In a case like this it is better to concentrate on the two lines which produce $f = 0$ and plan the required circuit on this information rather than the other six lines. Consider Table 3.7 and the following three examples.

x	y	z	f_1	f_2	f_3
0	0	0	1	1	0
0	0	1	1	0	1
0	1	0	1	1	1
0	1	1	0	1	0
1	0	0	1	1	1
1	0	1	1	0	1
1	1	0	1	1	1
1	1	1	0	1	0

Table 3.7

Example 9. Two lines for $f_1 = 0$ are

$$\bar{x}.y.z + x.y.z = (\bar{x} + x).y.z$$
$$= y.z \qquad (\bar{x} + x = 1)$$

Therefore, if we want f_1 then we require

$$f_1 = \overline{y.z}$$
$$\text{i.e.}\quad f_1 = \bar{y} + \bar{z} \qquad \text{(DeMorgan's laws)}$$

This circuit consists of switches $\bar{y}$, $\bar{z}$ connected in parallel.

Example 10. Two lines for $f_2 = 0$ are

$$\bar{x}.\bar{y}.z + x.\bar{y}.z = (\bar{x} + x).\bar{y}.z = \bar{y}.z$$
$$\therefore\quad f_2 = \overline{\bar{y}.z}$$
$$= \bar{\bar{y}} + \bar{z} \qquad \text{(DeMorgan's laws)}$$
$$= y + \bar{z}$$

This circuit consists of switches y, $\bar{z}$ connected in parallel.

Example 11. Three lines for $f_3 = 0$ are

$$\bar{x}.\bar{y}.\bar{z} + \bar{x}.y.z + x.y.z = \bar{x}.\bar{y}.\bar{z} + (\bar{x} + x).y.z$$
$$= \bar{x}.\bar{y}.\bar{z} + y.z$$
$$\therefore\quad f_3 = \overline{\bar{x}.\bar{y}.\bar{z} + y.z}$$

The application of DeMorgan's laws needs a little care at this stage so it helps to recall that $\overline{A + B} = \bar{A}.\bar{B}$. Think of $A = \bar{x}.\bar{y}.\bar{z}$ and $B = y.z$

$$\therefore\quad f_3 = (\overline{\bar{x}.\bar{y}.\bar{z}}).(\overline{y.z})$$

Recall that $\overline{P.Q} = \bar{P} + \bar{Q}$ and $\overline{P.Q.R} = \bar{P} + \bar{Q} + \bar{R}$, so that

$$\overline{\bar{x}.\bar{y}.\bar{z}} = \bar{\bar{x}} + \bar{\bar{y}} + \bar{\bar{z}} = x + y + z$$
$$\text{and } \overline{y.z} = \bar{y} + \bar{z}$$
$$\therefore\quad f_3 = (x + y + z).(\bar{y} + \bar{z})$$

The circuit for f_3 follows in Fig. 3.18.

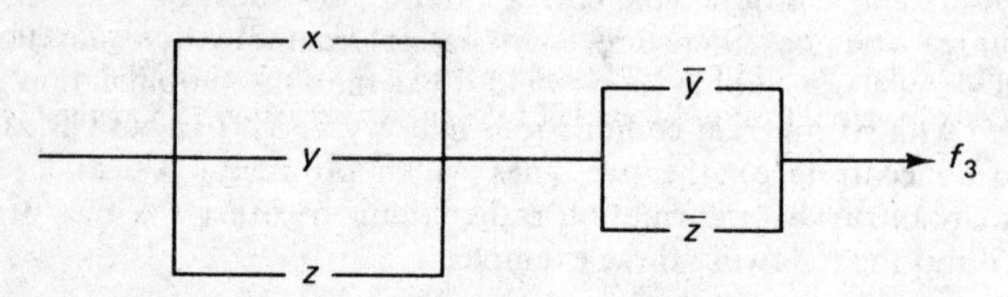

Fig. 3.18

There is an alternative method for finding the negation of a flow function which is carried out on the circuit diagram itself. We can appreciate the basic elements of the method by referring to the use of DeMorgan's laws which we

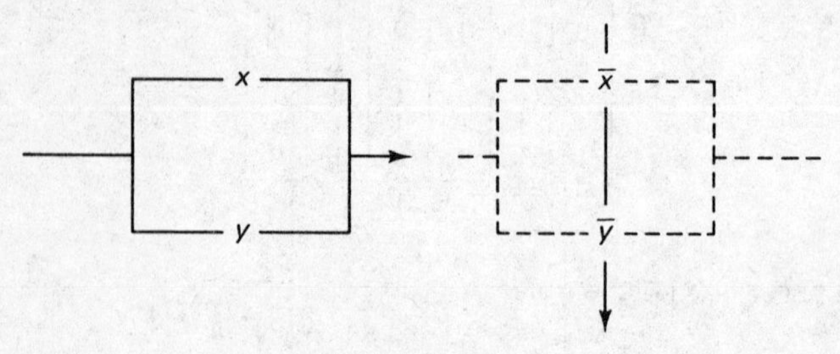

Fig. 3.19

have practised elsewhere. Recall that if we start with $x + y$ the negation becomes

$$\overline{x + y} = \bar{x}.\bar{y}$$

That is, a connection in parallel is converted to a connection in series and, as we see in Fig. 3.19, we insert the series connection onto the original diagram by reading down the page and negating the separate switches.

Again, if we start with $x.y$ the negation becomes

$$\overline{x.y} = \bar{x} + \bar{y}$$

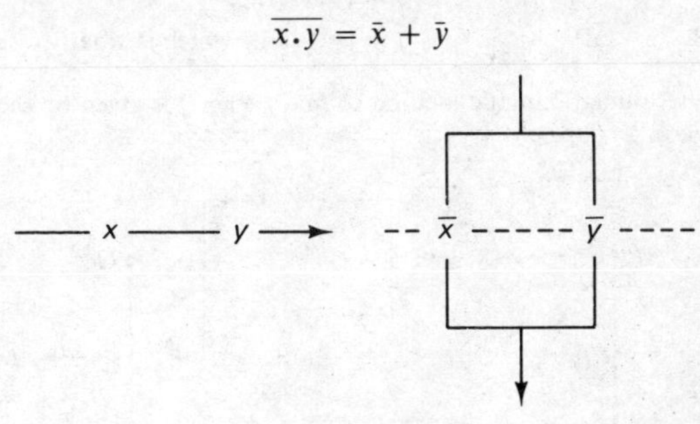

Fig. 3.20

That is, a connection in series is converted to a connection in parallel and, as we see in Fig. 3.20, we insert the parallel connection onto the original diagram by reading down the page and negating the separate switches. It only remains to combine these two basic changes in order to convert the general circuit.

Example 12. Obtain a circuit for f given the flow function $\bar{f} = \bar{x}.(y + z)$.

Solution. The circuit for $\bar{f}$ is drawn in dotted lines in Fig. 3.21 and negated circuits are drawn in unbroken lines in Fig. 3.22 with the original switches negated.

Realigning the circuit to read from left to right, we have Fig. 3.23 as the final circuit.

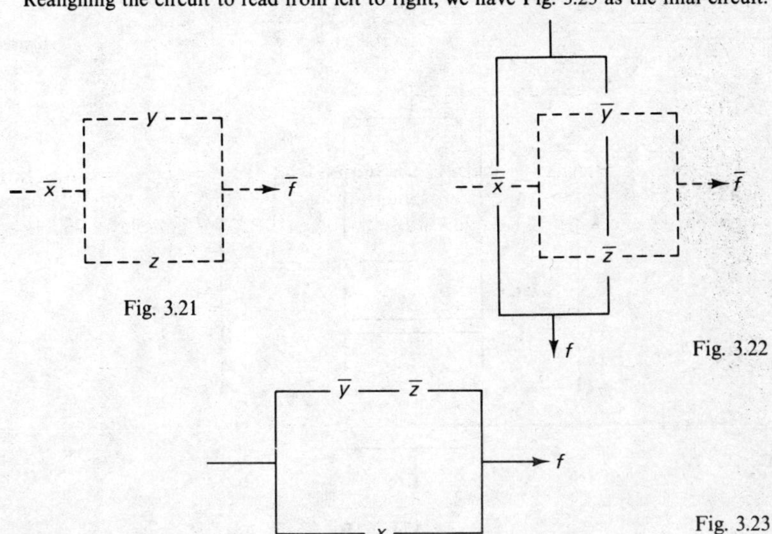

Fig. 3.21

Fig. 3.22

Fig. 3.23

As a somewhat unnecessary check on this result, consider carrying out the process again on Fig. 3.23 (left to the reader) or by applying DeMorgan's laws to the flow function of the circuit in Fig. 3.23.

$$\therefore \quad f = \bar{y}.\bar{z} + x$$
$$\therefore \quad \bar{f} = \overline{\bar{y}.\bar{z} + x}$$
$$= \overline{\bar{y}.\bar{z}}.\bar{x} \quad \text{(DeMorgan's Laws)}$$
$$= (\bar{\bar{y}} + \bar{\bar{z}}).\bar{x}$$
$$\therefore \quad \bar{f} = (y + z).\bar{x} \text{ which is what we started with.}$$

Answer

Example 13. Use a diagrammatic method to find f when $\bar{f}$ is given by the circuit in Fig. 3.24 (dotted lines).

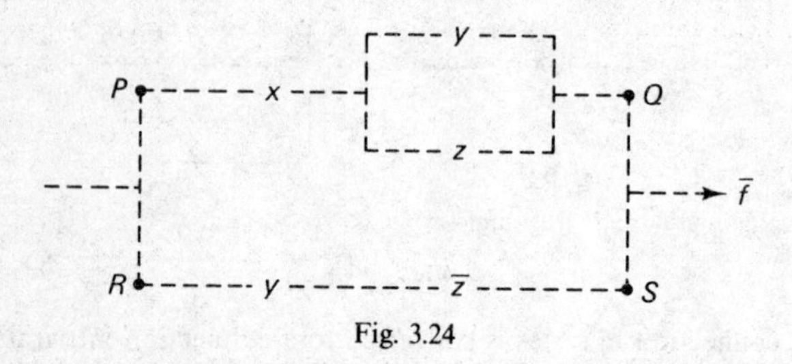

Fig. 3.24

Solution

There are two circuits PQ and RS which are in parallel so that after negation they will produce circuits in series.

The circuit PQ is similar to Fig. 3.21 and circuit RS is similar to Fig. 3.20. We merely put these two results together in Fig. 3.25. The final form is given by

$$\bar{\bar{f}} = f = (\bar{y}.\bar{z} + \bar{x}).(\bar{y} + z)$$
$$= \bar{y}.\bar{z} + \bar{x}.\bar{y} + \bar{x}.z$$
$$\therefore \quad f = \bar{y}.(\bar{z} + \bar{x}) + \bar{x}.z$$

Answer

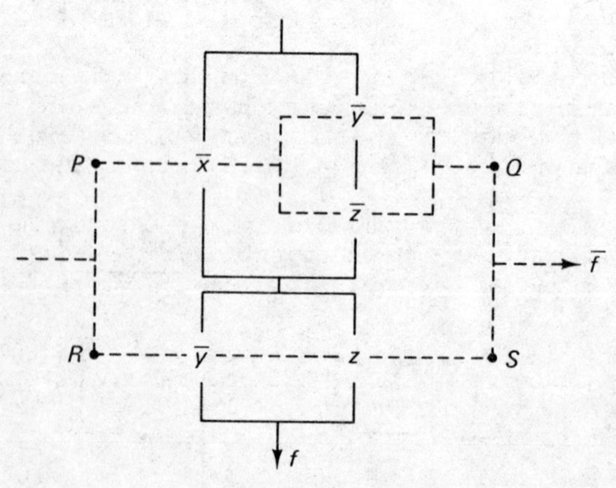

Fig. 3.25

Exercise 3.4

1. Use DeMorgan's laws to expand the following functions:

   (i) $\overline{\bar{x} + y}$
   (ii) $\overline{x + \bar{y}}$
   (iii) $\overline{\bar{x} + \bar{y}}$
   (iv) $\overline{x + y + z}$

2. Using Table 3.7 draw a circuit for f when:

   (i) $f = 0$ on lines 1, 2, 3, 4, 5, 6 only,
   (ii) $f = 0$ on lines 1, 2 only.

3. Using Table 3.7 a function f is 1 on lines 2, 4, 6, 7, 8 only. Obtain a circuit for f.
4. Draw the circuit for $f = x + y + z$ and construct the circuit for $\bar{f}$.
5. Using worked Example 12 above with Fig. 3.21, write down f given by $\bar{f} = x.(y + z)$.
6. Using worked Example 13 above with Fig. 3.24, write down f when $\bar{f}$ is given by the circuit of Fig. 3.24 with

   (i) the x switch removed,
   (ii) both y switches removed.

7. Use a diagrammatic method to find f when $\bar{f} = (x + y).(\bar{x} + z)$.

Gates

We have seen how a simple circuit of two switches x,y connected in series can represent the conjunction of two sentences by $x.y$ and we have adopted the notation of using . to represent AND. The wiring in series can be miniaturised and placed in a single unit called an **AND gate**. The notation used to indicate this replacement is shown in Fig. 3.26 (i). Thus the diagram indicates that x,y,z

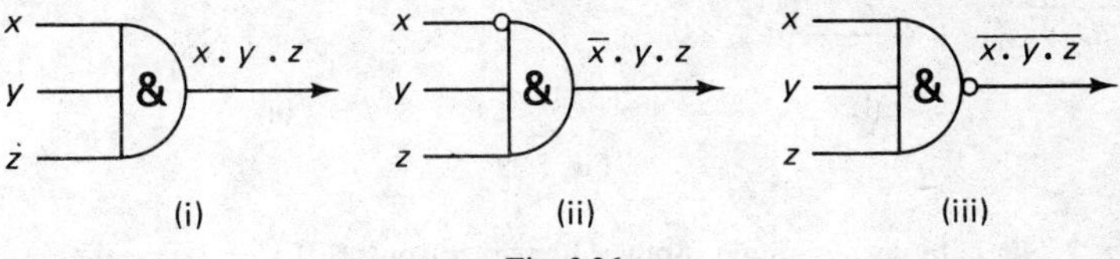

Fig. 3.26

are input to the AND gate and the resulting output is $x.y.z$ (x and y and z). The AND gate may also incorporate a negating device whose use is indicated in Fig. 3.26 (ii) by the small circle at the relevant input point and in this case giving an output of $\bar{x}.y.z$. A negater for the whole output is indicated in Fig. 3.26 (iii).

A similar situation holds for using + to represent OR so that the wiring for a parallel connection is contained in the single unit called an **OR gate**. The notation used to indicate this replacement is shown in Fig. 3.27 (i). The & of

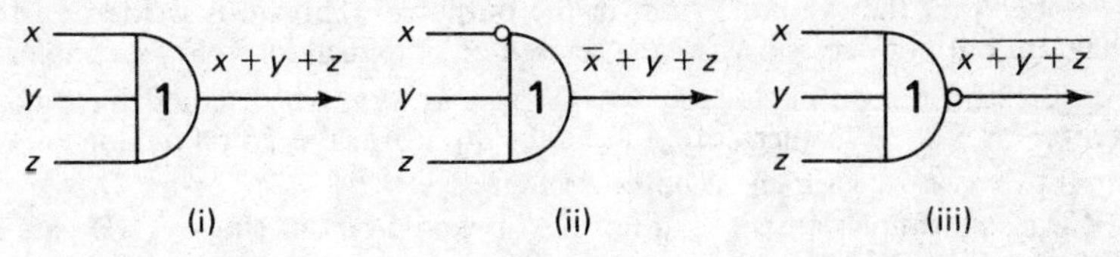

Fig. 3.27

the AND gate is replaced by the number 1 for the OR gate. The negaters are indicated as for AND gates and the corresponding outputs are shown in Fig. 3.27 (ii) and (iii). The number of inputs allowed will depend upon the design of the gate but in our short discussion we shall allow a maximum of three inputs.

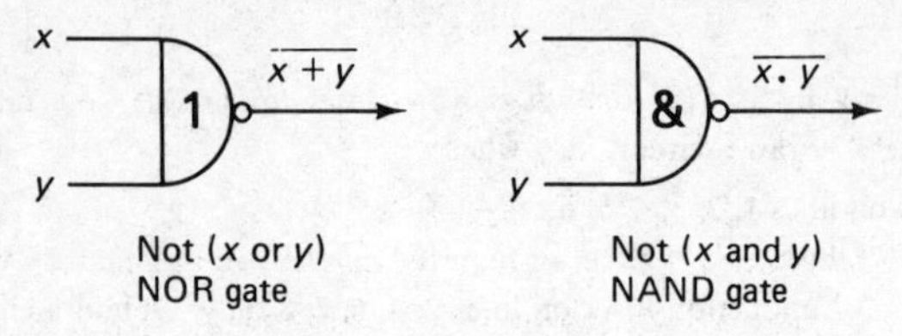

Fig. 3.28

Two other types of gates are **NOR** and **NAND** gates. Their meaning is fairly obvious since NOR is an abbreviation of 'NOT OR' and NAND is an abbreviation of 'NOT AND'. The notation diagrams for NOR and NAND gates are given in Fig. 3.26 (iii) and 3.27 (iii), but are repeated here in Fig. 3.28. Since all logic circuits may be constructed by using NOR gates exclusively or NAND gates exclusively one can appreciate the practical importance of these devices. Two circuits using NOR and NAND gates are shown in Fig. 3.29.

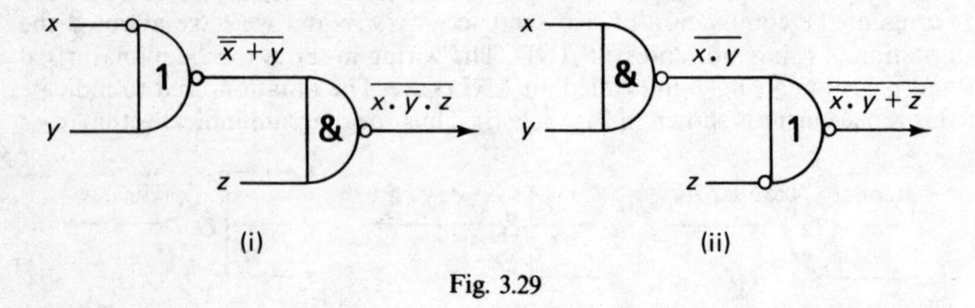

Fig. 3.29

A NOR gate with a single input x has an output $\bar{x}$. If $x = 0$ (i.e. there is no input) then the output is 1; thus a single NOR gate may be used as a negater. as shown in Fig. 3.30.

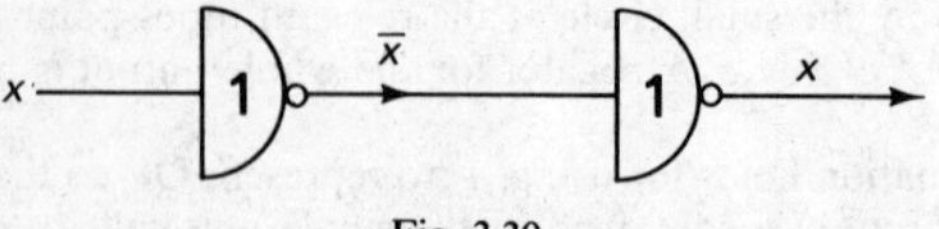

Fig. 3.30

Notice that the symbolic form of the output of the gate is written on the diagram but not necessarily in any revised or simplified form. For example, in Fig. 3.29 (ii) the output of the NAND gate is $\overline{x.y}$, which could have been written as $\bar{x} + \bar{y}$; furthermore, the final output, i.e. the flow function of the circuit, is $\overline{\overline{x.y} + \bar{z}}$, which could be written as $x.y.z$.

Clearly, a simpler circuit for Fig. 3.29 (ii) would be one single AND gate as illustrated in Fig. 3.31.

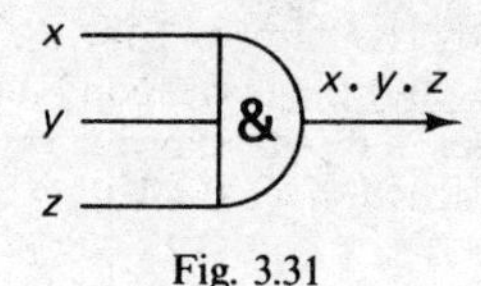

Fig. 3.31

Example 14. Explain how to analyse the flow function to see how many gates are necessary to obtain

$$f = (x + \bar{x}.\bar{y}).(y + \bar{x}.\bar{y})$$

Solution. We may suggest a number of required gates by noting that we need

two AND gates, one each for $\bar{x}.\bar{y}$ and ( ).( )
two OR gates, one for each (+) and (+)

The suggested circuit is now drawn in Fig. 3.32.

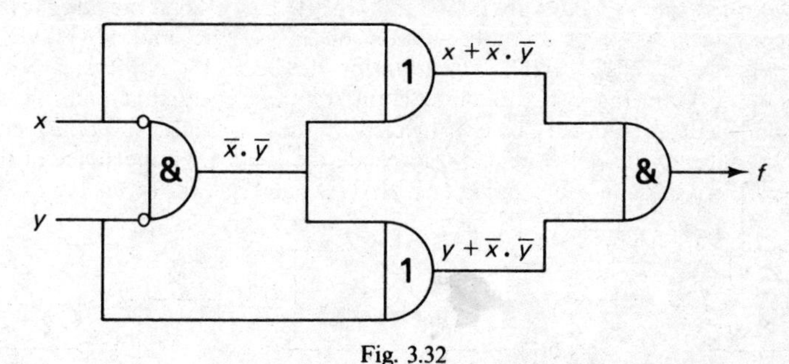

Fig. 3.32

Alternatively, we can simplify the proposition first by using the distributive law

$$\begin{aligned} f &= (x + \bar{x}.\bar{y}).(y + \bar{x}.\bar{y}) \\ &= x.y + \bar{x}.\bar{y} \qquad (x.\bar{x} = 0 \text{ and } y.\bar{y} = 0) \end{aligned}$$

which can be obtained by using only two AND gates with only one OR gate, as discussed in Example 16 below. *Answer*

As another alternative we can see

$$f = (x + \bar{y}).(\bar{x} + y)$$

with a different logic circuit using only one AND gate with two OR gates.

As a further alternative we may consider

$$\begin{aligned} f &= (x + \bar{y}).(\bar{x} + y) \\ &= \overline{(\bar{x}.y)}.\overline{(x.\bar{y})} \qquad \text{(DeMorgan's law)} \end{aligned}$$

and suggest a circuit using only two NAND gates with one AND gate, similar to Example 17 with $z = x$. *Answer*

Example 15. Obtain the same flow function as for the circuit in Fig. 3.33 by using NOR gates only. (Try to obtain f before reading on.)

Solution. We start by simplifying the flow function using the distributive law.

$$\begin{aligned} f &= \overline{(w + y).(\bar{x} + y)} \\ &= \overline{(w + y)} + \overline{(\bar{x} + y)} \qquad \text{(i) (DeMorgan's law)} \\ &= \bar{w}.\bar{y} + x.\bar{y} \qquad \text{(DeMorgan's law)} \\ f &= \bar{y}.(\bar{w} + x) \end{aligned}$$

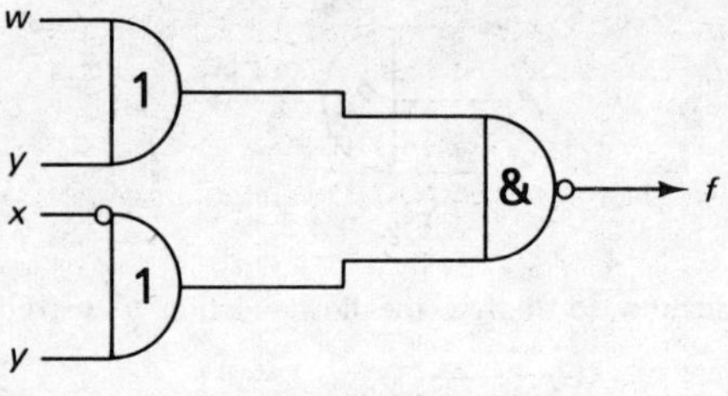

Fig. 3.33

Now the clue for the NOR circuit really lies in observing line (i), where we see the result expressed in terms of + only. This gives us the idea for the first two NOR gates, to obtain $\overline{w + y}$ and $\overline{\bar{x} + y}$ separately. After this we use two NOR gates in series, as displayed in Fig. 3.30, to produce the OR condition and the final result f as in Fig. 3.34.

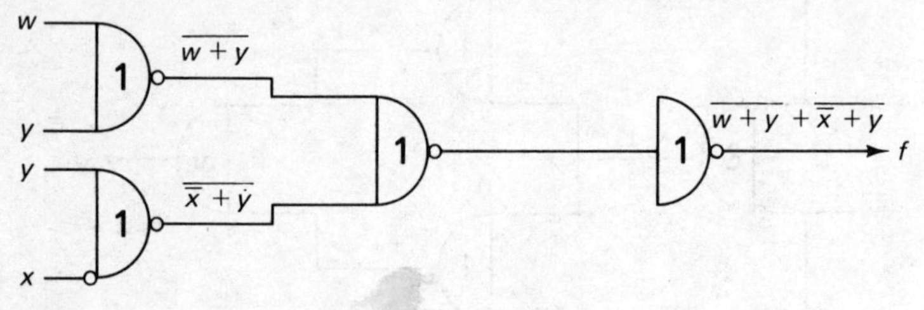

Fig. 3.34

Example 16. Draw a circuit using only NOR gates to represent the flow function $f = x.y + \bar{x}.\bar{y}$.

Solution. We have already seen in Examples 14 and 15 that by using appropriate inputs a NOR gate can be made to perform the AND operation. This depends upon DeMorgan's law that $\overline{\bar{x} + \bar{y}} = x.y$ so that a NOR gate with $\bar{x},\bar{y}$ as inputs is equivalent to an AND gate with x,y inputs. Examining the function f, we see $x.y$, $\bar{x}.\bar{y}$ as two such uses for NOR gates and the final circuit is seen in Fig. 3.35.

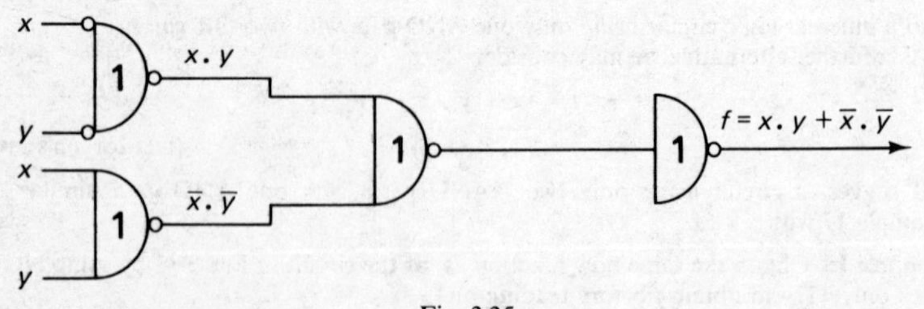

Fig. 3.35

Example 17. Use only AND and NAND gates to obtain a logic circuit to represent $(x \to y).(y \to z)$.

Solution. We first convert the function $f = (x \to y).(y \to z)$ to one involving . and + only

$$\therefore \quad f = (\bar{x} + y).(\bar{y} + z)$$

Since we shall require NAND gates to be used it follows that we must now see this expression in terms of . only, so we use DeMorgan's laws to get

$$f = \overline{(x.\bar{y})}.\overline{(y.\bar{z})}$$

and thereby see that we shall need two NAND gates and one AND gate to obtain f as shown in Fig. 3.36.

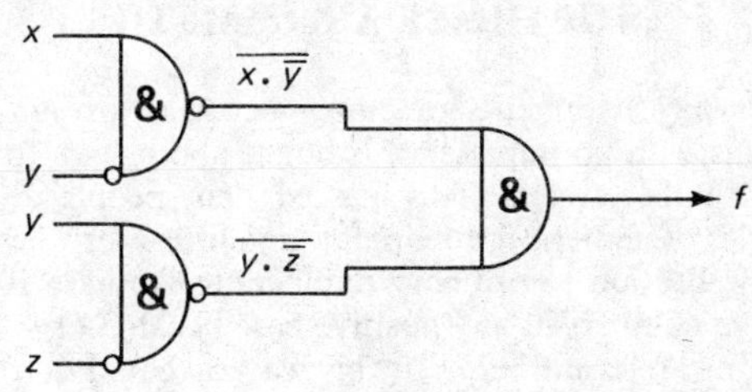

Fig. 3.36

Exercise 3.5

1. State the output of an OR gate receiving the following inputs:

   (i) x, y
   (ii) $\bar{x}, y$
   (iii) $\bar{x}, \bar{y}$
   (iv) $\bar{x}, y, \bar{z}$
   (v) $x.y, 1$
   (vi) $x.y, x$
   (vii) $x.y, y$
   (viii) $\bar{x}.\bar{y}, \bar{z}$

2. State the output of a NOR gate receiving the inputs of Question 1.
3. State the output of an AND gate receiving the inputs of Question 1.
4. State the output of a NAND gate receiving the inputs of Question 1, expressing each output as a disjunction by using DeMorgan's laws.
5. State how many gates are necessary to obtain the function

$$f = (x + \bar{z}).(y + x.z)$$

   Draw a circuit for f using AND as well as OR gates.
6. Use DeMorgan's laws to obtain the circuit for $f = x + \bar{y}.\bar{z}$ using NOR gates only.
7. Obtain a circuit for $f = x + \bar{y}.\bar{z}$ (Question 6) using NAND gates only.
8. Draw a circuit for $f = x.\bar{y} + \bar{x}.y$ using (i) NOR gates only, and (ii) NAND gates only.

4

MULTIBASE ARITHMETIC

Mistakes in everyday arithmetic sometimes occur as a result of treating elementary problems with an impatience brought about by a too shallow familiarity with the number system. This has led some people to believe that we might see the rules of numerical manipulation a little more clearly if instead of studying their application to ordinary numbers to the base 10 we shifted our interest to involve numbers in any positive base like 2, 3 or 4. (Negative bases are possible but not relevant here.) In this way we could do the same type of remedial exercise but with the possibility that the novelty of the new context might reveal our misunderstandings in a more meaningful light.

Multibase arithmetic has more than a remedial purpose since it enables us to look at some well-known puzzles and examine the binary number system which provides the basis for mathematical operations with computers.

We shall express any number in any base we please and these new numbers will be written in the usual shorthand by which both the position and size of a digit is used to indicate its numerical value. That is, just as 23 represents $(10 \times 2) + 3$ so a number like 8257.46 is an abbreviation for the resulting addition

$$8527.46 = (1000 \times 8) + (100 \times 5) + (10 \times 2) + (1 \times 7) + (\tfrac{1}{10} \times 4) + (\tfrac{1}{100} \times 6)$$

We sometimes describe this result by saying that the 8 is in the thousands column, the 5 is in the hundreds column, the 2 is in the tens column and the 7 is in the units column. The point is only used to separate the fractional part from the whole number part and the number continues with 4 in the tenths column and 6 in the hundredths column.

Thus, the individual digits may be considered to have values which vary according to their place in the number; hence the reference to 'place values'. The sequence of headings starting from the thousands column is shown in Fig. 4.1.

	10^3	10^2	10^1	10^0	10^{-1}	10^{-2}	10^{-3}
	1000	100	10	1	$\frac{1}{10}$	$\frac{1}{100}$	$\frac{1}{1000}$
(i)	2	5	6	3	0	0	0
(ii)		4	9	6	0	0	0
(iii)		8	7	3	4	9	

Fig. 4.1

To emphasise the composition of a denary number once again we note the third number in detail.

$$873.49 = (10^2 \times 8) + (10^1 \times 7) + (10^0 \times 3) + (10^{-1} \times 4) + (10^{-2} \times 9)$$

It is clear that we may recast any number we please by merely changing the value of the headings of columns. Thus, numbers expressed to base 8 will have columns with place values $8^3, 8^2, 8^1, 8^0, 8^{-1}, 8^{-2}, \ldots$ and, similarly, numbers expressed to base 6 will have columns with place values $6^3, 6^2, 6^1, 6^0, 6^{-1}, 6^{-2}, \ldots$

Notation

In order to distinguish numbers expressed in different bases we shall use a subscript notation as follows:

1. 23_{10} will be read as 'two three to base 10', i.e. the everyday twenty-three.
2. 23_8 will be read as 'two three to base eight'. We avoid reading this as 'twenty-three to base eight'. Note that the size of this number is given by the sum $(8^1 \times 2) + 3 = 19_{10}$. In this manner we convert 23_8 to base 10 so that we can write $23_8 = 19_{10}$. By inspection we can see that $23_{10} = (8^1 \times 2) + 7 = 27_8$.
3. 31_6 will be read as 'three one to base six' and again we see the size of the number in base 10 by writing $31_6 = (6^1 \times 3) + 1 = 19_{10}$. Again, notice that $31_6 = 23_8$.
4. Observe that the base subscript is always written as a denary number so that a hexadecimal number, i.e. a number to base 16, will be written like 139_{16} and expanded as

$$\begin{aligned} 139_{16} &= (16^2 \times 1) + (16^1 \times 3) + (16^0 \times 9) \\ &= 256 + 48 + 9 \text{ (all denary numbers)} \\ \therefore \quad 139_{16} &= 313_{10} \end{aligned}$$

Single digits may be written without base reference, e.g. $9_{10} = 9_{16} = 9$.
5. 32.32_4 will be read as 'three two point three two to base four' and we refer to 'the point' rather than 'the decimal point' since 'decimal' is associated with numbers to base 10. Expansion of this particular number gives

$$\begin{aligned} 32.32_4 &= (4^1 \times 3) + (4^0 \times 2) + (4^{-1} \times 3) + (4^{-2} \times 2) \\ &= 12 + 2 + \tfrac{3}{4} + \tfrac{2}{16} \text{ (all denary numbers)} \\ &= 14\tfrac{7}{8}\,{}_{10} \\ &= 14.875_{10} \end{aligned}$$

To summarise these initial ideas consider the reference table for octal numbers, i.e. numbers to base 8, in Fig. 4.2. The first octal number may be converted to base 10 as follows

	8^4	8^3	8^2	8^1	8^0	8^{-1}	8^{-2}	8^{-3}
	4096	512	64	8	1	$\frac{1}{8}$	$\frac{1}{64}$	$\frac{1}{512}$
(i)			3	2	3	5	0	1
(ii)		4	0	5	2	0	4	
(iii)			1	3	7	1	1	

Fig. 4.2

$$323.501_8 = (8^2 \times 3) + (8^1 \times 2) + (8^0 \times 3) + (8^{-1} \times 5) + (8^{-2} \times 0) + (8^{-3} \times 1)$$

$= 192 + 16 + 3 + \frac{5}{8} + \frac{1}{512}$ (all denary numbers)
$= 211\frac{321}{512}{}_{10}$ (all denary numbers)

Obviously the fractional part in some examples may be very tedious to assemble into one single fraction.

In Fig. 4.3 we give the reference table for binary numbers, i.e. numbers to the base 2. For binary numbers the only digits are 0, 1 so that a denary number of only modest size will require many 0, 1's when converted to base 2.

	2^5	2^4	2^3	2^2	2^1	2^0	. 2^{-1}	2^{-2}	2^{-3}	2^{-4}	2^{-5}
	32	16	8	4	2	1	. $\frac{1}{2}$	$\frac{1}{4}$	$\frac{1}{8}$	$\frac{1}{16}$	$\frac{1}{32}$
(i)	1	0	1	0	1	1	. 1	0	1		
(ii)		1	0	1	0	1	. 0	1			
(iii)			1	1	1	1	. 1	1	1		

Fig. 4.3

The first number is converted to base ten by the same method as the octal numbers shown previously

$101\,011.101_2 = 32 + 0 + 8 + 0 + 2 + 1 + \frac{1}{2} + 0 + \frac{1}{8}$ (all denary numbers)
$= 43\frac{5}{8}{}_{10}$
$= 43.625_{10}$

Conversion of numbers into denary numbers by using the reference table is by far the most straightforward general method, but there are one or two interesting shortcuts when converting from bases 2 to 4, 2 to 8, 2 to 16_{10}, 4 to 16_{10} and 3 to 9, as we shall see after the next Exercise.

Exercise 4.1

Convert the following numbers into denary numbers:

1. 32_8
2. 32_6
3. 32_4
4. 121_8
5. 137_8
6. 43_7
7. 11.4_5
8. 12.16_7
9. 53.14_6
10. 101.011_2
11. 34.34_5
12. 154.31_9

Conversion From Base 10

The general method of converting a number from any base to base 10 may be easily identified by looking at a particular conversion from, say, base 8 to base 10.

To find the powers of 8 in any number we divide repeatedly by 8 until we get a zero quotient. For our example consider step by step the conversion of 2375_{10} to an octal number as follows:

Conversion table

				8^4	8^3	8^2	8^1	8^0
Step (i)	8	2375						
Step (ii)	8	296 remainder 7	$\therefore\ 2375_{10} =$				296	7
Step (iii)	8	37 remainder 0	$\therefore\ 2375_{10} =$			37	0	7
Step (iv)	8	4 remainder 5	$\therefore\ 2375_{10} =$		4	5	0	7
		0 remainder 4	$\therefore\ 2375_{10} =$	0	4	5	0	7

The division process is now at an end because the last quotient is 0. Hence

$$2375_{10} = 4507_8$$

The method is the same for any conversion from base 10. To emphasise this we consider converting 2375_{10} to a hexadecimal number (i.e. base 16_{10}). We follow the same work plan as before:

Conversion table

				16^3	16^2	16^1	16^0
Step (i)	16	2375					
Step (ii)	16	148 remainder 7	$\therefore\ 2375_{10} =$			148	7
Step (iii)	16	9 remainder 4	$\therefore\ 2375_{10} =$		9	4	7
Step (iv)	16	0 remainder 9	$\therefore\ 2375_{10} =$	0	9	4	7

The division is now at an end because the last quotient is 0. Hence

$$2375_{10} = 947_{16}$$

One point should have become clear by now and that is that a whole number in any base remains a whole number when converted into any other base. In other words, the conversion will never introduce a fractional part if there was not a fractional part to begin with. Because we have other uses for the result we now expect the reader to convert 2375_{10} to a binary number using the same method and confirm by dividing 2375_{10} repeatedly by 2 that $2375_{10} = 100\ 101\ 000\ 111_2$.

If we group the digits of this binary number in triplets from the point it will look like

$$2375_{10} = 100\ 101\ 000\ 111_2$$

Now find the size of each triplet, i.e. $100_2 = 4$, $101_2 = 5$, $111_2 = 7$, and observe that we now have $2375_{10} = 100\ 101\ 000\ 111_2 = 4\ 5\ 0\ 7_8$. This shortcut conversion has used the relation $2^3 = 8$.

Similarly, by noting that $2^4 = 16$ we group the digits of this binary number in quartets from the point to get $2375_{10} = 1001\ 0100\ 0111_2$. Again we find the 'size' of each quartet, i.e.

$$1001_2 = 9,\ 0100_2 = 4,\ 0111_2 = 7$$

and observe this time that we now have

$$2375_{10} = 947_{16}$$

Do note that a number to the base 16_{10} will require a maximum of 16_{10} digits so we need to invent a few extras above the digits 0 to 9—in fact, we use the capital letters A to F to represent 10_{10} to 15_{10} in that order.

We can deduce a similar connection between numbers in base 3 and base 9. Since $9 = 3^2$ we shall group the digits of the base three numbers in pairs. Consider the number

$$\begin{aligned} 150_{10} &= 12120_3 \\ &= 01{,}21{,}20_3. \end{aligned}$$

Now find the 'size' of each pair

$$01_3 = 1,\ 21_3 = 7,\ 20_3 = 6$$
$$\therefore\quad 12120_3 = 176_9$$

The same method holds for the fractional parts of a number with bases related as in the above example, the grouping of the digits being measured from the point, with the appropriate addition of zeros to make up the last group if necessary.

Example 1. Express the binary number 1101.1011_2 to bases 4, 8 and 16_{10}.

Solution. To convert from base 2 to base 4 we group the digits in pairs from the point because $2^2 = 4$.

$$\therefore\quad 11, 01, .10, 11_2 = 31.23_4$$

To convert from base 2 to base 8 we group the digits in triplets from the point because $2^3 = 8$.

$$001, 101.101, 100_2 = 15.54_8$$

To convert from base 2 to base 16_{10} we group the digits in quartets from the point because $2^4 = 16_{10}$.

$$1101.1011_2 = \mathrm{D.B}_{16}$$

recalling $\mathrm{D} = 13_{10}$, $\mathrm{B} = 11_{10}$, in hexadecimal notation.

This last result shows the number to be $13\frac{11}{16}{}_{10}$ *Answer*

Example 2. Convert the number 1234_5 to base 7.

Solution. With no shortcuts available like those in Example 1 above we must convert our number into base 10_{10} and then into base 7. For the first stage we use the reference table for base 5 and so obtain the expansion

$$\begin{aligned} 1234_5 &= (5^3 \times 1) + (5^2 \times 2) + (5^1 \times 3) + (5^0 \times 4) \\ &= 125 + 50 + 15 + 4 \text{ (all denary numbers)} \\ &= 194_{10} \end{aligned}$$

For the second stage we divide 194_{10} repeatedly by 7 until we obtain a zero quotient.

Conversion table

				7^3	7^2	7^1	7^0
Step (i)	7	194					
Step (ii)	7	27 remainder 5	$\therefore\ 194_{10} =$			27	5
Step (iii)	7	3 remainder 6	$\therefore\ 194_{10} =$		3	6	5
		0 remainder 3	$\therefore\ 194_{10} =$	0	3	6	5

Hence

$$\begin{aligned} 1234_5 &= 194_{10} \\ &= 365_7 \end{aligned}$$
Answer

Finally in this section we deal with the general conversion of a number having a fractional part. The method to be used will be apparent when we examine in detail exactly what we are trying to do. Extending Example 2, we propose to convert 0.485_{10} to base 7. For the fractional part we must work left to right from the point into the columns headed $7^{-1}, 7^{-2}, 7^{-3}, \ldots$ Clearly the first move is to find how many sevenths there are in 0.485_{10}.

Dividing by $\frac{1}{7}$ is the same as multiplying by 7, and since

$$0.485_{10} \times 7 = 3.395_{10}$$

we now know that there are 3 whole sevenths in 0.485_{10}, which we remove, entering 3 in the 7^{-1} column and leaving $.395_{10}$ behind.

The next step is to find how many sevenths in 0.395_{10}. Again this is the same as multiplying by 7 and since

$$0.395_{10} \times 7 = 2.765_{10}$$

it follows that there are two forty-ninths for 0.485_{10}, which we remove, entering 2 in the 7^{-2} column and leaving $.765_{10}$ behind. Continuing in this way with repeated multiplication by 7 will produce the conversion to as many places as we please. We use the following work scheme for converting 0.485_{10} to base 7:

Conversion table

		7^{-1}	7^{-2}	7^{-3}	7^{-4}	7^{-5}
	0.485					
Step (i)	7					
	3 | .395	3				
Step (ii)	7					
	2 | .765	3	2			
Step (iii)	7					
	5 | .355	3	2	5		
Step (iv)	7					
	2 | .485	3	2	5	2	

We are now back where we started, i.e. multiplying 0.485 by 7. It follows that the results 3252 will recur over and over again. The recurrence may be represented by drawing a bar over those digits which recur so that the final answer may be written

$$0.485_{10} = 0.\overline{3252}_7$$

meaning that $0.485_{10} = 0.325232523252\ldots_7$.

Clearly, for conversions of numbers like 194.485_{10} to another base we must work on the whole number and fractional parts separately and add the two final results. Combining the result just obtained with that of Example 2, we have

$$194.485_{10} = 365.\overline{3252}_7$$

Approximation

We sometimes need to restrict a denary number to so many decimal places. Such a restriction means that we are offering an approximation to the original number. For example, an approximation to the number 29.76_{10} may be suggested by either 29.7_{10} or 29.8_{10}. Quoting 29.7_{10} means that we have lost 0.06_{10} from 29.76_{10}; this suggestion would be called 'rounding down'. Quoting 29.8_{10} means that we have gained 0.04_{10} on 29.76_{10}; this suggestion would be called 'rounding up'. Clearly, we shall take the suggestion 29.8_{10} because it is closer to the original number 29.76_{10}.

For another example consider offering a two-decimal place approximation to 6.805_{10}. Quoting 6.80_{10} means a loss of 0.005_{10} from 6.805_{10}, i.e. 'rounding down'. Quoting 6.81_{10} means a gain of 0.005_{10} on 6.805_{10}, i.e. 'rounding up'. In cases like this where both approximations are equally good we always choose to round up and write

$$6.805_{10} = 6.81_{10} \quad \text{(2 d.p.)}$$

We therefore obtain approximations to any number of decimal places by choosing to 'round up' if the next digit is 5 or more and rounding down if the next digit is less than 5.

Example 3

$$\begin{aligned} 3.3746_{10} &= 3.375_{10} \quad \text{(3 d.p.)} \\ 3.3746_{10} &= 3.37_{10} \quad \text{(2 d.p.)} \\ 3.375_{10} &= 3.38_{10} \quad \text{(2 d.p.)} \end{aligned}$$

Notice that the second result, 3.37_{10} (2 d.p.), must come from an inspection of the original number and *not* from 3.375_{10}, which is already an approximation to the original number.

Example 4

$$\begin{aligned} 2.2499_{10} &= 2.250_{10} \quad \text{(3 d.p.)} \\ 2.2499_{10} &= 2.25_{10} \quad \text{(2 d.p.)} \\ 2.2499_{10} &= 2.2_{10} \quad \text{(1 d.p.)} \end{aligned}$$

We must now consider how to obtain similar approximations to numbers in other bases. Note that we shall refer to approximations expressed to so many places and not decimal places since the numbers are no longer denary.

Approximating to numbers expressed to even bases will be similar to the method for base 10, i.e. we examine the 'next digit' to see if it is less than half the base.

Example 5

$$\begin{aligned} 2.234_8 &= 2.24_8 \quad \text{(2 p.)} \\ 2.234_8 &= 2.2_8 \quad \text{(1 p.)} \\ 2.24_8 &= 2.3_8 \quad \text{(1 p.)} \end{aligned}$$

Example 6

$$\begin{aligned} 1.523_6 &= 1.53_6 \quad \text{(2 p.)} \\ 1.523_6 &= 1.5_6 \quad \text{(1 p.)} \\ 1.53_6 &= 2.0_6 \quad \text{(1 p.)} \end{aligned}$$

Example 7

$$\begin{aligned} 1.1101_2 &= 1.111_2 \quad \text{(3 p.)} \\ 1.1101_2 &= 1.11_2 \quad \text{(2 p.)} \\ 1.1101_2 &= 10.0_2 \quad \text{(1 p.)} \end{aligned}$$

The principle of comparing with 'half the base' is the same for odd bases, but this time we must examine the 'next two digits' in the number, as shown in the following examples.

For example, 0.245_9 quoted as 0.2_9 (one place), means a loss of 0.045_9, but quoted as 0.3_9 (1 place) means a gain of 0.044_9 and therefore a better approximation, because this quotation is nearer the original.

We therefore have the following rules:

For base nine, if the next two digits form a number equal to or greater than 45, we add 1 to the digit in the previous place.

Thus 0.645_9 becomes 0.7_9 (one place) and 0.65_9 (two places)
again 1.2736_9 becomes 1.274_9 (three places)
or 1.27_9 (two places)
or 1.3_9 (one place)

For numbers to base seven, if the next two digits form a number equal to or greater than 34.

For numbers to base five, if the next two digits form a number greater than or equal to 23.

For numbers to base three if the next two digits form a number greater than or equal to 12.

Example 8

$2.145_9 = 2.15_9$ (2 p.) because the digits in the third and fourth places make 50, which is greater than 45 (round up).

$2.145_9 = 2.2_9$ (1 p.) because the digits in the second and third places make 45 (round up).

Example 9

$6.11336_7 = 6.114_7$ (3 p.) because the digits in the fourth and fifth places make 36, which is greater than 34 (round up).

$6.11336_7 = 6.11_7$ (2 p.) because the digits in the third and fourth places make 33, which is less than 34 (round down).

Example 10

$2.2112_3 = 2.22_3$ (2 p.) because the digits in the third and fourth places make 12 (round up).

$2.2112_3 = 2.2_3$ (1 p.) because the digits in the second and third places make 11, which is less than 12 (round down).

Exercise 4.2

Quote each of the following numbers to (3, 2, 1) places:

1. 8.83535_9
2. 7.7434_8
3. 5.63535_7
4. 2.2324_6
5. 1.4324_5
6. 2.3321_4
7. 2.2222_3
8. 1.01101_2
9. 16.8345_9
10. 11.1101_2

Obtain the following conversions:

11. 208_{10} to base 7.
12. 0.56_{10} to base 7.
13. 208.56_{10} to base 7.
14. 15.4_{10} to base 3.
15. 15.4_{10} to base 9.
16. 468.735_9 to base 3.
17. 11010.11011_2 to bases 4, 8 and 16.
18. 67.43_8 to bases 4 and 2.
19. 69.63_{10} to base 5.
20. 1132.13_4 to base 6.
21. 263.31_7 to base 5 working to 4 places.
22. 222.22_3 to base 8 working to 4 places.

Addition

Our remarks at the beginning of the chapter suggested that an examination of the structure of numbers in other bases would help to see denary numbers in a more meaningful light. This search for greater understanding continues with some attention to the usual operations of $+$, $-$, $\times$ and $\div$. The first operation of addition has already been used in the approximation work of the previous section. Here we wish to revise the addition of numbers in the context of the formal scheme which we use with denary numbers.

Example 11. Find the sum of 656.1_7 and 335.4_7.

Solution. With the rule that a subtotal of 7 in any column is replaced by 1 in the adjacent column to the left we may proceed with the addition:

	7^3	7^2	7^1	7^0	7^{-1}	
		6	5	6 .	1	
+		3	3	5 .	4	
		9	8	11 .	5	(Sub Addition in base ten)
		9	9	4 .	5	Use $11_{10} = 14_7$
		10	2	4 .	5	Use $9 = 12_7$ and $10_{10} = 13_7$
	1	3	2	4 .	5	= complete answer to base 7

The required sum is 1324.5_7.

Of course, we usually do the conversion of the subtotal mentally as we proceed with the addition, as illustrated in the next example.

Example 12. Find the sum of 678_9, 1756_9 and 387_9.

Solution. With the usual scheme of placing the digits in corresponding like columns we work from right to left in the usual manner. A subtotal of 9 in any column is replaced by 1 in the adjacent column to the left:

	9^3	9^2	9^1	9^0
		6	7	8
	1	7	5	6
+		3	8	7
	3	0	4	3
	2	2	2	

Step (i): First column, addition yields $21_{10} = 23_9$. Register 3 and carry 2.
Step (ii): Second column, addition yields $22_{10} = 24_9$. Register 4 and carry 2.
Step (iii): Third column, addition yields $18_{10} = 20_9$. Register 0 and carry 2.
Step (iv): Fourth column, addition yields 3. Register 3.
The required sum is 3043_9. *Answer*

Subtraction

For subtraction we shall use the method of decomposition in which any number may have part of the contents of one column converted (decomposed) into a column of lower denomination. Consider three different decompositions of the number 2314.51_6 in Fig. 4.4.

	6^3	6^2	6^1	6^0	6^{-1}	6^{-2}
(i)	2	3	1	4 .	5	1
(ii)	2	3	0	4 + 6 .	4	1 + 6
(iii)	2	2	0 + 6	3 + 6 .	4 + 6	1 + 6

Fig. 4.4

The original number in (i) may be reorganised into the arrangement (ii) or the arrangement (iii) by using the fact that a subtotal of 6 in any column is equal to an entry of 1 in the next column on the left—or, to put it another way, a subtotal of 1 in any column is equal to an entry of 6 in the next column on the right.

The reason for this manipulation will become clear in the following two examples.

Example 13. Find $323.5_6 - 154.2_6$.

Solution. The usual scheme of aligning the digits is followed and the idea is to break the overall subtraction into separate subtractions within each column.

	6^2	6^1	6^0	6^{-1}
	3	2	3 .	5
−	1	5	4 .	2
			.	

In the units column 3 − 4 is unsatisfactory so we make the decomposition shown in step (i):

		6^2	6^1	6^0	6^{-1}	
Step (i)		3	1	3 + 6 .	5	
	−	1	5	4 .	2	
				5 .	3	i.e. 3 + 6 − 4 = 5

In the next column 1 − 5 is unsatisfactory so we make the decomposition shown in step (ii):

		6^2	6^1	6^0	6^{-1}	
Step (ii)		2	1 + 6	3 + 6 .	5	
	−	1	5	4 .	2	
		1	2	5 .	3	i.e. 1 + 6 − 5 = 2

The subtraction in the final column is now completed and so we have

$$323.5_6 - 154.2_6 = 125.3_6$$

a result which also means that

$$323.5_6 = 154.2_6 + 125.3_6$$

Example 14. Find $6152_7 - 2456_7$.

Solution. We now perform the individual column subtractions and the necessary decompositions on the same layout without headings.

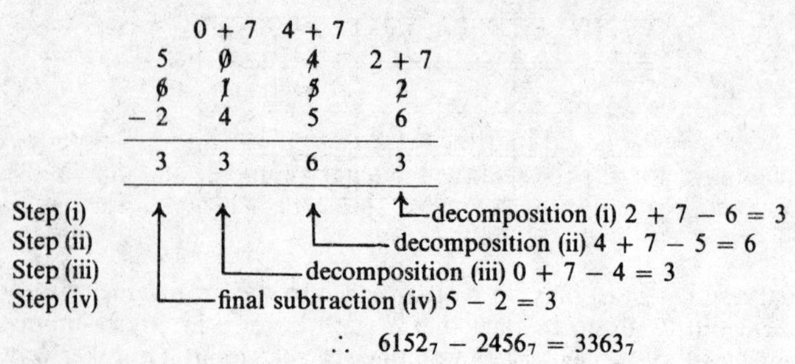

$$\therefore \quad 6152_7 - 2456_7 = 3363_7$$

a result which also means that

$$6152_7 = 2456_7 + 3363_7$$ *Answer*

Exercise 4.3

1. Find 13 − 5 in bases 9, 8, 7 and 6.
2. Find 21 − 14 in bases 9, 8, 7, 6, 5.
3. Find $435_6 + 242_6$ and $435_7 + 242_7$.
4. Find the sum of 231_5, 123_5, 312_5.
5. Find $1000_5 - 100_5$.
6. Find $1100_7 - 110_7$.
7. Find $3456_8 - 1567_8$.
8. Find $2211_3 - 1122_3$.
9. Find $1111_2 + 11_2$.
10. Find $1010_2 + 1010_4 + 1010_{16}$. Express the answer to base 4.
11. Find $325.14_6 \pm 243.54_6$.
12. Find $763.65_8 \pm 477.66_8$.

Multiplication

Multiplication is really successive addition but this procedure would be very tedious for a problem such as $1245_6 \times 23_6$ in which the first number might be written down 23_6 times and the sum of the numbers then found. Our method will be to carry out the multiplication completely in the base of the numbers we are multiplying, using the same layout scheme as for denary numbers.

Before reaching this stage we need to recall one or two standard results. Looking at

$$123_{10} \times 10_{10} = 1230_{10}$$

we ask: does the same result hold in any base (i.e. moving the point one place to the right)?

Consider

$$\begin{aligned} 243_6 \times 10_6 &= \{(6^2 \times 2) + (6^1 \times 4) \times (6^0 \times 3)\} \times 6^1 \\ &= (6^3 \times 2) + (6^2 \times 4) + (6^1 \times 3) \\ \therefore \quad 243_6 \times 10_6 &= 2430_6 \end{aligned}$$

The answer does take the expected form thus,

$$345_7 \times 10_7 = 3450_7,\ 2001_9 \times 10_9 = 20010_9 \text{ and so on.}$$

Likewise, $123_7 \times 20_7 = 2460_7$, $1222_3 \times 20_3 = 102210_3$, because a multiplication by 20 is the same as multiplication by 2 × 10. Similar reasoning shows that $243_6 \times 100_6$ results in moving the point two places to the right to give the answer 24300_6, since

$$10_6 \times 10_6 = 100_6 \text{ and so } 243_6 \times 100_6 = 243_6 \times 10_6 \times 10_6$$
$$= 2430_6 \times 10_6$$
$$= 24300_6$$

We have therefore established that the results of multiplying by any power of 10 in any base is pictorially the same as for denary numbers and this enables us to carry out multiplication by following the same scheme as for denary numbers.

Example 15. Multiply 254_6 by 121_6.

Solution. The result will be a sum of all the following submultiples

$$(254_6 \times 100_6), (254_6 \times 20_6), (254_6 \times 1)$$

which we assemble in the usual denary scheme:

		254
		× 121
$254_6 \times 100_6$	=	25400
$254_6 \times 20_6$	=	5520
$254_6 \times 1$	=	254
	Total	40014_6

N.B. For $254_6 \times 2$ we have multiplied mentally in denary and converted as we proceed. Thus $4 \times 2 = 8 = 12_6$; register 2, carry 1; $5 \times 2 = 10_{10} = 14_6$ and 1 carried gives 15_6, and so on.

$$\therefore \quad 254_6 \times 121_6 = 121_6 \times 254_6 = 40014_6$$ *Answer*

Example 16. Find the product of 342_5 and 214_5.

Solution. The product is $342_5 \times 214_5$.

		342
		× 214
$342_5 \times 200_5$	=	123400
$342_5 \times 10_5$	=	3420
$342_5 \times 4$	=	3023
	Total	140343_5

N.B. For $342_5 \times 4$ we have multiplied mentally in denary and converted as we proceed. Thus $2 \times 4 = 8 = 13_5$; register 3, carry 1; $4 \times 4 = 16_{10} = 31_5$ plus 1 carried makes 32_5; register 2 and carry 3; etc. The final result is

$$342_5 \times 214_5 = 214_5 \times 342_5 = 140343_5$$ *Answer*

Exercise 4.4

1. Find 234×100 in bases 9, 8, 7, 6, 5.
2. Find 3.4×1010 in bases 9, 8, 7, 6, 5.
3. Show that $11 \times 11 = 121$ in bases 3 to 10 inclusive.
4. In which bases is $12 \times 12 = 144$ true?
5. Find $1.52_8 \times 0.01_8$.
6. Find $1.43_5 \times 0.1_5 \times 0.1_5 \times 0.1_5$.
7. Examine the result $9 + 9 = 18$; $9 \times 9 = 81$ in base 10 and note the comparable result in base 9, namely $8 + 8 = 17$; $8 \times 8 = 71$. Write down the comparable results in bases 5, 6, 7.

8. Convert all the following denary numbers to binary numbers and find the sum S in base 2 when $S = 1 + 2 + 2^2 + 2^3 + 2^4 + 2^5 + 2^6 + 2^7$. Show that $S = 2^8 - 1$. Find the value of S when the last term of the right-hand side is 2^9.
9. Examine the results $1 \times 2 = 02_3$; $2 \times 3 = 12_4$; $3 \times 4 = 22_5$; and continue the pattern up to $A_{12} \times B_{12} = 92_{12}$. Explain the pattern of the results.
10. Express each of the following numbers as a product of its prime factors: 21_7, 36_8, 42_6, 302_5.
11. Find the least common multiple of the following three numbers: 82_5, 410_5, 240_5.

Division

This is the most difficult of the operations, particularly in unfamiliar bases, because it requires a ready knowledge of tables in order to assess the comparative sizes of the numbers which occur in the division. There is a great deal to be said for converting the problem to a familiar denary exercise and then reconverting the answer back to the original base.

For a simple example

$$\begin{aligned} 246_7 \div 21_7 &= 132_{10} \div 15_{10} \\ &= 8.8_{10} \\ &= 11.\overline{5412}_7 \end{aligned}$$

However, it can be helpful to grapple with the interplay between tables and the separate steps of a division. We prepare for the next examples by writing down the relevant tables:

$$\begin{array}{llll} 33_5 \times 1 &= 33_5 & 25_7 \times 1 &= 25_7 \\ 33_5 \times 2 &= 121_5 & 25_7 \times 2 &= 53_7 \\ 33_5 \times 3 &= 204_5 & 25_7 \times 3 &= 111_7 \\ 33_5 \times 4 &= 242_5 & 25_7 \times 4 &= 136_7 \\ 33_5 \times 10_5 &= 330_5 & 25_7 \times 5 &= 164_7 \\ 33_5 \times 11_5 &= 413_5 & 25_7 \times 6 &= 222_7 \\ 33_5 \times 12_5 &= 1001_5 & 25_7 \times 10_7 &= 250_7 \end{array}$$

We shall start by reminding ourselves of the method of long division in denary numbers.

Example 17. Divide 5812_{10} by 18_{10}.

Solution. The usual method reads as follows with all numbers denary.

(i) $58 \div 18$ is in between 3 and 4 (i.e. we know $18 \times 3 = 54$ and $18 \times 4 = 72$).
(ii) $18 \times 3 = 54$; register 3, subtract 54 from 58 then 'bring down the next digit, 1'.
(iii) $41 \div 18$ is between 2 and 3.
(iv) $18 \times 2 = 36$; register 2, subtract 36 from 41 then 'bring down the last digit, 2'.
(v) $52 \div 18$ is between 2 and 3.
(vi) $18 \times 2 = 36$; register 2, subtract 36 from 52.

$$\begin{array}{r} 322 \\ 18\,\overline{)\,5812} \\ \underline{54} \\ 41 \\ \underline{36} \\ 52 \\ \underline{36} \\ 16 \end{array}$$

The final result is $5812_{10} \div 18_{10} = 322_{10}$ remainder 16_{10}. *Answer*

Now consider the same problem in base 5.

Example 18. Divide 141222_5 by 33_5.

Solution. Use the same layout as in Example 17.

(i) $141_5 \div 33_5$ is between 2 and 3 (from tables).
(ii) $33_5 \times 2 = 121_5$; register 2, subtract 121 from 141 then 'bring down the next digit, 2'.
(iii) $202_5 \div 33_5$ is between 2 and 3 (from tables); register 2, subtract 121 from 202 then 'bring down the next digit, 2'.
(iv) $312_5 \div 33_5$ is between 4 and 10_5 (from tables).
(v) $33_5 \times 4 = 242_5$; register 4, subtract 242 from 312 then 'bring down the last digit, 2'.
(vi) the last step is shown.

$$\begin{array}{r} 2242 \\ 33\,\overline{)\,141222} \\ \underline{121} \\ 202 \\ \underline{121} \\ 312 \\ \underline{242} \\ 202 \\ \underline{121} \\ 31 \end{array}$$

The final result is $141222_5 \div 33_5 = 2242_5$ remainder 31_5. *Answer*

Example 19. Divide 13461_7 by 25_7.

Solution

(i) $134_7 \div 25_7$ is between 3 and 4 (from tables).
(ii) $25_7 \times 3 = 111_7$; register 3, subtract 111 from 134, 'bring down the next digit, 6'.
(iii) $236_7 \div 25_7$ is between 6 and 10_7.
(iv) $25_7 \times 6 = 222_7$; register 6, subtract 222 from 236, 'bring down the last digit, 1'.
(v) $25_7 \times 4 = 136_7$; register 4, subtract 136 from 141 to get the remainder.

$$\begin{array}{r} 364 \\ 25\,\overline{)\,13461} \\ \underline{111} \\ 236 \\ \underline{222} \\ 141 \\ \underline{136} \\ 2 \end{array}$$

$$13461_7 \div 25_7 = 364_7 \text{ remainder } 2$$ *Answer*

and again we see that a multiplication table is essential.

Exercise 4.5

Find the results in each of the following:

1. $220_3 \div 2$
2. $331_6 \div 10_3$
3. $10111_2 \div 10_2$
4. $3333_7 \div 2$
5. $144_7 \div 13_7$
6. $3555_8 \div 32_8$
7. $400_9 \div 10_9$
8. $4103_7 \div 100_7$
9. $20103_5 \div 33_5$
10. $31412_5 \div 33_5$
11. $3116_7 \div 25_7$
12. $6601_7 \div 25_7$
13. Calculate $\dfrac{1}{25_7}$ to 3 places
14. Simplify the following product given that all the numbers are in base 6:

$$\frac{100}{25} \times \frac{54}{12} \times \frac{4}{30}$$

15. Which of the following numbers are even?

$$11_7;\ 1001_3;\ 232_5;\ 31317171_9$$

Fractions

We have seen that certain fractions have sets of recurring digits when expressed in some bases, but are expressed exactly in other bases. For example, in denary

$$\tfrac{1}{9} = 0.\dot{1}_{10}$$

but in base 9 $\frac{1}{9} = 0.1_9$
or in base 3 $\frac{1}{9} = 0.01_3$

Similarly, $\frac{1}{7} = 0.\overline{142857}_{10}$ in denary but in base 7, we have

$$\tfrac{1}{7} = 0.1_7$$

It is instructive to start by suggesting a recurring set of digits and then try to find the fraction it represents.

Example 20. In an unknown base the fraction p is represented by $0.\overline{132}$. Find the possible values of the fraction p.

Solution. We have been told that $p = 0.132\ 132\ 132\ 132\ \ldots$ and we realise immediately that the base must be greater than 3.

Multiplication by 1000 gives

$$\begin{aligned} 1000p &= 132.132\ 132\ 132\ 132\ \ldots \\ \text{i.e. } 1000p &= 132 + 0.\overline{132} \\ \text{i.e. } 1000p &= 132 + p \\ \therefore\quad (1000\text{–}1)p &= 132 \\ p &= \frac{132}{1000\text{–}1} \end{aligned}$$

If we express p to base 10 the result will be that

$$\frac{132_{10}}{999_{10}} = 0.\overline{132}_{10}$$

If we express p to base 7 the result will be that

$$\frac{132_7}{666_7} = 0.\overline{132}_7$$ *Answer*

The same type of problem can be worked into a different context of summing an infinite number of quantities by finding a number to which we can get as close as we please by taking more and more terms.

Example 21. Find the numerical result of continuing to sum an increasing number of the following denary terms:

$$\frac{1}{3} + \frac{1}{9} + \frac{1}{27} + \frac{1}{81} \ldots$$

Solution. If we express these terms to base 3 we see immediately that

$$\frac{1}{3} = 0.1_3, \frac{1}{9} = \frac{1}{3^2} = 0.01_3, \frac{1}{27_{10}} = \frac{1}{3^3} = 0.001_3 \text{ and so on}$$

Summing the terms in base 3 we have

$$\begin{aligned} 0.111111\ldots_3 &= 0.\bar{1}_3 = p \text{ (say)} \\ \therefore\quad 10_3 p &= 1.\bar{1}_3 \\ 10_3 p &= 1 + p \\ p &= \frac{1}{2} \end{aligned}$$

Returning to the original denary question we have deduced that if we continue to add more and more terms in the series

$$\frac{1}{3} + \frac{1}{3^2} + \frac{1}{3^3} + \frac{1}{3^4} \ldots$$

we shall get closer and closer to $\frac{1}{2}$. *Answer*

This type of series is known as a geometric progression and in Example 21 we say we have found the sum of an infinite number of terms of the progression.

Exercise 4.6

Find the denary fractions which are represented by the following recurrences (1 to 4):

1. $0.\overline{132}_5$
2. $0.\overline{132}_4$
3. $1.\overline{10}_2$
4. $0.\overline{25}_7$

Convert each of the following denary series to recurrences in an appropriate base and hence find the 'sum' of each series:

5. $\frac{1}{10} + \frac{1}{100} + \frac{1}{1000} + \ldots$
6. $\frac{1}{5} + \frac{1}{5^2} + \frac{1}{5^3} + \frac{1}{5^4} + \ldots$
7. $\frac{1}{7} + \frac{1}{7^3} + \frac{1}{7^5} + \frac{1}{7^7} + \ldots$
8. $\frac{1}{4} + \frac{1}{4^4} + \frac{1}{4^7} + \frac{1}{4^{10}} + \ldots$

Puzzles

Bachet's weights problem has always been a popular outlet for recreations in multibase arithmetic.

We have seen that in the binary scale it is possible to enter denary quantities from 0 to 15_{10} inclusive using only the first four columns headed 8, 4, 2 and 1 respectively. This observation may be worded in terms of weighing by suggesting that we only need four masses of 8 g, 4 g, 2 g and 1 g in order to weigh any integral masses up to 15_{10} g. Thus, for a mass of 13_{10} g we write the binary equivalent $13_{10} = 1101_2$, which indicates that we need 8 g + 4 g + 1 g in one scale pan to balance 13_{10} g in the other pan. This particular problem is very straightforward since the given masses are all placed in the same scale pan for each weighing.

Far more interesting is Bachet's related problem of finding the minimum number of masses required to weigh integral masses from 1 g to 40_{10} g inclusive.

It is easy to see that we can continue with the binary representation and suggest that masses 32_{10} g, 16_{10} g, 8 g, 4 g, 2 g, 1 g will solve the problem, but a moment's reflection shows that these six masses will serve for loads up to 63_{10} g, which suggests an extravagance.

If we permit masses to be placed in either scale pan we find that 27_{10} g, 9 g, 3 g and 1 g will be enough. The reader will no doubt recognise the ternary numbers of base 3 here. The question is how do we systematically discover which masses should be put in which scale pan? For example, $22_{10} = 211_3$ suggests that we need two 9 g masses if they are all placed on one scale pan.

The secret of the exercise is to see,

	2 as $10_3 - 1$	i.e.	$2 = 3 - 1$
	20_3 as $100_3 - 10_3$	i.e.	$6 = 9 - 3$
and	200_3 as $1000_3 - 100_3$	i.e.	$18_{10} = 27_{10} - 9$

Hence $22_{10} = 211_3 = 1011_3 - 100_3$.

In similar fashion we see that

$$\begin{aligned} 25_{10} = 221_3 &= 200_3 + 20_3 + 1 \\ &= 1000_3 - 100_3 + 100_3 - 10_3 + 1 \\ &= 1000_3 - 10_3 + 1 \\ &= 1001_3 - 10_3 \end{aligned}$$

and so we weigh 25_{10} g by having 27_{10} g + 1 g on one side and 3 g on the other. The interesting point about Bachet's solution here is that it is the least number of different masses which can be used. Another solution is 1 g, 3 g and four 9 g masses and another is 1 g, 27_{10} g and four 3 g masses.

Another famous weighing problem concerns 12 identical coins. One of the coins is a fake having a weight which may be less or greater than the weight of a true coin, we do not know which. The problem is to use a simple balance with no given weights to find in three weighings not only the fake coin but also to determine whether it is heavier or lighter than a true coin.

It must be obvious that with no given weights the best we can do is to divide the coins into sets and compare their weights with each other. Taking three sets of four coins each we can weigh the first set (0) against the third set (2) and if they balance we know the fake is in the second set (1). If (0) and (2) do not balance then we do not know whether one side goes down because it contains an overweight fake or because the other side contains an underweight fake. But, if we take a note of which coins are in the 'down pan' then perhaps by changing the coins within the sets we shall be able to isolate the fake.

It follows that we need a number code which enables us to record not only which coins are placed in which scale pan but also whether the fake is over or under weight. So far this suggests that we try giving each coin a ternary number of three digits ranging from 000 to 222, where 0, 1 and 2 means placement in pan 0, box 1 and pan 2 of Fig. 4.5. Such a coding will require four 0's, 1's and 2's in each of the three places of the ternary numbers for the 12 coins. Fortunately such an arrangement is easy to find.

Thus a coin numbered as 012 will indicate that it is placed in pan 0 for the first weighing, in the box for the second weighing and in pan 2 for the third weighing.

Now suppose that pan 0 falls, the pans balance and pan 2 falls in the first, second and third weighings respectively. This will mean that an overweight fake has a number 012 but an underweight fake has a number 210. The sum of these two numbers is always 222_3 and since any of the 12 coins may be the fake it follows that each coin must be represented by two numbers rather than one, i.e. an overweight number and an underweight number, both of which must add up to 222_3, meaning that the two numbers are complementary with respect to the number 222_3, e.g. 112_3 is the complement of 110_3 with respect to 222_3. Finally, since no coin can be left in the same set of four coins throughout the three weighings it is clear that no coin can be numbered as 000, 111 or 222. This leaves 24 possible numbers out of the 27 between 000 and 222, inclusive.

We now choose the numbering arrangement shown in the following table with overweight numbers being listed in column H and underweight numbers being listed in column L. Details of the three weighings are given thereafter.

Number Denary	H *Ternary*	L *Ternary*	*Complement Denary*
1	001	221	25
3	010	212	23
4	011	211	22
5	012	210	21
14	112	110	12
15	120	102	11
16	121	101	10
17	122	100	9
18	200	022	8
19	201	021	7
20	202	020	6
24	220	002	2

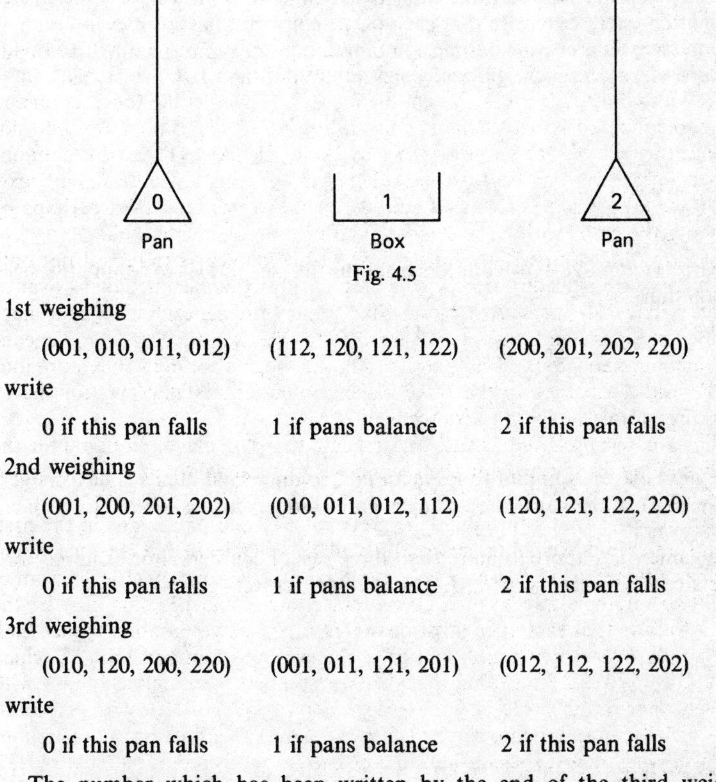

Fig. 4.5

1st weighing

(001, 010, 011, 012) (112, 120, 121, 122) (200, 201, 202, 220)

write

0 if this pan falls 1 if pans balance 2 if this pan falls

2nd weighing

(001, 200, 201, 202) (010, 011, 012, 112) (120, 121, 122, 220)

write

0 if this pan falls 1 if pans balance 2 if this pan falls

3rd weighing

(010, 120, 200, 220) (001, 011, 121, 201) (012, 112, 122, 202)

write

0 if this pan falls 1 if pans balance 2 if this pan falls

The number which has been written by the end of the third weighing identifies the fake coin, being under L or H corresponding to light or heavy.

Supposing the coin 010 had been a light coin, then the weighings would have registered as follows:

(i) pan 2 falls; register 2
(ii) pans balance; register 1
(iii) pan 2 falls; register 2

This gives coin numbered 3 as light. If coin 010 had been a heavy coin then the weighings would have registered the number 010.

Supposing the number registered at the end of the three weighings had been 202, then this coin descended each time it was weighed so the fake coin was numbered 20_{10} and was overweight.

The game of **Nim** is for two players playing alternately and can be controlled by any one who can add binary numbers.

A simple form of the game consists of starting with three separate sets of counters. The players take turns in removing one or more counters which are all taken from any one single set. The player who removes the last counter is the winner.

The secret of winning is as follows. Represent the number of counters in each set by binary numbers and add the columns without 'carrying'. Remove sufficient counters to ensure that each of the column totals is an even number and you have what is called a safe combination. For example, with 10_{10}, 7 and 3 counters in each set the binary representation of the 'state of play' is

$$\begin{array}{lcccc} & 1 & 0 & 1 & 0 \\ & & 1 & 1 & 1 \\ & & & 1 & 1 \\ \hline \text{No carry totals} = & 1 & 1 & 3 & 2 \end{array}$$

Clearly we remove 110_2 counters from the first set and leave behind the safe combination

$$\begin{array}{lccc} & 1 & 0 & 0 \\ & 1 & 1 & 1 \\ & & 1 & 1 \\ \hline \text{No carry totals} = & 2 & 2 & 2 \end{array}$$

The next player is bound to leave an odd column total after his next move. If at the start of the game the arrangement is already a safe combination then do not go first!

The game rules apply to more than three sets of counters. For example, for five sets containing 14_{10}, 10_{10}, 7, 2, 4 counters respectively

$$\begin{array}{lcccc} & 1 & 1 & 1 & 0 \\ & 1 & 0 & 1 & 0 \\ & & 1 & 1 & 1 \\ & & & 1 & 0 \\ & & 1 & 0 & 0 \\ \hline \text{No carry total} = & 2 & 3 & 4 & 1 \end{array}$$

Remove 11_2 from the last set or 101_2 from the third set to gain a safe combination.

Exercise 4.7

1. Using no more than the four masses 27 g, 9 g, 3 g, 1 g show how to weigh each of the following masses: (i) 11 g, (ii) 22 g, (iii) 26 g, (iv) 35 g, (v) 32 g.
2. An 81 g mass is added to the set of masses in question 1. Show how to weigh the following masses: (i) 45 g, (ii) 59 g, (iii) 72 g.

Determine whether the following are winning denary combinations for Nim.

3. 13, 5, 3
4. 15, 9, 7
5. 13, 10, 7
6. 14, 9, 6, 2
7. 5, 4, 3, 2, 1

8. In the 'fake coin' example in this chapter, the following registrations were made: (i) 101, (ii) 201, (iii) 022. What was the denary label of each coin and was it over or under weight?

5

FINITE SYSTEMS

Clock Arithmetic

We can produce a finite mathematical system by defining an operation of addition on a set of numbers $S = \{0, 1, 2, 3, 4, 5\}$ arranged round a one-handed clock-face like Fig. 5.1. Using this device, we obtain results such as

$$3 + 5 = 2, \ 4 + 4 = 2, \ 2 + 4 = 0$$

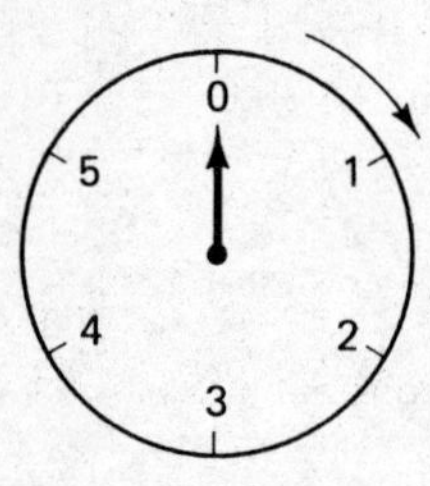

Fig. 5.1

but how?

Each separate addition starts with the hand at 0 and as each number is read so the hand is turned clockwise through that number of spaces until the sum is given by the final position of the hand. To find $3 + 5$ we start at 0, rotate clockwise through 3 spaces and then onwards through 5 more spaces, and the hand is now pointing at an answer of 2.

Clockwork addition is clearly not the same as ordinary addition so we should really invent a new symbol like $*$ and write $3 * 5 = 2$, but where the context makes the meaning obvious we continue to use the common 'plus' sign.

Some further results are

$$5 + 2 = 1, \ 3 + 3 = 0, \ 4 + 5 = 3$$

With six numbers in S there are 36 possible combinations for the addition of two numbers at a time so we can record all the possible results in Table 5.1.

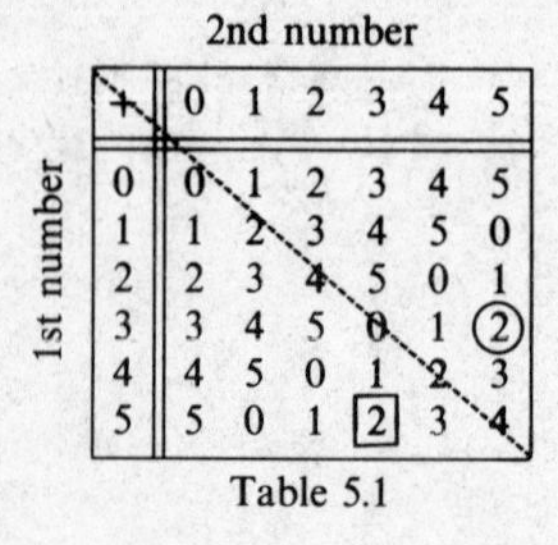

2nd number (columns); 1st number (rows)

+	0	1	2	3	4	5
0	0	1	2	3	4	5
1	1	2	3	4	5	0
2	2	3	4	5	0	1
3	3	4	5	0	1	(2)
4	4	5	0	1	2	3
5	5	0	1	[2]	3	4

Table 5.1

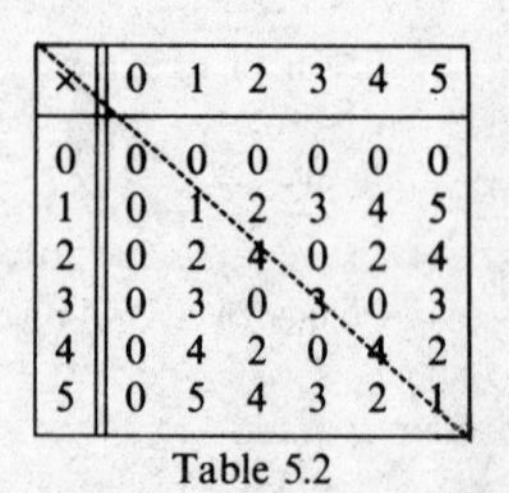

×	0	1	2	3	4	5
0	0	0	0	0	0	0
1	0	1	2	3	4	5
2	0	2	4	0	2	4
3	0	3	0	3	0	3
4	0	4	2	0	4	2
5	0	5	4	3	2	1

Table 5.2

Since all the results belong to S the binary operation $+$ is therefore closed on S. A typical result $3 + 5 = 2$ is ringed on Table 5.1.

The first element is 3 followed by 5 as the second element. The result for $5 + 3 = 2$ is shown in a square on the table.

In compiling the table it will be noticed that clock addition is commutative, since we may substitute any of the numbers in S into the equation

$$a + b = b + a$$

This makes the table symmetrical about the leading diagonal.

We can use Table 5.1 to prove that clock addition is associative, i.e.

$$(a + b) + c = a + (b + c)$$

for all a, b, c belonging to S (page 17). For example

$$\begin{aligned} (3 + 4) + 5 &= 1 + 5 = 0 \\ 3 + (4 + 5) &= 3 + 3 = 0 \\ \therefore\quad (3 + 4) + 5 &= 3 + (4 + 5) = 3 + 4 + 5 \end{aligned}$$

which shows that the result is independent of the position of the bracket. Similarly, $2 + 3 + 4 = (2 + 3) + 4 = 5 + 4 = 3$. It is a tedious process to check all possible combinations but at least such proofs are always available in a finite system.

Considering multiplication as repeated addition gives the following typical results:

$$\begin{aligned} 4 \times 3 &= (4 + 4) + 4 = 2 + 4 = 0 \\ 3 \times 4 &= (3 + 3) + (3 + 3) = 0 + 0 = 0. \quad \therefore \quad 4 \times 3 = 3 \times 4 \\ (4 \times 2) \times 5 &= 2 \times 5 = 4 \\ 4 \times (2 \times 5) &= 4 \times 4 = 4. \quad \therefore \quad (4 \times 2) \times 5 = 4 \times (2 \times 5) \qquad \text{(i)} \end{aligned}$$

Compiling Table 5.2 shows multiplication to be commutative and by confirming all results such as (i) we could prove that multiplication is also associative, as expected from the comparable results for addition.

We should note that all these results could be obtained by performing ordinary addition or multiplication and converting the result into clock form with division by 6 and registering the remainder as the answer.

Example 1. By ordinary addition: $3 + 5 + 5 = 13$
Division by 6 leaves a remainder of 1
By clock addition: $3 + 5 + 5 = 1$

Example 2. By ordinary multiplication: $3 \times 5 \times 5 = 75$
Division by 6 leaves a remainder of 3
By clock multiplication: $(3 \times 5) \times 5 = 3 \times 5 = 3$

The repetition of results in 'packets of 6' suggests the more formal title of 'modular 6 arithmetic' or more usually 'arithmetic mod 6' for our example of clock arithmetic.

Exercise 5.1

Use Tables 5.2 and 5.1 to find the following:

1. $5 + 4; 4 + 5$
2. $2 + 4 + 3; 3 + 4 + 2$
3. $1 + 2 + 3 + 4; 4 + 3 + 2 + 1$

4. $(3 + 4) \times 2; 3 + (4 \times 2)$
5. $(5 \times 5) + 4; 5 \times (5 + 4)$
6. $3 \times (5 + 4); (3 \times 5) + (3 \times 4)$
7. $5 + (5 \times 4); (5 + 5) \times (5 + 4)$
8. $(3 + 4) \times (2 \times 5); (3 \times 4) + (2 + 5)$

Find the values of n so that:

9. $n + 3 = 1$
10. $5 + n = 2$
11. $n^2 = 4$
12. $n(n + 1) = 0$
13. With S as the set of odd numbers are the operations of ordinary addition and multiplication closed on S? Consider the same question for E the set of even numbers.

Compile the addition and multiplication tables for arithmetic mod 3 and answer the following questions:

14. $(2 \times 2 \times 2) + 2$
15. $(1 + 2 + 2) \times 2$
16. $2 \times (1 + 2)$
17. Find n if (i) $n^2 = 1$ and (ii) $n^2 = 2$.
18. Find n if $2n + 1 = 2$.

Subtraction

The results of clock subtraction could easily be obtained by rotating the single hand anticlockwise in Fig. 5.1, e.g. finding $2 - 5$ by starting at 0, moving clockwise through two spaces followed by moving anticlockwise through five spaces to record the final answer of 3. Thus

$$2 - 5 = 3$$

We must beware of wrongly concluding from the results $2 - 5 = 3 = 5 - 2$ that subtraction is commutative. All possible cases need to be explored and the results $5 - 1 = 4$ and $1 - 5 = 2$ immediately show that clock subtraction is not a commutative operation on the set $\{0, 1, 2, 3, 4, 5\}$.

Our aim in this section is to define subtraction in terms of addition and we shall subsequently describe subtraction as the inverse operation to addition.

Additive Inverses

In arithmetic mod 6 we have

$$0 + 0 = 0, 1 + 0 = 1, 2 + 0 = 2, 3 + 0 = 3, 4 + 0 = 4 \text{ and } 5 + 0 = 5$$

to show that the addition of 0 to each number of the set leaves the number unchanged. Describing each side of these equations as identical suggests that we call 0 the identity element with respect to addition in arithmetic mod 6. Formally, this is merely a particular example of a more general statement for any algebra in which we find that

$$a * e = a = e * a$$

for every element a in the set S (this includes $a = e$). We say that e is the **identity element** for the operation $*$ defined on the set S.

Having established the existence of an identity element for $+$ in arithmetic mod 6 we now examine how we may combine the elements of our set $\{0, 1, 2,$

3, 4 5} so as to produce this identity. We have already compiled the addition table 5.1 and seen that

$$0 + 0 = 0 = 0 + 0$$
$$1 + 5 = 0 = 5 + 1$$
$$2 + 4 = 0 = 4 + 2$$
$$3 + 3 + 0 = 3 + 3$$

Each component of these pairs is therefore said to be the **additive inverse** of the other component with respect to the operation $+$. That is, 5 is the additive inverse of 1 and vice versa, all because of their sum being equal to the identity element 0.

The practical use for this line of thinking is seen in our attempt to solve an equation like $n + 4 = 3$ without defining a new operation of subtraction. We proceed as follows:

Add the additive inverse of 4 to both sides:

$$\therefore \quad n + 4 + 2 = 3 + 2$$

Simplify with the aid of the associative law:

$$\therefore \quad n + 0 = 3 + 2$$

Observe that 0 is the identity element for $+$

$$\therefore \quad n = 3 + 2$$

We see therefore that using such a process avoids the need to consider $n + 4 = 3$ as an exercise in subtraction, and this appeals to mathematicians' urge for efficiency and so they define a meaning for subtraction in terms of addition.

Definition 5.1. In modulo arithmetic the result of $a - b$ is that number c such that $a = b + c$.

The practical significance of this definition is to interpret

$$a - b + b = b + c$$

as making use of $b - b = 0$ the identity element, so that instead of subtracting a number we add its additive inverse.

Example 3. Obtain the additive inverse of 4 and hence find $2 - 4$.

Solution. The additive inverse of 4, arithmetic mod 6, is 2 because

$$4 + 2 = 0$$
$$\therefore \quad 2 - 4 = 2 + 2 = 4$$

Example 4. Find $2 - (2 - 4) - 5$.

Solution. We evaluate brackets first, i.e. $(2 - 4) = (2 + 2) = 4$.

$$\therefore \quad 2 - (2 - 4) - 5 = 2 - 4 - 5$$

(When no brackets are inserted expressions are read from left to right. Notice that $(2 - 4) - 5 \neq 2 - (4 - 5)$.)

$$\therefore \quad 2 - 4 - 5 = 2 + 2 + 1 = 5 \qquad \textit{Answer}$$

Example 5. Find $((2 - 3) - 4) - 5$ by using the additive inverse wherever possible.

Solution. We may write

$$((2 - 3) - 4) - 5 = ((2 + 3) + 2) + 1$$
$$= 2 + 3 + 2 + 1 = 2 \qquad \textit{Answer}$$

Exercise 5.2

The following questions are in arithmetic mod 6. Write in additive form and then simplify, using Table 5.1.

1. $1 - 5$
2. $5 - 1$
3. $2 - (1 - 5); 2 - 1 + 5$
4. $(2 - 1) - 5$
5. $((3 - 5) - 4) - 2$
6. $0 - 4$
7. $(0 - 4) - 5$
8. $(0 - 4) - (0 - 5)$
9. $(3 - 4) - (4 - 3)$
10. $0 - (4 - 2) - (2 - 4)$
11. Find the identity element for addition in arithmetic mod 7 and state the additive inverse for each of the numbers 0, 1, 2, 3. Hence find each of the following: (i) $2 - 5$, (ii) $3 - 4$, (iii) $0 - 2$, (iv) $1 - 6$.

Multiplicative Inverses

Just as we can define subtraction in terms of addition so we can define division in terms of multiplication. Inspection of Table 5.2 reveals familiar results

$$0 \times 1 = 0 = 1 \times 0$$
$$1 \times 1 = 1 = 1 \times 1$$
$$2 \times 1 = 2 = 1 \times 2, \text{ etc.}$$

to show that 1 is the identity element with respect to multiplication. We also note that $1 \times 1 = 1$ and $5 \times 5 = 1$ are the only such products and it is not possible, for example, to have $2 \times ? = 1$ in arithmetic mod 6.

Definition 5.2. The multiplicative inverse of an element a is that element which multiplies a to give the multiplicative identity.

The inverse of a with respect to multiplication is written a^{-1} (read as 'a inverse') so that

$$a \times a^{-1} = 1 = a^{-1} \times a$$

For ordinary multiplication on the set of real numbers the inverse of 4 is $4^{-1} = \frac{1}{4}$ because $4 \times \frac{1}{4} = 1$; similarly, the inverse of 5 is $5^{-1} = \frac{1}{5}$ because $5 \times \frac{1}{5} = 1$; finally 0 does not have an inverse. Thus all the inverses belong to the set (i.e. the multiplicative inverse of a real number is a real number). On the set of whole numbers $\{0, 1, 2, 3, \ldots\}$ only 1 has an inverse with respect to multiplication, which belongs to the set of whole numbers.

Definition 5.3. The result $a \div b$ is the unique number c such that $a = c \times b$.

This means that if we multiply both sides of $a = c \times b$ by b^{-1} we have

$$a \times b^{-1} = c \times b \times b^{-1} = c \times 1 = c$$

i.e.

$$a \div b = a \times b^{-1} = c$$

and the operation of division has changed into multiplication by an inverse. In other words, instead of dividing by a number we multiply by its multiplicative inverse. Recall that a binary operation must always produce a unique result whenever that result exists, i.e. whenever $a \times b^{-1}$ exists it must be unique, otherwise division will not be a binary operation on the set.

In Table 5.2 we saw that it is not possible to multiply 2 by any number in the set to gain an answer of 1, i.e. $2 \times 2^{-1} = 1$ is not possible and so 2^{-1} does not exist. We can see now that it is not possible to speak of $4 \div 2$ because 2^{-1} does not exist, i.e. we cannot express $4 \div 2$ as 4×2^{-1}, and yet the answer appears to be 2 because $4 = 2 \times 2$. However, our multiplication table 5.2 also shows that $4 = 5 \times 2$; consequently the definition 5.3 reveals two possible values for c in the equation $4 = c \times 2$ but since c is not therefore unique it follows that no meaning can be given to $4 \div 2$. It is this shift from division to multiplication by inverses that enables us to preserve division as a binary operation. Thus we can obtain $4 \div 5$ because this means $4 \times 5^{-1} = 4 \times 5 = 2$ and the answer is unique.

Example 6. $2 \div 3 = 2 \times 3^{-1}$ does not exist because 3^{-1} does not exist in arithmetic (mod 6).

Example 7. $2 \div 5 = 2 \times 5^{-1} = 2 \times 5 = 4$ in arithmetic (mod 6).

Example 8. Solve the equation $3n = 3$ in arithmetic (mod 6). Now this does not mean that $n = 3 \div 3$ although this may be a method of obtaining n if 3^{-1} existed but, we know that 3^{-1} does not exist. However, using $S = \{0, 1, 2, 3, 4, 5\}$ as our replacement set and inspecting Table 5.2 we see that $n = 1$, 3 and 5. This is one of the reasons why we rely upon replacement sets rather than methods of division when solving equations.

Exercise 5.3

Use Tables 5.1 and 5.2 to evaluate the following:

1. 1^{-1}
2. 2^{-1}
3. $3 \div 5$
4. $(5 + 1) \div 3$
5. $(2 + 2 + 2) \div 2$
6. $4 \times (2 \div 4)$
7. $4 \times (2 \div 5)$
8. $2 - 5^{-1}$
9. $((3 + 2)^{-1}) \times 5$

10. Solve: $4n = 2$
11. Solve: $2n = 4$
12. Solve: $2n + 1 = 3$
13. Solve: $3n + 5 = 2$
14: Solve: $5n = 2$

Modular Arithmetic

The modal clock restricts the arithmetic to consider only the numbers which appear on its face. An arithmetic (mod m) can deal with the set of whole numbers but restricts all the results to a subset of the whole numbers, by interpreting the results of a binary operation as a remainder after division by m.

Thus arithmetic (mod 6) would examine a number like 73 and write $73 \equiv 1$ (mod 6) because 1 is the remainder after dividing 73 by 6.

We read this last statement as '73 is congruent to 1, mod 6'. Since $25 \equiv 1$ (mod 6) it is clear why we write

$$73 \equiv 25 \pmod 6$$

Similarly $$20 \equiv 62 \pmod 6, \text{ and so on}$$

This will mean in turn that the solution of any equation like

$$n \equiv 2 \pmod 6$$

with the whole numbers for the replacement set will have a solution set $\{2, 8, 14, 20, 26, 32, \ldots\}$. Since an arithmetic (mod m) depends on ordinary addition prior to the division by m we correctly expect the new binary operations of + and × to be both commutative and associative. But we must not make the rash conclusion that because subtraction and division have been defined as the inverse operations to addition and multiplication they too are commutative—they certainly are not and neither are they associative.

One of the difficulties in dealing with modular arithmetic is the presence of zero divisors which we do not have in ordinary arithmetic.

Definition 5.4. The non-zero elements a, b with the product $a \times b = 0$ are called zero divisors.

In ordinary arithmetic if $a \times b = 0$ then we know that a or b is 0 (inclusive or). This result provides the basis of factorisation and deduction of roots in algebra in as much as the expression $(x - 2)(x - 3) = 0$ allows the only conclusion that $x - 2 = 0$ or $x - 3 = 0$ and the solutions $x = 2$ or $x = 3$ are thereby obtained. Regrettably, some readers misunderstand the significance of this result and believe incorrectly that $(x - 2)(x - 3) = 10$ implies that $x - 2 = 10$ or $x - 3 = 10$.

Thus in ordinary arithmetic $a \times b = 0$ implies that a or b is 0, but in modular arithmetic $a \times b = 0$ may well be satisfied by non-zero values for a and b.

In arithmetic (mod 6) just studied, we have seen that with $2 \times 3 = 0$ then 2 and 3 are zero divisors, and with $3 \times 4 = 0$ then 3 and 4 are zero divisors. This is a troublesome difficulty which has caused us to rely upon replacement sets for solving equations instead of the usual methods of attempting to divide and multiply both sides of the equation—for example

$$2n = 4 \pmod 6$$

dividing by 2 to get

$$n = 2$$

when we are not entitled to divide by 2 because 2^{-1} does not exist.

Alternatively, multiplying both sides by 3 to gain

$$2n \times 3 = 4 \times 3 = 0$$
$$\therefore \quad 0 = 0$$

correct but useless. Both examples demonstrate how careful we must be.

Not all arithmetics (mod m) are so awkward. Those with m a prime number are well behaved and have no zero divisors. It is when m is non-prime that we have these difficulties. Tables 5.3 and 5.4 show the results for arithmetic (mod 8) on the set of whole numbers.

+	0	1	2	3	4	5	6	7
0	0	1	2	3	4	5	6	7
1	1	2	3	4	5	6	7	0
2	2	3	4	5	6	7	0	1
3	3	4	5	6	7	0	1	2
4	4	5	6	7	0	1	2	3
5	5	6	7	0	1	2	3	4
6	6	7	0	1	2	3	4	5
7	7	0	1	2	3	4	5	6

Table 5.3

×	0	1	2	3	4	5	6	7
0	0	0	0	0	0	0	0	0
1	0	1	2	3	4	5	6	7
2	0	2	4	6	0	2	4	6
3	0	3	6	1	4	7	2	5
4	0	4	0	4	0	4	0	4
5	0	5	2	7	4	1	6	3
6	0	6	4	2	0	6	4	2
7	0	7	6	5	4	3	2	1

Table 5.4

Again we notice that the commutativity of both operations is made visible by the symmetry of the tables about the leading diagonals. (Note that we used the commutativity of the operations to build the table in the first place.)

The identity element for addition (mod 8) is again 0, and the additive inverses are gained from equations like

$$2 + 6 = 0 \text{ implies } (-2) = 6 \text{ and } (-6) = 2$$
$$4 + 4 = 0 \text{ implies } (-4) = 4$$

The identity element for multiplication is again 1 and the multiplicative inverses are gained from equations like

$$5 \times 5 = 1 \text{ indicating that } 5^{-1} = 5$$

and

$$7 \times 7 = 1 \text{ indicating that } 7^{-1} = 7$$

Example 9. Obtain the additive inverses of 2 and 3 and hence simplify $1 - 2 - 3$ in arithmetic (mod 8).

Solution. Since $2 + 6 = 0$, the additive inverse of 2 is 6. Similarly, the additive inverse of 3 is 5

$$\therefore \quad 1 - 2 - 3 = 1 + (6 + 5)$$
$$= 1 + 3 = 4$$

Answer

Example 10. Obtain the zero divisors of arithmetic (mod 8).

Solution. We look for zero in the body of the table 5.4 and note the following results:

$$2 \times 4 = 0 \text{ implies that 2 and 4 are zero divisors}$$
$$4 \times 6 = 0 \text{ implies that 4 and 6 are zero divisors}$$

Thus 2, 4 and 6 cannot have multiplicative inverses in (mod 8).

Example 11. Solve the equation $6n = 3$ in arithmetic (mod 8).

There is no solution to this equation since the replacement set $\{0, 1, 2, 3, 4, 5, 6, 7\}$ for n does not yield $6n = 3$.

Answer

Exercise 5.4—The following questions are in arithmetic (mod 8)

1. Which elements have additive inverses?
2. Which elements have multiplicative inverses?
3. Is division a closed binary operation?

Simplify the expressions in Questions 4–9.

4. $0 - 2 - 7$
5. $(0 - 2) \times (0 - 7)$
6. $3 \times (6 + 5)$
7. $(0 - 3) \times (6 + 5)$
8. $(3 \times 3 \times 3) \div 7$
9. (i) $4 \div 2$ (ii) $2 \div 4$ (iii) $6 \div 3$ (iv) $3 \div 6$ (v) $7 \div 5$ (vi) $5 \div 7$
10. Solve the equations
    (i) $2n = 6$ (ii) $2n = 4$ (iii) $4n = 4$ (iv) $4n = 0$ (v) $6n = 4$ (vi) $6n = 2$

Fractions

Fractions are not necessary in modular arithmetic defined on the set of whole numbers since the operation of division is defined in terms of multiplication by inverses which always belong to the set.

Definition 5.5. The fraction written $\frac{a}{b}$ will represent $a \times b^{-1}$ and will exist only if b^{-1} exists.

Thus $\frac{2}{3}$ exists in arithmetic (mod 8) (Table 5.4, page 94) but not in arithmetic (mod 6) (Table 5.2, page 88). Of course, we have already discussed such awkward results in division but it is of some interest to see what happens if we try to manipulate fractions in the manner of ordinary arithmetic. A typical peculiarity in arithmetic (mod 8) is $\frac{2}{6}$, which is meaningless because 6^{-1} does not exist yet a 'cancelling down by 2' (quite wrongly) suggests $\frac{2}{6} = \frac{1}{3}$ where the right-hand side does exist.

Alternatively, suppose we start in arithmetic (mod 8) with $\frac{1}{3}$ which is acceptable and multiply this by 1

$$\tfrac{1}{3} \times 1 = \tfrac{1}{3}\text{(acceptable)}$$

Now write 1 as $\frac{2}{2}$ and consider

$$\tfrac{1}{3} \times \tfrac{2}{2} = \tfrac{1}{3} = \tfrac{2}{6}\text{(not acceptable)}$$

The error, of course, is in assuming that we may suggest $1 = \frac{2}{2}$ when 2 is a zero divisor in arithmetic (mod 8), i.e. $\frac{2}{2}$ in arithmetic (mod 8) is as meaningless as $\frac{0}{0}$ in ordinary arithmetic.

Now consider the addition and multiplication of fractions in modular arithmetic.

Example 12. Simplify $\frac{2}{5} + \frac{3}{7}$ in arithmetic (mod 8).

Solution. We know that 5^{-1} and 7^{-1} exist and that these fractions may be rewritten

$$\begin{aligned} 2 \times 5^{-1} + 3 \times 7^{-1} &= (2 \times 5) + (3 \times 7) \\ &= 2 + 5 = 7 \end{aligned}$$

As an aside consider the silly remark that each of the fractions is less than a half yet their sum is seven! It is not possible to say that some numbers are greater than others. For example, $1 - 6 = 1 + 2 = 3$ (mod 8) and it is unreasonable to see the ordinary size comparisons between 1 and 6.

Again we have $18 \equiv 2$ (mod 8) and $33 \equiv 1$ (mod 8). Is 33 less than 18? It follows that all feelings for comparisons of size of numbers in modular arithmetic must be set aside.

Let us see what happens if we follow the usual procedures for adding two fractions in arithmetic (mod 8).

i.e. $$\frac{2}{5} + \frac{3}{7} = \frac{2 \times 7 + 3 \times 5}{5 \times 7} \quad \text{(i.e. converting to a common denominator)}$$

$$= \frac{6 + 7}{3} = \frac{5}{3}$$

But $$\frac{5}{3} = 5 \times 3^{-1} = 5 \times 3 = 7$$

and we have the same answer as before. *Answer*

Example 13. Simplify $\frac{6}{7} \times \frac{2}{3}$ in arithmetic (mod 8).

Solution. (i) Cancellation leads to

$$\frac{4}{7} = 4 \times 7^{-1} = 4$$

(ii) Interpretation leads to

$$(6 \times 7^{-1}) \times 2 \times 3^{-1} = (6 \times 7) \times 2 \times 3 = (2 \times 2) \times 3 = 4 \times 3 = 4$$

(iii) Multiplication leads to

$$\frac{6 \times 2}{7 \times 3} = \frac{4}{5} = 4 \times 5^{-1} = 4 \times 5 = 4$$

Although this example is not a general proof the fact is that as long as the fractions exist then we may manipulate them with the methods of ordinary arithmetic.

Exercise 5.5

Solve the following 10 questions in both arithmetic (mod 6) and (mod 8) if possible:

1. $\frac{1}{3} + \frac{1}{5}$
2. $1 - \frac{4}{3}$
3. $5 \times \left(\frac{1}{3} + \frac{1}{5}\right)$
4. $\frac{n}{3} = 2$
5. $n = \frac{4}{2}$ and $2n = 4$
6. $n(n - 1) = 0$
7. $(n - 1)(n - 2) = 0$
8. $n(n - 1) = 4$
9. $n^2 + 4 = 0$
10. $n^n = 4$
11. Obtain the addition and multiplication tables for arithmetic (mod 5).
    (i) State the identity elements for addition and multiplication.
    (ii) State the multiplicative inverses where they exist.
    (iii) Are there any zero divisors in this arithmetic?
    (iv) Simplify the following: (a) $\frac{1}{3} + \frac{1}{4}$, (b) $1 - \frac{2}{4}$, (c) $\frac{1}{3} - \frac{1}{2}$, (d) $\frac{1}{3} \times \frac{1}{2} \times \frac{1}{4}$.
    (v) Solve the equation $kn = 1$ for $k = 1, 2$ and 3.
12. The days of the week may be considered as elements of arithmetic (mod 7). Christmas Day is on a Tuesday one year; on which day of the week will it be the following year if the next year is (i) not a leap year, (ii) is a leap year?
13. February 29th is on Friday; on what day will the next February 29th fall?

A More Formal Discussion

Formal statements are usually devised to display the generality of mathematical ideas. Having obtained some practical experiences with modular arithmetic the shift of this experience into the context of a formal discussion of modular arithmetic will possibly enable us to appreciate the wider implications of these ideas.

Any two elements of a non-empty set may be brought together or combined according to any rules we care to define for the set and these rules alone will determine the resulting outcome. For mathematical purposes we shall require at least two virtues in such rules and they are that:

(i) Anyone starting with the same two elements—a and b, say—must obtain the same result c; in other words, all results are uniquely determined.

(ii) All the results must belong to the set we started with, thus enabling us to apply the rule again.

As we described earlier (page 16), combining two elements at a time suggests the word binary and the notion of uniqueness of result, when it exists in mathematics, is carried in the word operation. Since we have not yet made up our mind what to call the binary operation we shall settle for representing it by $\ast$ and refer to its use with elements a, b of the set by writing

$$a \ast b = c$$

and reading this as 'a star b equal to c' and noting that possibly this may not be the same as $b \ast a$. In other words, there is an order to be followed when executing the operation $\ast$. Our comment in (ii) above is symbolised by $c \in S$ where S is the set from which we drew a and b, i.e. all $a, b \in S$.

Definition 5.6. A binary operation $\ast$ defined on a set S is a rule whereby to each pair of elements a, b taken in order, there corresponds exactly one element c. Thus

$$a \ast b = c$$

If $c \in S$ for all $a, b \in S$ we say that the binary operation $\ast$ is closed with respect to S.

Just as for $\cap$, $\cup$, $\wedge$, $\vee$, we shall be interested to know if $\ast$ is commutative or associative, so we will enquire if $a \ast b = b \ast a$ for all $a, b \in S$ and also if $(a \ast b) \ast c = a \ast (b \ast c)$ for all $a, b \in S$ (pages 17, 89). Consider the following examples:

Example 14. Let $S =$ {real numbers} and define the following combination rule $\ast$ for all $a, b \in S$

$$a \ast b = \tfrac{1}{2}(a + b)$$

where $+$ represents ordinary addition. We make the following typical observations:

(i) Since $\frac{1}{2}(a + b)$ belongs to S we have therefore defined a closed binary operation on S.

(ii) Knowing that $a + b = b + a$ in ordinary addition tells us that

$$a \ast b = b \ast a$$

We therefore say that $\ast$ is a commutative operation on S.

(iii) We note the obvious particular results

$$a \ast a = a; \; b \ast b = b$$

(always realise that it is possible to take $a = b$)

$$\begin{aligned} a \ast 0 &= \tfrac{1}{2}a \\ a \ast (-a) &= 0 \\ 1 \ast b &= \tfrac{1}{2}(1 + b) \end{aligned}$$

(iv) We should always examine how the result of combining three elements is obtained. Again with brackets

$$\begin{aligned} (a \ast b) \ast c &= \tfrac{1}{2}(a + b) \ast c \\ &= \tfrac{1}{2}\{\tfrac{1}{2}a + \tfrac{1}{2}b + c\} \\ a \ast (b \ast c) &= a \ast \tfrac{1}{2}(b + c) \\ &= \tfrac{1}{2}\{a + \tfrac{1}{2}b + \tfrac{1}{2}c\} \end{aligned}$$

and these results are not the same unless $a = c$. Check with a numerical substitution and put $a = 4$, $b = 12$, $c = 28$

$$\begin{aligned} (a \ast b) \ast c &= 8 \ast 28 = 18 \\ a \ast (b \ast c) &= 4 \ast 20 = 12 \end{aligned}$$

i.e. the operation $\ast$ is not associative on S.

Example 15. Let $S =$ {real numbers} and define the closed binary operation $\ast$ by $a \ast b = a^2b$ for all $a, b \in S$, where ordinary multiplication $a \times a \times b$ is intended when writing a^2b. Again, without being asked, we would immediately make the following type of observations:

(i) With $a * b = a^2b$ and $b * a = b^2a$ we see at once that since a^2b is not always equal to b^2a this binary operation is not commutative on S (we could illustrate this numerically by considering $a = 2, b = 7$).

(ii) We note the results

$$a * a = a^3, \quad b * b = b^3$$
$$a * 0 = 0, \quad 0 * b = 0$$
$$a * 1 = a^2, \quad 1 * a = a$$
$$1 * 1 = 1, \quad 1 * 0 = 0$$

(iii)
$$(a * b) * c = (a^2b) * c = a^4b^2c$$
$$a * (b * c) = a * (b^2c) = a^2b^2c$$

Check numerically $a = 2, b = 3, c = 5$

$$(a * b) * c = (12) * 5 = 720$$
$$a * (b * c) = 2 * (45) = 180$$

The operation $*$ is not associative on S.

It must not be assumed on the basis of this example that if $*$ is not commutative then it is not associative (see Question 1, Exercise 5.6).

In Chapter 2 we saw that it was possible to replace a descriptive definition by a table of results of all possible combinations of the variables. In the above examples this replacement was not possible because there were an infinite number of elements in the set. Consider the next example in which the set only has two elements.

Example 16. With $S = \{a, b\}$, a binary operation $*$ is defined by the entries in Table 5.5. List all the possible combinations and their results, and comment on the possible commutativity and associativity of $*$ on S.

2nd element (columns); 1st element (rows)

$*$	a	b
a	a	a
b	a	b

Table 5.5

Solutions. The instruction on the table about which is the first and second element respectively must be obeyed. Reading our results direct from the table we have

(i)
$$a * a = a, \ a * b = a$$
$$b * a = a, \ b * b = b$$

We note that the binary operation $*$ is closed on S.

(ii) With $a * b = a = b * a$ as the only result obtainable for different elements we see that $*$ is commutative on S.

(iii)
$$\begin{aligned} a * (b * c) &= \text{either } a * (b * a) \text{ or } a * (b * b) \\ &= \text{either } a * a \qquad \text{or } a * b \\ &= \text{either } a \qquad \text{or } a \\ &= a \end{aligned}$$

Also
$$(a * b) * c = a * c = a$$

Notice that c must be either a or b because these are the only elements in S.

Hence $*$ is associative on S. *Answer*

Exercise 5.6

In each of the following questions 1 to 5 the binary operation $*$ is defined on $S = \{\text{real numbers}\}$ for all $a, b \in S$. Determine whether or not $*$ is (i) commutative, (ii) associative on S.

1. $a * b = a$
2. $a * b = a - b$; where $-$ means ordinary subtraction.
3. $a * b = a + b - 1$; where $+$, $-$ mean ordinary addition and subtraction.
4. $a * b = 2(a + b)$; where $+$ means ordinary addition.
5. $a * b = \sqrt{a^2 + b^2}$; where $+$ means ordinary addition.
6. Suggest two numbers which satisfy Table 5.5 with $*$ as ordinary multiplication.
7. Suggest two numbers which satisfy Table 5.5 with $*$ as ordinary multiplication and the entry $a * a = b$.
8. Compile a table like Table 5.5 to register the results of adding odd and even numbers.
9. Two binary operations $*$ and $\circ$ are defined on the set of real numbers and are given by

$$a \circ b = ab + 2a + 2b + 2, \; a * b = a + b + 2.$$

Show that $\circ$ is distributive over $*$, i.e. show that

$$a \circ (b * c) = (a \circ b) * (a \circ c).$$

Other Finite Systems

In the previous section we saw how it was possible to use a binary operation to impose an algebraic structure on a set, and so create what we call a mathematical system, i.e. a set with one or more binary operations defined on the set. Immediately our binary operation has been defined we explore the applications of the operation so as to discover its characteristics. We ask is it commutative?, is it associative?, are there any special results arising from its application? When the set S is infinite such exploration may be tedious and demanding but if the set is finite the table of values may be rapidly constructed and the structure of the finite systems adequately exposed for our further enquiries in order that we may classify the systems by the properties they possess. In other words, we may start with two different finite systems which have two different binary operations but which produce the same structure in their table of values so that the algebraic manipulations in one system are completely duplicated in the other. With true mathematical efficiency we ask why study both systems when one system represents them both?

Definition 5.7. A mathematical system consists of a set S together with one or more binary operations defined on the set.

Definition 5.8. A mathematical system with only one closed binary operation $*$ defined on set S is called a groupoid and represented by writing $(S, *)$.

As a minor example consider Table 5.5 and Question 6 in the previous exercise. This table represents ordinary multiplication on the set $\{0, 1\}$ and also the wider idea of representing the results of multiplying odd and even numbers, i.e. a represents the set of even numbers, b represents the set of odd numbers and the result $a * b = c$ is registered as either a member of a or of b.

$\cup$	A	B
A	A	A
B	A	B

Table 5.6

$\vee$	T	F
T	T	T
F	T	F

Table 5.7

$\times$	0	1
0	0	0
1	0	1

Table 5.8

Two more Tables 5.6 and 5.7 demonstrate the same structure as shown in these examples.

For Table 5.6 the operation is union $\cup$ on the set A and a subset B. For Table 5.7 the operation is disjunction $\vee$ on statements true, false.

Thus, four different finite systems are seen to have the same algebraic structure and a mere change in mathematical symbols enables us to switch from one to the other. Hence the arithmetic (mod 2) multiplication Table 5.8 may represent them all.

A practical way of obtaining a finite system without using numbers is to consider the equivalence of rotations about three mutually perpendicular lines. The scheme to be followed is given in Fig. 5.2 and we shall consider rotating the letter P about the perpendicular lines h and v and a line e erected perpendicular to the page through the intersection of v and h.

The elements of the set $S = \{I, V, H, E\}$ are rotations of the letter P as follows:

H: Rotate P through 180° about line h. (1 to 4, 4 to 1, 2 to 3, 3 to 2)
V: Rotate P through 180° about line v. (1 to 2, 2 to 1, 4 to 3, 3 to 4)
E: Rotate P through 180° about line e. (1 to 3, 3 to 1, 2 to 4, 4 to 2)
I: Leave P where it is.

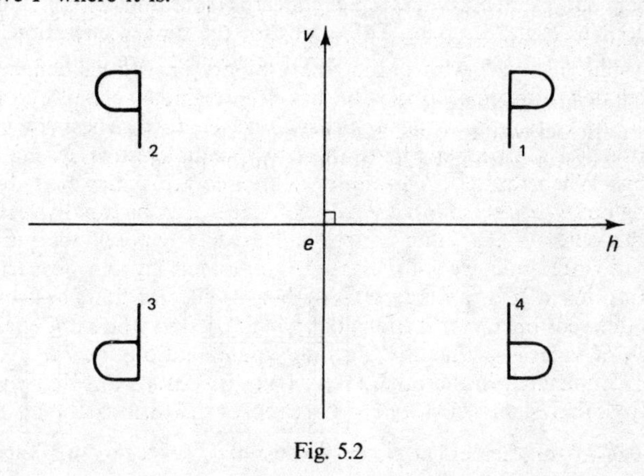

Fig. 5.2

The binary operation $\ast$ is defined as 'followed by' so that $H \ast V$ means 'do H followed by V'. Suppose we start with P_1; then the combination $H \ast V$ takes P_1 to b_4 and then to d_3, an end result which we could have obtained by using E only.

Hence, we have the equivalence $H \ast V = E$ and it is clear that

$$H \ast V = V \ast H = E$$

Similarly, starting with ꟼ$_2$ say, and finding the equivalent single rotation to $V \ast V$, shows that ꟼ$_2$ goes to P_1 and then back to ꟼ$_2$ so we may as well leave ꟼ$_2$ where it is.

$$\therefore \quad V \ast V - I$$

The completion of Table 5.9 is now self evident and we conclude that $\ast$ is a closed commutative operation on the set $S = \{I, V, H, E\}$.

$\ast$	I	V	H	E
I	I	V	H	E
V	V	I	E	H
H	H	E	I	V
E	E	H	V	I

Table 5.9

A typical test for associativity might be to explore $V \ast H \ast E$. Thus

$$(V \ast H) \ast E = E \ast E = I$$
$$V \ast (H \ast E) = V \ast V = I$$
$$\therefore \quad (V \ast H) \ast E = V \ast H \ast E = V \ast (H \ast E)$$

In fact, the operation $\ast$ is associative on S. (We shall return to Table 5.9 in a later chapter.)

In a sense, once the table has been compiled we really do not care any more where it came from because we have erected an algebraic structure independent of the practical content which produced it, and it is this abstraction of mathematical detail which is so satisfying.

A further but elementary example concerns rotations of an equilateral triangle PQR in Fig. 5.3, giving Table 5.10 for the binary operation $\ast$, 'followed by', on the set $S = \{I, A, C\}$ where the elements are defined as

I: Leave the triangle where it is.
A: Rotate anticlockwise through 120° about 0, in the plane of PQR.
C: Rotate clockwise through 120° about 0, in the plane of PQR.

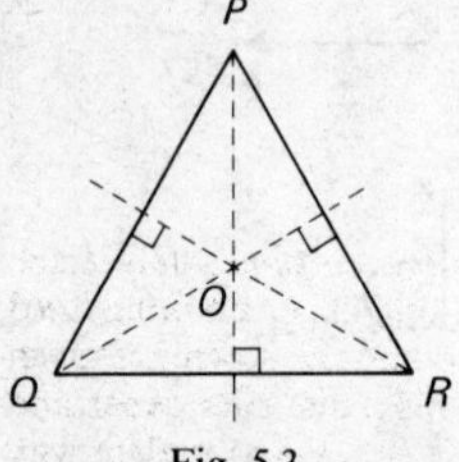

Fig. 5.3

$\ast$	I	A	C
I	I	A	C
A	A	C	I
C	C	I	A

Table 5.10

+	0	1	2
0	0	1	2
1	1	2	0
2	2	0	1

Table 5.11

As outlined in the previous example, $A \ast C$ means: rotate anti-clockwise through 120° 'followed by' rotate clockwise through 120°. This will leave the triangle where it was and so $A \ast C = I$. Clearly $C \ast A = I$, $A \ast A = C$ and so on.

Completion of Table 5.10 shows that the binary operation $\ast$ is closed, commutative and associative on S. Comparison with the addition table for arithmetic (mod 3) in Table 5.11 shows that the structure of the tables is the same so that all the rotations of $\triangle PQR$ may be controlled by Table 5.11. In other words, both systems have the same structure.

Exercise 5.7

1. Using Table 5.9, simplify the following:

(i) $V \ast H \ast H$, (ii) $V \ast H \ast V$, (iii) $V \ast V \ast V$,
(iv) $H \ast V \ast H$, (v) $H \ast E \ast V$, (vi) $E \ast H \ast H$

2. In Table 5.9:

   (i) Which is the identity element?
   (ii) State the inverses of each element with respect to $\ast$.
   (iii) Show that $\ast$ is a closed binary operation on each of the sets $\{I, V\}$, $\{I, H\}$, $\{I, E\}$

3. Use Fig. 5.2 and the same $\ast$, 'followed by', binary operation on the set $S = \{I, A, B, C\}$ where the elements are given by:

   I: Leave P where it is.
   A: Rotate through 90° anticlockwise about e.
   B: Rotate through 180° anticlockwise about e.
   C: Rotate through 90° clockwise about e.

   Complete a table for $\ast$ and state the characteristics of $\ast$, including inverse elements. Identify if possible another system with the same structure.

4. The binary operation $\ast$ is defined on $S = \{1, 2, 3, 4\}$ by $a \ast b = (a \times b) \pmod 5$ where $\times$ is ordinary multiplication.
   Compile a table for $\ast$ on S.

   (i) Is $\ast$ commutative or associative?
   (ii) State the inverses of each element with respect to $\ast$.

5. Two binary operations $\ast$ and o are defined on the set $\{a, b, c\}$ by Tables 5.12 and 5.13. Discuss the characteristics of $\ast$ and o and give all possible inverses. Simplify the following and draw a conclusion from the results.

$\ast$	a	b	c
a	c	b	a
b	a	c	b
c	b	a	c

Table 5.12

o	a	b	c
a	a	b	c
b	b	c	a
c	c	a	b

Table 5.13

   (i) $a \ast (b \text{ o } c)$, (ii) $(a \ast b) \text{ o } (a \ast c)$,
   (iii) $c \text{ o } (b \ast a)$, (iv) $(c \text{ o } b) \ast (c \text{ o } a)$

6. The game of paper (P), stone (S), knife (K) is played by two players alternately. Each makes one hand represent P: flat hand, S: clenched fist, K: one finger pointing, held behind back. On the word go, both players show their hand and the scoring is given by: paper wraps stone and wins, stone blunts knife and wins, knife cuts paper and wins. Complete a table for the 'conjunction $\ast$' on the set $\{P, S, K\}$, and explain why $\ast$ is not an operation on $\{P, S, K\}$.

Groups

Some of the mathematical systems in the previous sections, even when different in size, still had the same characteristics. For example, in arithmetic (mod 7) and (mod 5) each element had an inverse with respect to both $+$ and $\times$ whereas in arithmetic (mod 6) and (mod 4) this was not the case. In the formal discussion we saw that some operations may be associative but not commutative and vice versa. Clearly the time has come to attempt to classify these systems in some way and although we do not wish to go too far in this direction the idea of a group is of some mathematical importance.

Definition 5.9. A group $(S, \ast)$ is a set of elements S, together with a binary operation $\ast$ defined on S, satisfying the following four conditions for any and every element a, b, c in S.

1. S is closed under the operation $\times$ (i.e. the result $a \times b$ is unique, includes all cases $a = b$ and always belongs to S).

2. The associative law holds for $\times$ (i.e. $(a \times b) \times c = a \times (b \times c)$ for all the elements in S, including all cases when a, b, c are not distinct).

3. For the operation $\times$ there exists an identity element e in S such that $a \times e = a = e \times a$, for ever $a \in S$.

4. Every element a of S has an inverse a^{-1} in S with respect to $\times$ such that

$$a^{-1} \times a = e = a \times a^{-1}$$

If $\times$ is also commutative then we shall call $(S, \times)$ a commutative group.

We have already encountered conditions 1 and 2. For 3 and 4 we emphasise the two-sided nature of the requirement. Question 5, Exercise 5.7 showed a one-sided identity for $\times$ which is not enough for condition 3 here.

An inspection of Table 5.9 shows that the operation $\times$, 'followed by' on the set $\{I, V, H, E\}$ forms a system $(S, \times)$ which is a group. Again the systems given by Tables 5.10 and 5.11 satisfy all the four conditions for a group. By comparison we see that $\{a, b, c\}$ and o in Question 5, Exercise 5.7, form a group, but $\{a, b, c\}$ and $\times$ do not form a group.

Isomorphisms

Seeing the same structure in two groups simply by changing the names of the elements and the operation is given a special name of identification.

Definition 5.10. Two groups

$S_1 = \{a_1, a_2, a_3, a_4, \ldots\}$ with operation $\times$,

and $S_2 = \{b_1, b_2, b_3, b_4, \ldots\}$ with operation o

are said to be **isomorphic** if a one-to-one correspondence (matching, page 3) can be established between the two sets S_1 and S_2 in which $a_1 \to b_1$, $a_2 \to b_2$, $a_3 \to b_3$, etc., so that if $a_k \times a_n = a_m$ then $b_k \circ b_n = b_m$.

We can illustrate this definition with some examples we have seen before.

Example 17. With $S_1 = \{I, A, C\}$ and operation $\times$ (Table 5.10) and
$S_2 = \{0, 1, 2\}$ and operation + (Table 5.11)

we match S_1 and S_2 with $I \to 0$, $A \to 1$, $C \to 2$, and see a typical connection between the two groups in noting

$$A \times C = I \text{ with } 1 + 2 = 0$$
$$A \times A = C \text{ with } 1 + 1 = 2$$

Now we can say that $(S_1, \times)$ and $(S_2, +)$ are isomorphic.

Example 18. The two groups $(S_1, +)$ arithmetic (mod 4), $S_1 = \{0, 1, 2, 3\}$ and
$(S^2, \times)$ arithmetic (mod 5), $S_2 = \{1, 2, 3, 4\}$

are isomorphic.

+	0	1	2	3
0	0	1	2	3
1	1	2	3	0
2	2	3	0	1
3	3	0	1	2

Table 5.14

×	1	2	3	4
1	1	2	3	4
2	2	4	1	3
3	3	1	4	2
4	4	3	2	1

Table 5.15

×	1	3	4	2
1	1	3	4	2
3	3	4	2	1
4	4	2	1	3
2	2	1	3	4

Table 5.16

Tables 5.14 and 5.15 show the algebraic structure of the groups. All we have to do is produce the required matching of the two sets S_1 and S_2. This is revealed as follows:

(i) Clearly the identity elements must correspond, so we need $0 \to 1$.
(ii) Since $2 + 2 = 0$ in S_1 and $4 \times 4 = 1$ in S_2 we must have $2 \to 4$.
(iii) Since $1 + 1 = 2$ in S_1 and $3 \times 3 = 4$ in S_2, we must have $1 \to 3$.
(iv) Finally, $3 \to 2$.

The complete one-to-one correspondence from S_1 to S_2 is

$$0 \to 1,\ 1 \to 3,\ 2 \to 4,\ 3 \to 2$$

If we now realign the row and columns of Table 5.15 to give a pictorial correspondence between Tables 5.16 and 5.14 the isomorphism becomes obvious.

Subgroups

Sometimes within a group $(S, \ast)$ there is a subset of S which also forms a group under $\ast$. This subgroup must contain the identity element at least. A subgroup with the identity element as its only element is called a **trivial subgroup**.

Examination of Table 5.15 shows that the subset $\{1, 4\}$ together with the operation $\times$ forms a subgroup of $(S, \times)$ as defined by the table. Indeed, the structure of this subgroup can be seen in Table 5.17 below.

If we go back to Table 5.1 we see the subgroup $\{0, 3\}$ with $+$ which is now displayed in Table 5.18.

×	1	4
1	1	4
4	4	1

Table 5.17

+	0	3
0	0	3
3	3	0

Table 5.18

Sometimes associativity is very tedious to prove and we try to fall back on short-cut tests when we suspect that an operation may not be associative on a set. An interesting theorem by Lagrange states that:

If a finite group of order m has a subgroup of order n then n must be a factor of m.

This means that if we are inspecting a table of order five and we discover a subgroup of order two then we know that the table cannot represent a group because 2 is not a factor of 5. Notice that all the subgroups we have obtained so far have been of orders which are factors of the order of the parent group. We must be warned against the obvious mistake in assuming that the presence of subgroups in a table implies that the table represents a group, as we see in Table 5.19 below of a system of order six for the operation $\ast$ on $S = \{0, 1, 2,$

∗	0	1	2	3	4	5
0	0	1	2	3	4	5
1	1	2	0	4	5	3
2	2	0	1	5	3	4
3	3	4	5	0	1	2
4	4	5	3	2	0	1
5	5	3	4	1	2	0

Table 5.19

∗	0	1	2	3	4
0	0	1	2	3	4
1	1	0	3	4	2
2	2	4	0	1	3
3	3	2	4	0	1
4	4	3	1	2	0

Table 5.20

3, 4, 5} which contains subgroups {0, 4}, {0, 1, 2}, i.e. subgroups of order 2 and 3 which are both factors of 6, but this does not mean that $(S, \ast)$ is a group. Indeed, $(S, \ast)$ cannot be a group because the operation $\ast$ does not obey the associative law for as we readily see,

$$(3 \ast 5) \ast 4 = 2 \ast 4 = 3$$
$$3 \ast (5 \ast 4) = 3 \ast 2 = 5$$

so

$$(3 \ast 5) \ast 4 \neq 3 \ast (5 \ast 4)$$

Table 5.20 will be used in the next exercise.

Exercise 5.8

1. There are two subgroups in the group given by Table 5.1 (page 88). Identify these subgroups given that one has two elements and the other has three elements.
2. Identify the three subgroups in the group given by Table 5.9.
3. Prove that any two groups with only two elements are isomorphic.
4. Examine Table 5.3 for a subgroup of two and four elements.
5. Prove that Tables 5.14 and 5.15 represent groups.
6. Show that from Table 5.4 the set {1, 2, 3, 4, 5, 6, 7} does not form a group with respect to $\times$ for arithmetic (mod 8). Identify three subgroups of two elements each and one subgroup of four elements.
7. Do either of the tables in Question 5, Exercise 5.7, represent a group?
8. Establish whether or not Tables 5.10 and 5.13 represent isomorphic groups.
9. Find any subgroups in Table 5.20.
10. Using Table 5.20, find:

    (i) $4 \ast (1 \ast 2)$
    (ii) $(4 \ast 1) \ast 2$

    Show that Table 5.20 does not represent a group.

6

RELATIONS

We have seen how to select and describe various subsets of a universal set by using a characteristic property of the elements; now we wish to go a stage further to see how the individual elements of a set may be related to each other.

In Chapter 1 we found the need to express a progressive sequence of terms as a matter of trust, writing something like 2, 4, 8, 16, ... and leaving this as sufficient evidence to enable all the other terms to be generated. The link between successive terms as we read from left to right might be stated as '2 is a half of 4', '4 is a half of 8', ...'16 is a half of 32', and so on. Thus each term is related to the next one by the phrase

'is a half of'

Suppose now that we examine the elements of the set $S = \{2, 4, 6, 8, 10, 12, 14, 16, 18, 20\}$ and try to create a chain of elements by making up different links.

Conditioned as we are by the order in which the elements have been written the obvious relation is given by the phrase 'is two less than' and the result of defining this on the set S is to produce (2, 4), (4, 6), (6, 8) as typical ordered pairs in this binary relation. (Binary because it relates two elements at a time.)

The following is a list of alternative relations which may be defined on the set S

Relation	*Ordered pairs*
'is a quarter of'	(2, 8), (4, 16)
'is ten less than'	(2, 12), (4, 14), (6, 16), (8, 18), (10, 20)
'is ten more than'	(12, 2), (14, 4), (16, 6), (18, 8), (20, 10)
'is twice'	(20, 10), (16, 8), (12, 6), (8, 4), (4, 2)

Thus each relation on the set S produces its own set of ordered pairs. There is a certain misleading neatness about these examples inasmuch as the elements only ever appear once in the ordered pairs. A relation like 'is less than' is now seen to have a much greater number of ordered pairs belonging to it: here are some of them

(2, 4), (2, 6), (2, 8), (2, 10), ... (4, 6), (4, 8), (4, 10), ... (6, 8), (6, 10). ...

and so on, with the dots not only relieving the fatigue but also suggesting the absent pairs.

Given any set S of random elements readers are likely to make different choices of relations but we shall all need a common notation with which to indicate the direction of our lines of thought and show that we are thinking in terms of 4, 8, 12 and not 12, 8, 4 say. So we shall agree to represent the relation of one element to another according to the following definition:

Definition 6.1. A relation on a set S is represented by a set of ordered pairs (a, b) where $a, b \in S$. If a connective phrase is used to describe the relation then this will be represented by R and we shall write $a \text{ R } b$ to indicate that elements a and b are so related. We note that the pairs (a, b), (b, a) will not necessarily be pairs in the same relation R.

Example 1. Identify the ordered pairs which belong to the relation R:'is three less than' on the set $S = \{1, 2, 4, 5, 7, 10, 26, 29, 30\}$.

Solution. We examine each element of S, subtract three and see if the answer is an element of S. By examining the elements in this manner we see that R is represented by the set of ordered pairs

$$(1, 4), (2, 5), (4, 7), (7, 10), (26, 29)$$ *Answer*

Example 2. Obtain the ordered pairs of the relation R:'has the same prime factors as' on the set $S = \{6, 12, 15, 18, 30, 45, 50, 60, 75\}$.

Solution. The prime factors involved are 2, 3, 5, 7.

Taking each element in turn we start with $6 = 2 \times 3$ and look for another member of the set which is also a product of only 2s and 3s. Thus 6 R 12, 6 R 18, 12 R 18, i.e. ordered pairs (6, 12), (6, 18), (12, 18) belong to R.

Of course 12 has the same prime factors as 6, etc. so the ordered pairs (12, 6), (18, 6), (18, 12) also belong to R.

Interestingly enough, we may reason that 6 has the same prime factors as 6 so (6, 6) belongs to R. And similarly so does (12, 12), (18, 18). Without listing all the remaining ordered pairs in R we can see that R partitions the set S into the following disjoint subsets.

$$A = \{6, 12, 18\}$$
$$B = \{15, 45, 75\}$$
$$C = \{30, 60\}$$
$$D = \{50\} \text{ so that } S = A \cup B \cup C \cup D$$ *Answer*

Example 3. The relation R on a set S is given by the following ordered pairs: (9, 3), (16, 4), (25, 5), (64, 8), (100, 10), (a, b). Give a descriptive title for R, and suggest possible values for a, b.

Solution. This is very much a trial and error exercise in which we hope that our practical experience will bring forth suggestions which fit the ordered pairs in the binary relation R.

We 'see' that R:'is the square of', i.e. 64 is the square of 8 and so on. Hence $b^2 = a$.

Answer

Exercise 6.1

1. Give the ordered pairs for the following relations defined on the set $S = \{1, 2, 3, 4, 6, 8, 9, 12, 16, 64\}$:
   (i) 'is half of'
   (ii) 'is a quarter of'
   (iii) 'is eight times as great as'
   (iv) 'is three more than'
   (v) 'gives an odd number when multiplied by'
   (vi) 'is twenty less than'
2. Give the ordered pairs for the following relations defined on the set $S = \{1, 2, 3\}$:
   (i) 'is greater than'
   (ii) 'is less than'
   (iii) 'is equal to'
3. The relation R on a set is given by the following set of ordered pairs $\{(2, 10), (-1, -5), (0, 0), (a, 30), (3, b), (x, y)\}$. Assuming that an arithmetical pattern may be deduced from the first three pairs in the set, find a, b and an equation connecting x to y.

4. Write down the set of all ordered pairs which it is possible to form in the set $S = \{0, 1, 2\}$.
5. Obtain the ordered pairs of the relation R:'has the same prime factors as' in the set $S = \{6, 12, 21, 22, 36, 44, 63, 88\}$ and show that this relation partitions S into three disjoint subsets.

Relation Diagrams

The pictorial reference to lines of thought in establishing a relation between two elements is more than a passing remark. Diagrams like Fig. 6.1 and 6.2

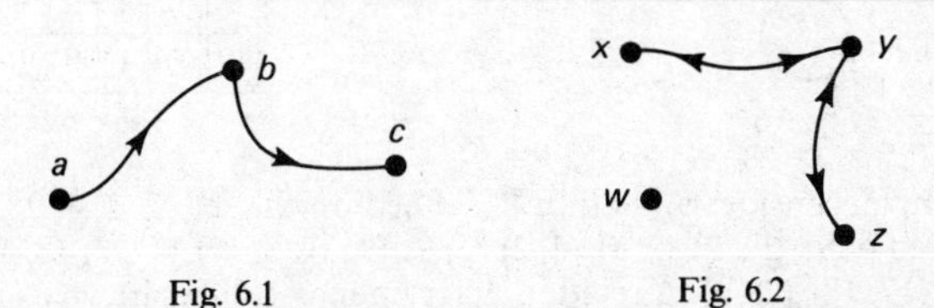

Fig. 6.1 Fig. 6.2

can convey information almost without defining their meaning. Thus with $S = \{a, b, c\}$ and a relation R the diagram shows, by joining a to b and using the arrowhead, that a is related to b and similarly b is related to c. The formal representation of these remarks is a R b and b R c or we can state that the ordered pairs (a, b) and (b, c) belong to R.

A typical relation which this diagram fits is R:'is a third of' so that the numerical suggestion might be

$$a = 2, b = 6, c = 18$$

or

$$a = 10, b = 30, c = 90$$

A relation such as R:'is less than' would not fit the diagram because although a is less than b *and* b is less than c is indicated it now follows that these two statements together will imply that a is less than c, in which case we should join a to c.

For Fig. 6.2 consider the relation R:'results in an odd sum when added to' on the set of four whole numbers $S = \{w, x, y, z\}$. (*N.B.* As we shall see, this relation will not fit the diagram.) The diagram indicates that $x + y$ is an odd number, and shows that we need two arrowheads on the connection because $y + x$ is the same odd number as $x + y$. A little thought will indicate that of x, y, z either two must be odd numbers or two must be even numbers.

To put the meaning of the diagram more formally the relation R consists of the set of ordered pairs $\{(x, y), (y, x), (y, z), (z, y)\}$. As an indication of the thoughts such diagrams can provoke we now wonder why w is not connected to one or more of x, y and z because since w, x, y, z are whole numbers then w is either odd or even and so must be in relation to one of x, y or z.

Suppose we consider a relation R:'results in a multiple of 5 when added to' on S, Now this would be acceptable because we could have

$$x = 2, y = 8, z = 12, w = 1$$

Relations are completely and more conveniently defined by a set of ordered pairs drawn from the set on which the relation is defined, because it is not always possible to offer a description of the relation in a descriptive form like 'is greater than' or 'results in an even sum when added to'.

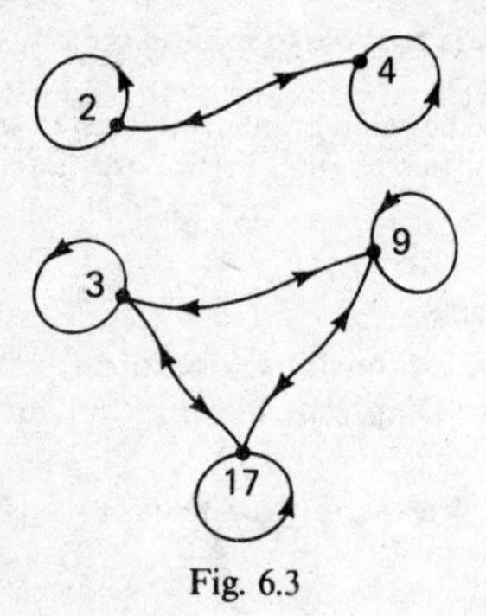

Fig. 6.3

2nd element

1st element	2	3	4	9	17
2	1	0	1	0	0
3	0	1	0	1	1
4	1	0	1	0	0
9	0	1	0	1	1
17	0	1	0	1	1

Table 6.1

For example, consider a relation R defined on the set {2, 3, 4, 9, 17} by the set of ordered pairs {(2, 2), (2, 4), (4, 2), (3, 9), (9, 3), (3, 17), (17, 3), (3, 3), (4, 4), (9, 17), (17, 9), (17, 17), (9, 9)}. A relation diagram may still be drawn as in Fig. 6.3. The loops attached to each element, e.g. 2, register the fact that 2 R 2 and (2, 2) belong to the relation. The two arrows on each connection indicate that the relation reads both ways, e.g. (3, 17) and (17, 3) each belong to R.

Clearly, as a foolproof representation of any relation we choose to invent on a set, the list of suggested ordered pairs is superior to all others. For a set S with, say, 5 elements there are $5^2 = 25$ different ordered pairs to choose from and any random selection defines a relation. It just so happens that on occasion we may 'see' a descriptive phrase to cover such a random selection. For example, for Fig. 6.3 we may describe R: 'gives an even result when added to', a rather clumsy substitute for the clear elegance of a list of ordered pairs. Note that (3, 3) belongs to R because $3 + 3 = 6$.

Another way to keep an account of the ordered pairs is to list the connections in what is called an **incidence matrix**. This is a table of 0, 1 entries in which a connection is indicated by inserting 1 in the appropriate square. The incidence matrix for the relation diagram of Fig. 6.3 is given in Table 6.1 (the entries are normally made without a grid—compare Tables 6.1 and 5.14). The double-arrowed connection for the binary relation is the same as commutativity for a binary operation and results in a matrix or table which is symmetrical about the leading diagonal. Thus, given the incidence matrix of Table 6.1 we could easily draw the corresponding relation diagram or list the ordered pairs of the relation. The main point to note at this stage is the complete abstraction of defining any relation on a set by the quotation of ordered pairs. There is no need whatever to search for linguistic finesse.

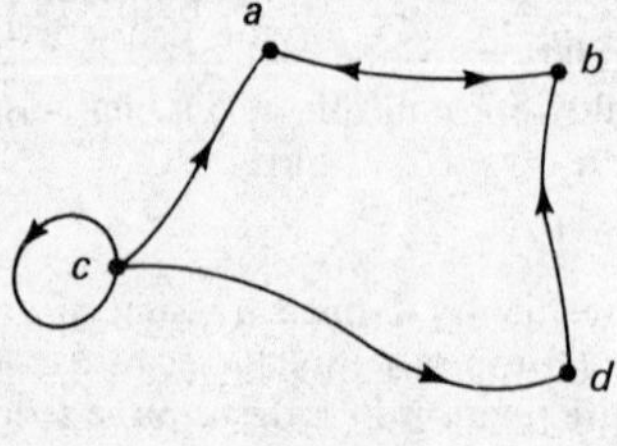

Fig. 6.4

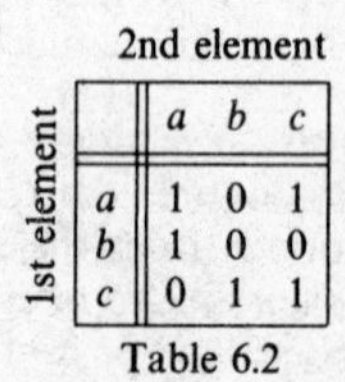

2nd element

1st element	*a*	*b*	*c*
a	1	0	1
b	1	0	0
c	0	1	1

Table 6.2

Thus Table 6.2 or Fig. 6.4 may both represent their respective relations although we might just as well have listed ordered pairs in the first place—that is, $\{(a, b), (b, a), (d, b), (c, d), (c, c), (c, a)\}$ for Fig. 6.4 and $\{(a, a), (a, c), (b, a), (c, b), (c, c)\}$ for Table 6.2.

Exercise 6.2

1. Draw a relation diagram for the relation R: 'results in an even sum when added to' on the set

$$\text{(i) } S = \{3, 7, 8, 10\}$$
$$\text{(ii) } S = \{2.6, 1.4, 4, \sqrt{3}\}$$

2. Without drawing a relation diagram write down the ordered pairs in the relation given by Table 6.2 with the last row reading 0, 0, 0.
3. Write down the ordered pairs in the relation given by Fig. 6.4 with all arrows reversed.
4. Given that *a*, *b*, *c*, *d* are whole numbers explain why the relation given by Fig. 6.4 cannot be:

   (i) 'is equal to'
   (ii) 'is greater than'
   (iii) 'has a sum divisible by 3 when added to'

5. What is indicated by Table 6.2 not being symmetrical about the leading diagonal? Would the incidence matrix for Fig. 6.4 be symmetrical?
6. Draw the incidence matrix for Fig. 6.4. What would the matrix represent if all the entries were changed to their reflections in the leading diagonals?

Types of Relations

Our work with relations in mathematics is not concerned with whether one relation is equal to another relation, i.e. has the same set of ordered pairs, but whether the two relations are of the types which are identified below.

Reflexive Relations

Informally, this is the type of relation which produces loops on *every* element in the relation diagram, or has a 1 entered in every position on the leading diagonal of the incidence matrix. Fig. 6.4 does not represent a reflexive relation because only the element *c* has a loop, Fig. 6.3 does represent a reflexive relation because every element has a loop. Formally we write:

Definition 6.2. A relation R on a set S is said to be reflexive if (a, a) belongs to R for every $a \in S$.

Some descriptive examples of reflexive relations on the set of whole numbers are:

(i) 'is equal to', (ii) 'is a multiple of', (iii) 'has a common factor with'

For $S = \{1, 2, 3\}$, (1, 1), (2, 2) and (3, 3) must all belong to R before R is called reflexive.

Symmetric Relation

Informally this is the type of relation which carries two arrowheads in each and every connection between different elements drawn in the relation diagram, and produces an incidence matrix which is symmetrical about the leading diagonal.

Fig. 6.3 represents a symmetric relation but Fig. 6.4 does not because it does not have two arrowheads on *every* connection between different elements. Formally we write:

Definition 6.3. A relation R on a set S is said to be symmetric if (b, a) belongs to R whenever (a, b) belongs to R.

This is sometimes written as 'a R b implies b R a', i.e. any interchange of the two elements of the ordered pairs in R leaves R unchanged.

Some descriptive examples of symmetric relations on the set of whole numbers are:

(i) 'is equal to', (ii) 'has a factor in common with', (iii) 'has an even sum when added to'

For $S = \{1, 2, 3\}$, $R = \{(1, 2), (2, 3), (3, 2), (3, 3)\}$ we would know that R was *not* symmetric because of the absence of the ordered pair (2, 1).

Anti-symmetric Relation

Informally this is the type of relation for which no connection may carry two arrowheads, and in the incidence matrix no entry of 1 is reflected in the leading diagonal.

Unfortunately it is not possible to describe this relation as being unsymmetric or not-symmetric because we do not have a two-state value here like true–false as we did in Chapter 2. The following example illustrates this point.

For $S = \{1, 2, 3\}$ and $R = \{(1, 2), (2, 3), (3, 2)\}$ we have suggested a relation which is not symmetric because (2, 1) is not an element of R, and is not anti-symmetric because both (2, 3) and (3, 2) belong to R. Formally we write:

Definition 6.4. A relation R on a set S is said to be anti-symmetric if whenever (a, b), $a \neq b$, belongs to R the ordered pair (b, a) does not belong to R.

This is sometimes written as $a \mathrel{R} b \Rightarrow b \not\mathrel{R} a$.

Some descriptive examples of anti-symmetric relations on the set of people:

(i) 'is heavier than', (ii), 'is taller than', (iii) 'is twice as old as'

Emphasis must be placed on the significance of the word 'whenever' in the last two definitions. Thus for:

(i) R symmetric: If we have (a, b) then we must also have (b, a).
(ii) R anti-symmetric: If we have (a, b) then we must *not* have (b, a).

Now suppose (a, b) does not belong to R in the first place. This will mean that it is possible to have a relation which is both symmetrical and anti-symmetrical, e.g. $\{(1, 1), (2, 2), (3, 3)\}$ on $\{1, 2, 3\}$.

Transitive Relation

Informally this is the type of relation which completes a triangle of connections between three different elements. Thus Fig. 6.1 would need to have a connected to c to produce a transitive relation. A numerical illustration is useful here. Suppose we have $14 > 10$ and $10 > 3$, then these two pairs together tell us that $14 > 3$. In algebra this is expressed by

$$x > y \text{ and } y > z \text{ together, imply that } x > z$$

In other words, we can transfer or pass over from x to z, hence the word transitive (passes over, transfer).

Formally we write:

Definition 6.5. A relation R on a set S is said to be transitive if (a, c) belongs to R whenever (a, b) and (b, c) belongs to R. (Again notice the use of the word whenever.)

Some descriptive examples of transitive relations on the set of real numbers are

(i) 'is equal to', (ii) 'is greater than', (iii) 'is less than'

For the set $S = \{1, 2, 3\}$ and $R = \{(1, 2), (2, 3)\}$ the relation R cannot be transitive because the ordered pair (1, 3) is absent from R.

Equivalence Relation

Definition 6.6. An equivalence relation on a set is a relation which is

(i) reflexive,
and (ii) symmetric,
and (iii) transitive.

Obviously an equivalence relation is highly valued in having so many virtues. Some descriptive examples of an equivalence relation on the set of whole numbers are

(i) 'is equal to', (ii) 'has the same prime factors as', (iii) 'has an even sum when added to'

One very important feature of an equivalence relation defined on a set S is that it divides (or partitions) S into disjoint subsets so that each element of S belongs to one and only one of the subsets and thereby corresponds to a sorting process of the elements of S. Aside from this aspect of the equivalence relation it is clear that given R as a large set of ordered pairs the search to establish whether R is an equivalence relation could be tedious. Strangely enough, a descriptive phrase for R may shorten the search for identification if we think of the equivalence relation as a sorting process and ask ourselves 'does the descriptive phrase sort all the elements of S into disjoint subsets?' This is the standard purpose of a questionnaire, for example:

Question	*Relation*	*No. of disjoint subsets*
(i) State male or female.	'is the same sex as'	2
(ii) Married or single.	'in the same marital state as'	2
(iii) Date of birth.	'born on the same day as'	366

Order Relations

We are here referring to a relation on a set S which will enable all the elements of S to be listed in an order, i.e. a total ordering of S. The most obvious relation which comes to mind is the relation 'is less than' on $S = \{\text{whole numbers}\}$.

For a total ordering of S by R we require R to be:

(i) anti-symmetric,
and (ii) transitive,
and (iii) complete,

i.e. every pair of distinct elements must be in R so that we have either $a \mathrm{R} b$ or $b \mathrm{R} a$.

Exercise 6.3

State whether the following relations are reflexive (R), symmetric (S) antisymmetric (A), transitive (T) and state which ones are equivalence relations (E) or order relations (O):

	Relation	*Set*
1.	'is equal to'	set of whole numbers
2.	'is not equal to'	"
3.	'is greater than'	"
4.	'is less than'	"
5.	'is a factor of'	set of natural numbers
6.	$a \mathrm{R} b$ if $a + b$ divisible by 3	"
7.	$a \mathrm{R} b$ if $a - b$ divisible by 3	"
8.	$a \mathrm{R} b$ if ab an even number	"
9.	$a \mathrm{R} b$ if ab an odd number	"
10.	$a \mathrm{R} b$ if $(a + b) > 4$	$\{1, 2, 3, 4, 5, 6\}$

11. For the set $S = \{1, 2, 3, 4\}$ relation $\mathrm{R} = \{(1, 1), (1, 2), (2, 3), (3, 3), (4, 4)\}$. Classify R.
12. For the set $S = \{1, 2, 3, 4\}$ the relation $\mathrm{R} = \{(1, 1), (2, 2), (3, 3), (1, 3), \ldots\}$, where some elements are considered to be missing. Include the minimum number of ordered pairs so as to make R:
    (i) reflexive
    (ii) symmetric
    (iii) antisymmetric
    (iv) transitive
    (v) an equivalence relation.
13. For the set $S = \{\text{straight lines in a plane}\}$ classify the relations R_1 and R_2 defined on S. R_1:'parallel to'; R_2:'perpendicular to'. (Thus $a \mathrm{R} b$ means 'line a is parallel to line b.)

Cartesian Product

Relations which have been classified in the previous section were defined on one set, each relation being given by a set of ordered pairs. It is appropriate to think that we have related the elements of a set S to the same set S so that in the ordered pair (a, b) belonging to R we know that $a \in S$ and $b \in S$. Now we can consider relating the elements of set A to a different set B.

Since we are speaking of a relation from set A to set B it follows that any ordered pair (a, b) in a relation R will be drawn from these two sets by taking $a \in A$ and $b \in B$.

If $n(A) = 10$ and $n(B) = 5$ we shall be able to invent R from a maximum choice of $10 \times 5 = 50$ different ordered pairs. Starting with this set of 50 different ordered pairs it becomes clear that any relation from set A to set B is going to consist of a subset of this grand total.

Definition 6.7. The cartesian product of two non-empty sets A, B in that order, is the set of all ordered pairs (a, b) for all $a \in A$ and $b \in B$.

This cartesian product will be written $A \times B$. (We note that $A \times B$ may not be the same as $B \times A$.)

For example, if $A = \{1, 2, 3\}$, $B = \{p, q\}$ then the cartesian product $A \times B$ is given by the set of six ordered pairs.

$$A \times B = \{(1, p), (1, q), (2, p), (2, q), (3, p), (3, q)\}$$

Any relation R from set A to set B is a subset of $A \times B$, e.g.

$$R_1 = \{(1, p), (3, q)\},\ R_2 = \{(1, p), (2, p), (3, p)\}$$

In most cases we shall have $A = B$ and we shall refer to the cartesian product $A \times A$.

Definition 6.8. A binary relation from set A to set B is any subset of $A \times B$.

Notice the versatility of this definition: it does not depend on our ability to select an appropriate descriptive phrase, we merely select those ordered pairs which fit our creative needs.

As a further illustration consider relations R_1 and R_2 from set A to set B where

$$A = \{1, 3, 5, 7\} \text{ and } B = \{2, 4, 6\}$$

We list the set of 12 ordered pairs of $A \times B$

$$\begin{array}{llll} (1, 6), & (3, 6), & \underline{(5, 6)}, & (7, 6)^* \\ (1, 4)^*, & \underline{(3, 4)}, & (5, 4), & (7, 4) \\ \underline{(1, 2)}, & (3, 2), & (5, 2)^*, & (7, 2) \end{array}$$

and a suggested relation R_1 is underlined

$$\text{i.e. } R_1 = \{(1, 2), (3, 4), (5, 6)\}$$

and a further relation R_2 is asterisked

$$\text{i.e. } R_2 = \{(1, 4), (5, 2), (7, 6)\}$$

The individual elements such as a or b in (a, b) are called the **components** or **coordinates** of the ordered pair. Thus a is the first coordinate and b the second coordinate, and for example, as we have already seen, (1, 2) is different from (2, 1).

Definition 6.9. The set of first coordinates in the relation R is called the **domain** of R. (This is sometimes referred to as the domain of definition of R.) The set of second coordinates in the relation R is called the **range** of R.

When R is drawn from the cartesian product $A \times B$ the domain of R is a subset of A and the range of R is a subset of B.

Example 4. Identity the relation R:'is two less than' in the cartesian product $A \times B$ and $B \times A$ where $A = \{1, 2, 3, 4, 5\}$, $B = \{3, 5, 7\}$.

Solution. In $A \times B$ we have $R_1 = \{(1, 3), (3, 5), (5, 7)\}$; i.e. the range of R_1 is $\{3, 5, 7\}$.
In $B \times A$ we have $R_2 = \{(3, 5)\}$ i.e. the range of R_2 is $\{5\}$.
This shows the dependence of R on the domain available. *Answer*

The formal definition of a relation following that for a cartesian product $A \times B$ leads to the introduction of a universal notation which now follows. These definitions may be tiresome but they are part of a constant effort to standardise the notational presentation of mathematical ideas.

Notation: With the relation R defined from the set A to set B and consisting of a set of ordered pairs we either list the set of ordered pairs or write in the following set builder notation,

$$R = \{(a, b) \mid p\,(a, b),\ a \in A,\ b \in B\}$$

where p (a, b) is some proposition connecting a with b such as $a > b$, $b < a$, $b = 5a + 3$ and so on, meaning that R consists of *all* the ordered pairs (a, b) such that p (a, b) is true and $a \in A$, $b \in B$.

For example, with

$$A = \{1, 2, 3\}, B = \{\text{whole numbers}\}$$
$$R = \{(a, b) \mid b = 5a + 3, a \in A, b \in B\} \text{ gives}$$
$$R = \{(1, 8), (2, 13), (3, 18)\}$$

that is, R consists of *all* the ordered pairs (a, b) such that $b = 5a + 3$, and $a \in A$, $b \in B$.

But it would be impossible to list all the ordered pairs in R given $A = \{\text{real numbers}\}$.

Example 5. List all the elements of R defined by

$$R = \{(a, b) \mid b = a^2, a \in A, b \in B\}$$

where $A = \{1, 2, 3, 4, 5, 6\}$, $B = \{\text{odd numbers}\}$

Solution. We already know that if a is even then so is a^2. Hence R = $\{(1, 1), (3, 9), (5, 25)\}$ and the range of R is $\{1, 9, 25\}$ *Answer*

Inverse Relations

The inverse relation may be considered as an undoing exercise like those referred to on page 22. Since relations have direction in the sense of ordered pairs we can see the obvious notion of undoing in the inverse relation R^{-1} since it is obtained by interchanging the coordinates in all the ordered pairs which belong to R.

Definition 6.10. The inverse relation to R is written R^{-1} and given by

$$R = \{(b, a) \mid (a, b) \in R\}$$

which means that R^{-1} consists of *all* ordered pairs (b, a) such that $(a, b) \in R$.

Example 6. The relation R from set A to set B where $A = \{-1, 0, 1\}$ and $B = \{0, 1, 2, 3, 4\}$ is given by

$$R = \{(a, b) \mid b = a + 1, a \in A, b \in B\}$$

Obtain the inverse relation R^{-1}.

Solution. By inspection:

$R = \{(-1, 0), (0, 1), (1, 2)\}$, i.e. domain of $R = \{-1, 0, 1\}$ and range of $R = \{0, 1, 2\}$.

Hence $R^{-1} = \{(0, -1), (1, 0), (2, 1)\}$, i.e. domain of $R^{-1} = \{0, 1, 2\}$ and range of $R^{-1} = \{-1, 0, 1\}$.

The domain and range of R become the range and domain of R^{-1} respectively.

Notice also that the set builder notation gives

$$R^{-1} = \{(b, a) \mid b = a + 1, a \in A, b \in B\}$$

i.e. a straightforward interchange of a and b. In most cases we shall have $A = B$. *Answer*

Exercise 6.4

1. With set $A = \{1, 2, 3, 4, 5, 6\}$ and set $B = A \cup \{7, 8, 9\}$ how many ordered pairs are there in the cartesian product $A \times B$ and $B \times A$?

   (i) Write down all the ordered pairs in $R = \{(a, b) \mid a = 3, b > 5, a \in A, b \in B\}$.
   (ii) Find R^{-1} for R in (i).

(iii) Which ordered pairs could possibly belong to both a new R^{-1} and R defined from set A to set B?
(iv) If $R = R^{-1}$ what can be said about the elements of R?

2. With $A = \{1, 2, 3\}$ write down the cartesian product $A \times A$.
   (i) Given that $R^{-1} = \{(1, 1), (1, 2), (1, 3)\}$ find R.
   (ii) List the ordered pairs in $R = \{(a, b) \mid a > b, a, b \in A\}$.
   (iii) Find the inverse relation of R in (ii).

3. The relation R on $A \times A$, $A = \{\text{positive integers}\}$ is given by

$$R = \{(x, y) \mid y = x^2, x, y \in A\}$$

List those elements of R for which (i) $x \leqslant 3$, (ii) $y \leqslant 26$.

4. The relation R on $A = \{\text{positive integers}\}$ is given by

   $R = \{(x, y) \mid x - y \text{ is an even number}, x, y \in A\}$
   (i) List all the elements of R for which $x \leqslant 4$, $y \leqslant 4$.
   (ii) Show that R is an equivalence relation.
   (iii) Into what classes does R partition the set A?

5. The relation R is defined on $A \times A$ with $A = \{\text{natural numbers}\}$. With (a, b) R (c, d) meaning that $\dfrac{a}{b} = \dfrac{c}{d}$ prove that R is an equivalence relation. (This is the relation which recognises that $\frac{1}{2} = \frac{2}{4} = \frac{3}{6}$, etc. or $\frac{5}{7} = \frac{10}{14} = \frac{15}{21}$, etc.)
6. Given that R is an equivalence relation on the set $A \times A$ prove that $R = R^{-1}$.

Graphs

The graph of a relation is a record of all its ordered pairs, and the compiling of such a record is called graphing the relation. The pictorial or graphic representation of a relation is most conveniently carried out on a piece of graph paper suitably scaled for size and carrying two perpendicular axes so that the posi-

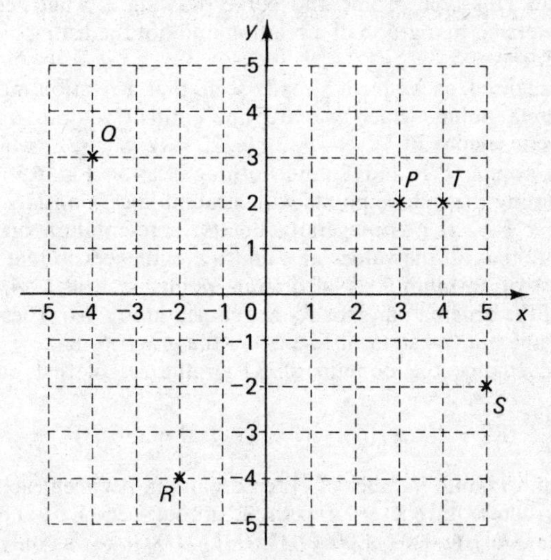

Fig. 6.5

tion of any point on the graph paper can be fixed by reference to its distance from each of these axes.

Briefly the representation can be 'seen' from the Fig. 6.5, which shows the arrangement that is required—namely, two perpendicular lines 0x, 0y intersecting at 0 and called the x axis (0x) and y axis (0y). In order to give some assistance to measuring the distance from the lines, they are graduated by the integers, as shown in the figure. We can appreciate the proposed representation by examining the position of the points P, Q, R, S. We always measure parallel to the x axis first and parallel to the y axis second. The position of P is given as 3 parallel to the x axis and 2 parallel to the y axis, which we may abbreviate by writing (3, 2). Thus the position of a point is quoted as an ordered pair with the convention that the first number is the x coordinate, i.e. measured parallel to the x axis, and the second number is called the y coordinate, i.e. measured parallel to the y axis. This means that the general point in Fig. 6.5 may be referred to as the point (x, y). It is clear by inspection that Q is the point $(-4, 3)$, R is the point $(-2, -4)$ and S is the point $(5, -2)$.

Since any relation R is a set of ordered pairs it follows that the elements of R may be translated into a corresponding set of points on the graph paper, so giving a pictorial graph of the relation. The arrangement of axes and points briefly outlined in Fig. 6.5 is referred to as the **cartesian plane** with 0x and 0y called the coordinate axes. For this reason we refer to the point $(-4, 2)$ as meaning a point having an x coordinate of -4 and a y coordinate of 2. For the arrangement on this Fig. 6.5 we can identify the grid points as representing the ordered pairs which are in cartesian products such as $\{0, 1, 2, 3, 4, 5\} \times \{0, 1, 2, 3, 4, 5\}$.

If we now consider the relation $R = \{(x, 2) \mid 3 \leqslant x \leqslant 4, x \in \text{(real numbers)}\}$ we see that although it is impossible to list all the elements in R we can suggest their presence when drawing the graph of R by joining P to T on Fig. 6.5 and saying that every position on the line segment PT represents an element in R and vice versa. This type of line and curve drawing is what people usually mean when referring to a graph of a relation and not the listing of elements as originally defined.

It must be realised, as we have already seen, that a relation may consist of a set of discrete points which we are not entitled to join up. Consider, for example, the relation $R = \{(-4, 3), (-2, -4), (5, -2)\}$, which is represented by the points Q, R and S (and nothing else) on Fig. 6.5. There is no question of joining these three points with straight lines. Consider the relation $R_1 = \{(x, y) \mid x = y, x, y \in \text{(integers)}\}$ and its representation on a cartesian graph. We note that all the values of x and y are integers so that the relation R_1 will consist of an infinite set of discrete points such as $(-4, -4)$, $(0, 0)$, $(3, 3)$. Some of the ordered pairs for R_1 are circled in Fig. 6.6. These are shown on a straight line but the straight line is *not* the graph of R_1.

Suppose we change the domain of R_1 to the set of real numbers and consider

$$R_2 = \{(x, y) \mid x = y, x, y \in \text{(real numbers)}\}$$

Now we have an infinite number of ordered pairs in between each one of the circled points representing R_1. For example, in between (1, 1) and (2, 2) consider (1.0001, 1.0001), (1.0002, 1.0002), (1.0003, 1.0003), ... as only three such points.

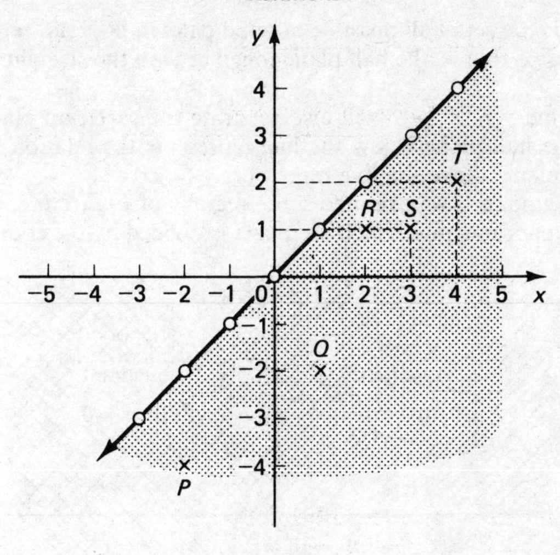

Fig. 6.6

The final result of inserting all these points is to draw a straight line through the lot and say that the straight line represents the relation R_2. It clearly makes sense to call this straight line $y = x$ or say that $y = x$ is the equation to the straight line.

Any straight line in the cartesian plane divides the plane into two half planes. In this example we may refer to half planes above and below the line without ambiguity. Looking at points like P, Q, R, which are below the line, take note of their coordinates $P(-2, -4)$, $Q(1, -2)$, $R(2, 1)$ and see that below the line, the y coordinate of a point is always less than its x coordinate, i.e. $y < x$.

Therefore, *all* the points in the shaded half plane in Fig. 6.6 belong to R_3, where

$$R_3 = \{(x, y) \mid y < x,\ x, y \in \text{(real numbers)}\}$$

In other words, the whole of the shaded half plane is the graph of the relation R_3.

Equations such as (i) $y = 3x + 2$, (ii) $y = -2x + 7$, (iii) $y = \frac{1}{2}x + 1$ and (iv) $y = \frac{1}{3}x - 2$ may be thought of as particular examples of the equation $y = mx + c$, where we can choose particular values for m and c. If we put $m = 3, c = 2$ we get (i); if we put $m = \frac{1}{3}, c = -2$ we get (iv). Any such equation represents a straight line and separates the cartesian plane into two half planes.

Since relations are sets it follows that two relations may have a non-empty intersection, so we may speak of $R_1 \cap R_2$ and $R_2 \cap R_3$, say. Starting from the one line $y = x$ we may suppose that in $\mathbb{R} \times \mathbb{R}$ (recall $\mathbb{R}$ = {real numbers}) there is

(i) a set R_2 represented by all the points on the line $y = x$,
and (ii) a set R_3 represented by all the points in the half plane $y < x$,
and (iii) a set R_4 represented by all the points in the half plane $y > x$.

$\therefore$ $R_3 \cup R_2$ is the set of all possible ordered pairs in $\mathbb{R} \times \mathbb{R}$, represented by the relation $y \leqslant x$, that is, the half plane together with the straight line boundary $y = x$.

A different line $y = 2x + 1$ will also separate the cartesian plane into two half planes. The half plane below the line represents the relation $y < 2x + 1$ while the half plane above the line represents $y > 2x + 1$.

For further guidance we examine some regions of Figure 6.7, in which we note the four quadrants into which the plane is divided by the coordinate axes.

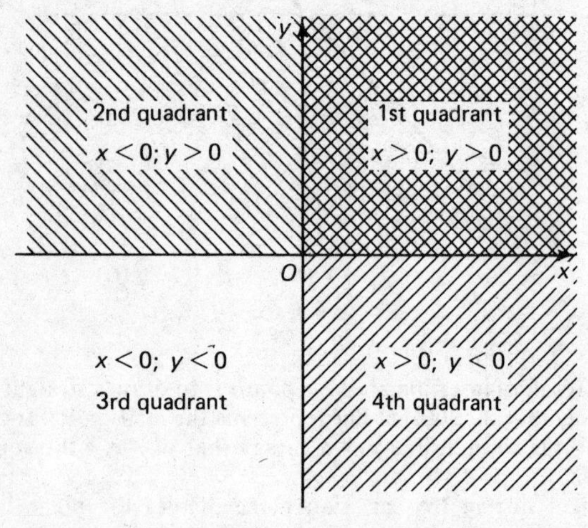

Fig. 6.7

Everywhere on the line 0x, the x axis, we have points whose y coordinate is 0 so we may write an equation for the x axis as $y = 0$. Similarly, we have $x = 0$ as the equation for the line 0y or the y axis. Everywhere above the line $y = 0$ we have $y > 0$, which we may call

$$R_1 = \{(x, y) \mid y > 0, x, y \in \mathbb{R}\}$$

Everywhere to the right of the line 0y we have $x > 0$, which we may call

$$R_2 = \{(x, y) \mid x > 0, x, y \in \mathbb{R}\}$$

All the points in the first quadrant are represented by ordered pairs which belong to $R_1 \cap R_2$ (i.e. $x > 0$ and $y > 0$). The points which lie in the other three quadrants may be thought of in terms of similar intersections.

The diagrams on page 121 now provide us with similar examples to the above.

Example 7. In Fig. 6.8 the equation of the straight line is $y = x + 2$; then all the points on this line have ordered pairs which satisfy the relation $R_1 = \{(x, y) \mid y = x + 2, x, y \in \mathbb{R}\}$. The method of drawing the line is simply to choose two values for x, e.g. 1 and 8, and work out the corresponding values 3 and 10 for y to give two points (1, 3), (8, 10) which are then joined by a line produced in each direction as far as possible. The shaded half plane represents the relation $R_2 = \{(x, y) \mid y < x + 2, x, y \in \mathbb{R}\}$.

Example 8. In Fig. 6.9 the equation of the straight line is $y = -2x + 2$. All the points on this straight line make up the relation $R_1 = \{(x, y) \mid y = 2 - 2x, x, y \in \mathbb{R}\}$.

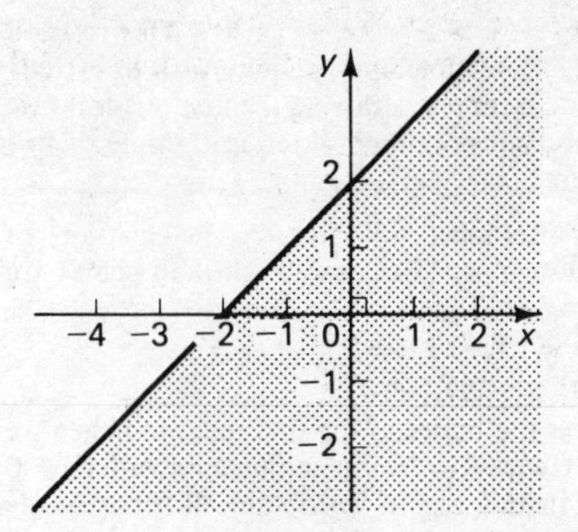

Fig. 6.8

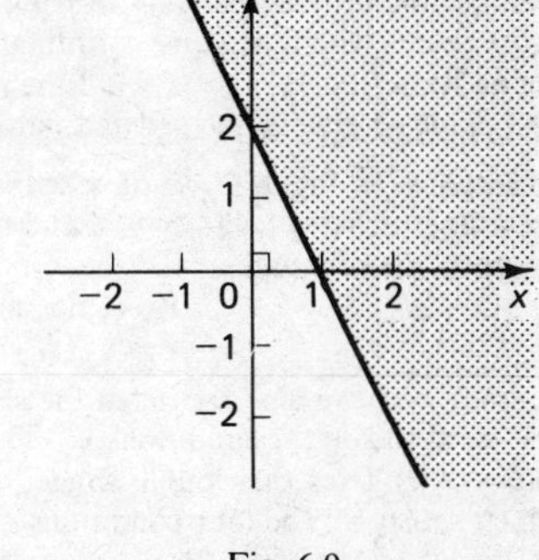

Fig. 6.9

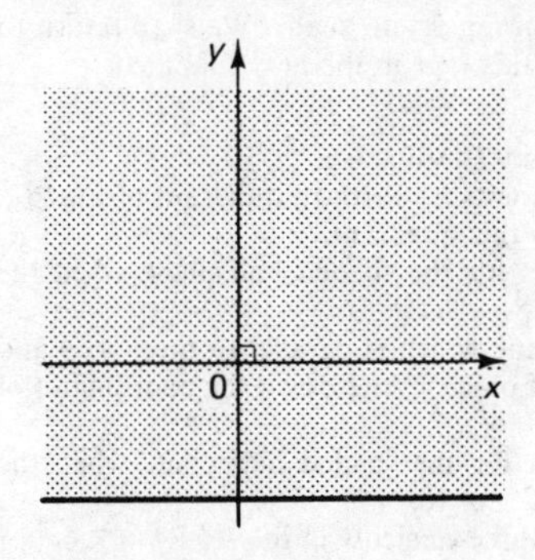

Fig. 6.10

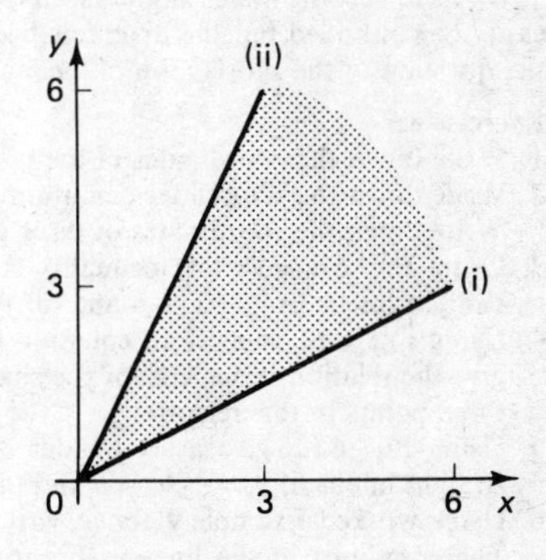

Fig. 6.11

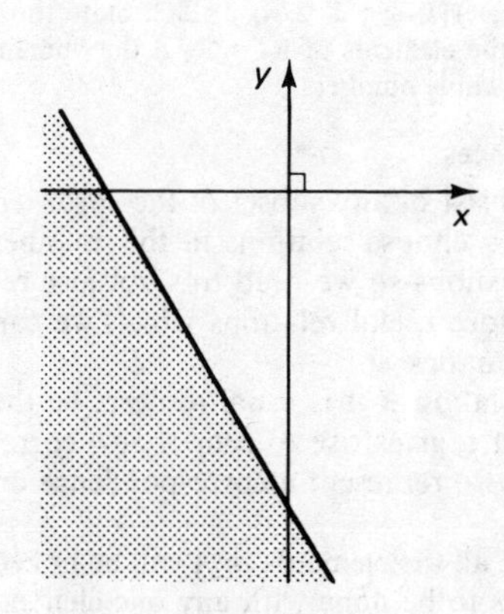

Fig. 6.12

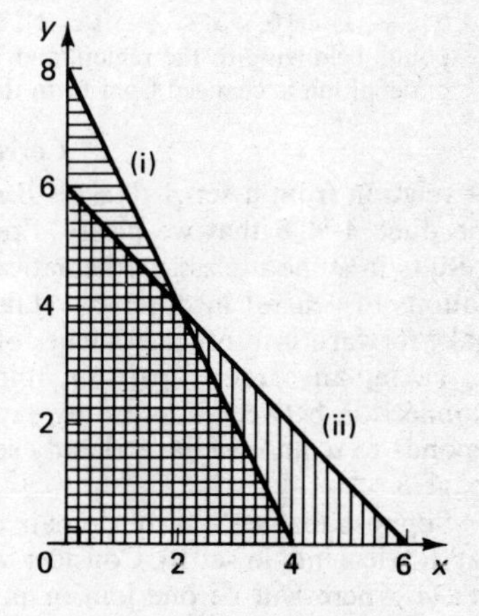

Fig. 6.13

The shaded half plane represents $R_2 = \{(x, y) \mid y > 2 - 2x, x, y \in \mathbb{R}\}$. This relation includes the ordered pairs $(-1, 5)$, $(3, -1)$, $(3, -3)$. Sometimes we may wish to exclude those points with a negative coordinate so aside from *only* drawing the first quadrant we define $R_3 = \{(x, y) \mid x \geqslant 0, y \geqslant 0, x, y \in \mathbb{R}\}$ and consider $R_2 \cap R_3$ to restrict the shaded half plane to give us the ordered pairs we want.

Example 9. In Fig. 6.13 we have two straight lines labelled (i) and (ii). The equations of these lines are (i) $y = -2x + 8$ and (ii) $y = -x + 6$. The two relevant half planes are given by the relations

$$R_1 = \{(x, y) \mid y < 8 - 2x, x, y \in \mathbb{R}\}$$

and

$$R_2 = \{(x, y) \mid y < 6 - x, x, y \in \mathbb{R}\}$$

Since we have also restricted the shading on the figure to the first quadrant then we also need to add the conditions $x \geqslant 0$, $y \geqslant 0$. Hence $R_1 \cap R_2$ in conjunction with ($x \geqslant 0$ and $y \geqslant 0$) gives the double shaded region. In this region we obtain all those points which satisfy all the four conditions

$$y < 8 - 2x, \; y < 6 - x, \; 0 \leqslant y, \; 0 \leqslant x$$

We have become increasingly selective in these three examples and this process could easily be continued but the diagrams become more messy to shade. We shall return to the question of the satisfaction of inequalities of this type in the next chapter.

Exercise 6.5

1. Write down the coordinates of the points S and T in Fig. 6.6.
2. Write down the set builder definition for all points in the third quadrant of Fig. 6.7. Write down the coordinates of three points in this quadrant.
3. Using Fig. 6.8, give the inequality statements for the shaded half plane when the straight line is (i) $y = x + 4$ and (ii) $y = x + 8$.
4. Using Fig. 6.10, suggest an equation for the line which is parallel to the x axis and give the relation statement for the shaded half plane. Write down the coordinates of three points in this region.
5. Using Fig. 6.12, give a set builder definition for the shaded half plane when the straight line is (i) $y = \frac{-5x}{3} - 5$ and (ii) $y = \frac{-5x}{3} - 10$.
6. Using worked Example 9 above, write down three elements in the set $R_1 \cap R_2$.
7. The equations of the lines in Fig. 6.11 are (i) $y = \frac{1}{2}x$ and (ii) $y = 2x$. Define the relation given by all the points in the shaded region, and write down three elements in this relation.
8. Draw the region of the cartesian graph which corresponds to $R_1 \cap R_2$ where $R_1 = \{(x, y) \mid 0 \leqslant x \leqslant 2, x, y \in \mathbb{R}\}$, $R_2 = \{(x, y) \mid 0 \leqslant y \leqslant 2, x, y \in \mathbb{R}\}$. State three points belonging to the region, and give all the elements of $R_1 \cap R_2$ if the domain of definition is changed from $\mathbb{R}$ to the set of whole numbers.

Correspondences

A relation from a set A to a set B may consist of any subset of the cartesian product $A \times B$ that we please. Freedom to choose relations in this manner results in some useless mathematical suggestions so we need to scrutinise relations in general in order to retain the more useful relations which we can take forward into other branches of mathematics.

Taking any ordered pairs (a, b) in our relation R it is usual to describe the connection between a and b by saying that a gives rise to b or that b corresponds to a, and we have already seen how to represent a correspondence on page 3.

Suppose that set A is the domain of R (i.e. all the elements in A will be linked up to elements in set B). Consider what has to be done with any one element $a \in A$. There will be one join or many joins as suggested by Fig. 6.14 (i), in which a is related to one and only one element in B, and by Fig. 6.14 (ii), in

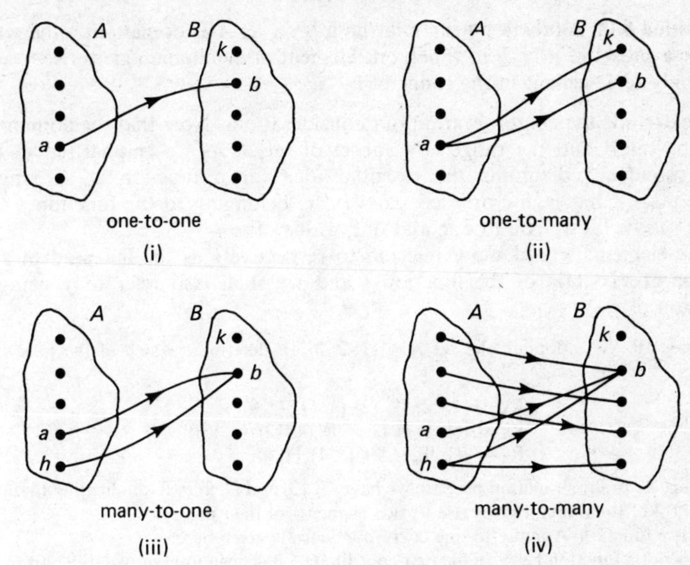

Fig. 6.14

which a is related to more than one element in B which for descriptive purposes is referred to as 'one to many' (i.e. many stands for more than one). In this situation, on some future occasion we shall have to make a choice: shall we use (a, k) or (a, b)? 'Many valued' choices like this are very troublesome in mathematics and we are about to exclude them.

If we look at the connections from the other end consider what to do *to* any $b \in B$. Ignore it—no problem there! Have one and only one connection to b as in Fig. 6.14 (i)? Or, we may have many connections coming *to* b as in (iii), i.e. many to one. Finally, we can mingle the situations of (ii) and (iii) so as to obtain (iv), many to many, to complete the possible correspondence diagrams from the set A to set B.

We can now summarise the descriptions of the four types of correspondences from set A to set B:

(i) A one-to-one correspondence. Each element of the domain gives rise to one and only one element of the range and each element of the range arises from one and only one element of the domain. (In Chapter 1 we said that the domain matched the range.)

(ii) A one-to-many correspondence: a correspondence in which at least one element of the domain gives rise to two or more elements of the range.

(iii) A many-to-one correspondence: a correspondence in which two or more elements in the domain give rise to the same element in the range.

(iv) A many-to-many correspondence features both (ii) and (iii).

Functions

The convenience of a one-to-one or many-to-one correspondence in which the elements of the range are always uniquely determined is worthy of a special name. We call it a function.

Definition 6.11. A function f is a relation from a set A as domain into the set B and is a subset of $A \times B$ in which one element of the domain gives rise to one and only one element in the range of f.

We exclude the empty set from our considerations. Note that the domain of f is the set A and the range is a subset of set B. As we emphasised in the correspondence diagrams the essential idea for a function is its single-valuedness—that is, for ordered pairs (x, y) belonging to the function f one value of x will give rise to one and only value of y.

The elements x and y are referred to respectively as the **independent** and **dependent variables** of the function f and we shall also refer to y being a function of x.

Example 10. With domain the set $A = \{1, 2, 3, 4\}$, determine which of the following relations are functions:

$$R_1 = \{1, 12), (2, 13), (3, 41), (2, 9)\}$$
$$R_2 = \{(1, -1), (2, -1), (4, 13), (3, 13)\}$$
$$R_3 = \{(1, 0), (1, 3), (3, 1), (4, 0)\}$$

Solution. R_1 is not a function because we have (2, 13) and (2, 9) in R_1, indicating that one element 2 of the domain gives rise to two elements of the range.

R_2 is a function. A many-to-one correspondence is given here.

R_3 is not a function because the first coordinate 1 has been used in two different pairs. (Searching for a repeat of the first coordinate is the quickest test to find if we do not have a function, because while all functions are relations not all relations are functions.)

Answer

Example 11. The equation $y^2 = x$ is used to generate the ordered pairs (x, y) of a relation with {positive integers} as the domain. Show that the relation is not a function.

Solution. We must search for a value of x which leads to two (or more) values of y, observing that $y = \pm\sqrt{x}$.

By inspection $x = 9$ gives rise to $y = 3$ and $y = -3$, i.e. (9, 3) and (9, −3) belong to the relation, which thereby cannot be a function.

Do notice that had the equation $y = x^2$ been used then the resulting relation would have been a function.

If we look back to the diagrams on page 121 we can see that none of the shaded regions represents the graph of a function because there is always more than one value of y to correspond to any one value of x. But, every straight line other than the y axis does represent a function. We should note the function defined by the equation $y = -2$ in Fig. 6.10. Here the function contains elements such as (−1, −2), (0, −2), (1, −2), (3, −2), i.e. $(x, -2)$ for any $x \in \mathbb{R}$; a good example of a 'many-to-one' correspondence.

Answer

Example 12. Use the equation $y = x^2 + x + 3$ to define a relation with domain $A = \{-1, 0, 1, 2\}$ and confirm that this relation is a function f.

Solution. Substituting $x = -1, 0, 1, 2$ in turn we obtain the relation R given by $R = \{(-1, 3), (0, 3), (1, 5), (2, 9)\}$ and since no first coordinate is repeated the relation is a function f with domain A and range $\{3, 5, 9\}$.

We say that f is defined by or given by the equation $y = x^2 + x + 3$. *Answer*

Notation. When each of the ordered pairs (x, y) in a function f is obtained from a formula, equation or rule, we shall say that 'y is a function of x'. Sometimes the value of y will be written as $y = f(x)$ and read as 'f of x'. Thus $f(x)$ is the element in the range of f which corresponds to the element x in the domain. $f(x)$ is also called the image of x in the function f. In

Example 12 above we could use the abbreviation of referring to the function '$f(x) = x^2 + x + 3$ with domain $\{-1, 0, 1, 2\}$', meaning (i) function f defined by the equation $f(x) = x^2 + x + 3$ on domain $\{-1, 0, 1, 2\}$ or (ii) $f = \{(x, y) \mid y = x^2 + x + 3, x \in A, y \in \mathbb{R}\}$, $A = \{-1, 0, 1, 2\}$.

In short, f stands for a set of ordered pairs, $f(x)$ stands for a second co-ordinate in an ordered pair.

Exercise 6.6

1. Using the domain $A = \{-2, -1, 0, 1, 2\}$ obtain the ranges in $\mathbb{R}$ for each of the relations defined by the following equations. State whether the relations are functions.

   (i) $y = x + 5$ (iv) $y^3 = x$
   (ii) $y = x^2 - x$ (v) $y > x$
   (iii) $y^2 = x + 2$ (vi) $y < x$

2. State which of the following relations is a function, and suggest an equation which defines the function:

   (i) $\{(0, 1), (1, 0), (2, 0), (0, 2), (1, 2)\}$
   (ii) $\{(0, 1), (1, 1), (2, 1), (3, 1)\}$
   (iii) $\{(1, 2), (2, 3), (3, 4), (4, 5), (5, 6)\}$
   (iv) $\{(-1, 0), (-1, 1), (-1, 2), (-1, 3)\}$

3. A relation R is defined by the equation $y = 3$ for all values of x. Show that R is a function.
4. A relation is defined by the equation $x = 3$ for all values of y. Show that R is not a function.
5. Given $A = \{1, 2, 3, 4, 5, 6\}$ which of the following relations are functions in $A \times A$:

   (i) 'is equal to' (v) '$x + y$ is an even number'
   (ii) 'is less than' (vi) '$x - y$ is an even number'
   (iii) 'is greater than' (vii) 'is half of'
   (iv) 'is a factor of'

6. Which of the following sketches represent a function with the real numbers between P and Q as domain.

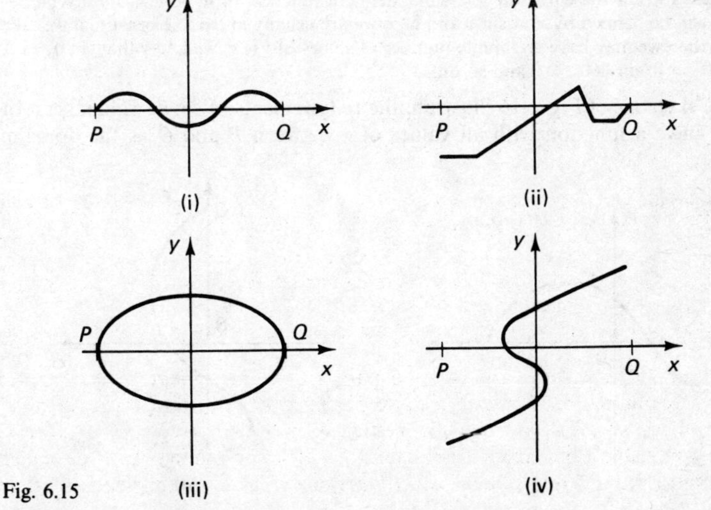

Fig. 6.15

Inverse Functions

The inverse of any relation R is obtained by interchanging the values of x and y in all the ordered pairs (x, y) which belong to R. As a simple example, with R given by $R = \{(1, 12), (2, 12), (5, 13), (0, 11)\}$, the inverse relation is given by $R^{-1} = \{(12, 1), (12, 2), (13, 5), (11, 0)\}$, so that the domain of R becomes the range of R^{-1}.

It is easy to see here that this relation R is a function but R^{-1} is not a function. We conclude therefore that the inverse of a function is not always a function.

Had we started with R^{-1}, i.e. a relation which is not a function, we would have obtained the inverse of R^{-1}, which is the function R.

Let us examine this inversion process through the equations which might define the function f. Thus we may have f defined by the ordered pairs satisfying $y = x^2$ in $\mathbb{R} \times \mathbb{R}$ so that the inverse relation is defined by the equation $x = y^2$ in $\mathbb{R} \times \mathbb{R}$, i.e. interchanging x and y. We have already seen that $y^2 = x$ does not define a function in $\mathbb{R} \times \mathbb{R}$.

Of course any relation is dependent on the sets used for the domain and range, an influence we can see very clearly here if we change the above examples to the following:

$$R = \{(x, y) \mid y = x^2, x, y \in \mathbb{R}^+\}$$
$$\therefore R^{-1} = \{(x, y) \mid y^2 = x, x, y \in \mathbb{R}^+\}$$

and the restriction to positive real numbers (recall that $\mathbb{R}^+ = \{$positive real numbers$\}$) means that when we take the square root in $y^2 = x$ to find y, only the positive root can be taken. Thus the restriction to $\mathbb{R}^+$ has retained the single-valuedness of R^{-1} and thereby, not only is R a function, but so is R^{-1}.

The presence of a single-valued function is very important in mathematics and its absence can lead to an irritating frequency of alternatives in some sections of elementary work like trigonometry, for example, whenever we are involved in a doing and undoing problem as in the next example.

Example 13. Let the equation $y = \sin x$ define a function in $\mathbb{R} \times \mathbb{R}$. The inverse relation will be defined by $x = \sin y$ and here we are usually in trouble because if we take $x = 0$ then we may have an infinite number of values of y (e.g. we know that $\sin 0° = 0$, $\sin 180° = 0$, $\sin 360° = 0$ and so on).

The sketches of Fig. 6.16 illustrate the requirements of single-valuedness. In (i) we have a function with all values of x between P and Q as the domain.

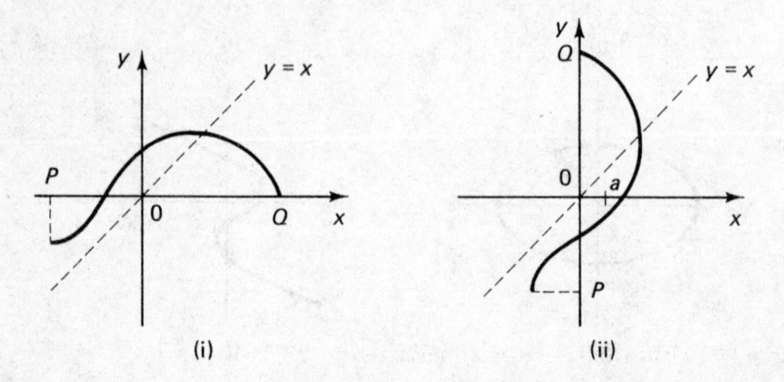

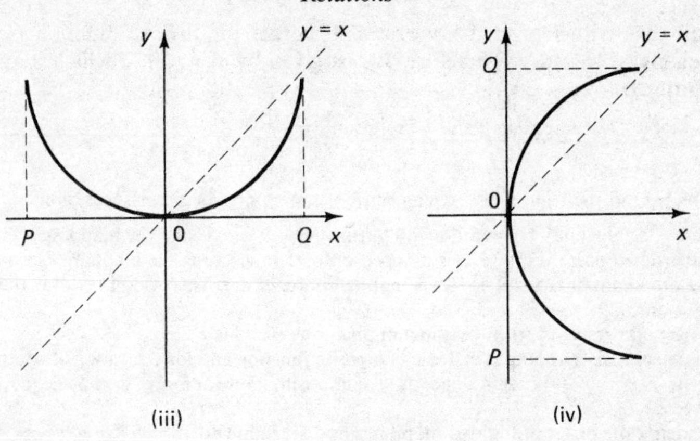

Fig. 6.16

Reflection of this graph in the line $y = x$ produces the inverse relation in (ii) which is not a function (page 173). (Interchanging x and y in (x, y) is equivalent to reflection in the line $y = x$.) For example, put $x = a$ and we get two values of y and we have lost single-valuedness.

In (iii) we have a function ($y = x^2$ is the equation we had in mind) and in (iv) we have the inverse relation which is not a function.

Composition of Functions

It is quite common in mathematics to combine functions so as to produce one overall function. For example, we might start with the element $x = 5$ and see this correspond to 11 in the function f, after which 11 is made to correspond to 8 in the function g, i.e. $(5, 11) \in f$ and $(11, 8) \in g$. This suggests that we might combine f and g in some way so as to connect 5 directly with 8 in one pair (5, 8).

The language we are using lends itself to speaking in terms of the correspondence diagrams of Fig. 6.14, page 123, but since the word correspondence is attached to relations in general we choose to use the word **mapping** for functions in particular. Thus any function is called a mapping and we may describe f as a mapping from set A to set B. (It is possible to refine the description even further so that if the range of f is a proper subset of B we say 'from set A *into* set B' whereas if the range is the set B then we say 'from set A *onto* set B'.)

Notation. It is easier to illustrate the notation by choosing particular examples.

If f is defined by the equation $y = x + 3$ then each value of x is mapped onto $x + 3$ (i.e. $x = 1$ gives rise to 4, $x = 10$ gives rise to 13, $x = x$ gives rise to $x + 3$). We describe this by writing

$$f: x \longmapsto x + 3$$

meaning that x maps onto $x + 3$.

The inverse function is

$$f^{-1}: x \longmapsto x - 3$$

(e.g. $x = 4$ gives rise to 1, $x = 13$ gives rise to 10).

(Recall that with f given by $y = x + 3$ we get the inverse function f^{-1} by interchanging x and y. Therefore, f^{-1} is given by $x = y + 3$ which may also be written $y = x - 3$.)

Example 14. Two mappings f and g are given by

$$f{:}x \longmapsto x + 4 \qquad g{:}x \longmapsto x^2$$

for $x \in \mathbb{R}$. Find the results of carrying out both mappings in either succession.

Solution. The idea may be identified by putting $x = 5$, say. Using the mapping g first we get the ordered pair (5, 25), i.e. 5 is mapped onto 25 in g. Using the mapping f second we get the ordered pair (25, 29), i.e. 25 is mapped onto 29 in f. The overall result is that 5 is mapped onto 29.

The general result is given by the mapping $x \longmapsto x^2 + 4$.

This successive mapping is called a **composite function** and for the case just completed we write $f \circ g{:}x \longmapsto x^2 + 4$ read as 'f composite g' meaning g first f second, or g followed by f.

Reversing the order of the two mappings and starting with $x = 5$ again we see that 5 is mapped onto 9 in f, and 9 is mapped onto 81 in g.

The overall result is that 5 is mapped onto 81 by $g \circ f$. Thus

$$g \circ f{:}x \longmapsto (x + 4)^2.$$

As expected $f \circ g$ is not the same mapping as $g \circ f$. *Answer*

Example 15. Show that the two mappings $f{:}x \longmapsto x + 3$, $g{:}x \longmapsto x - 2$, $x \in \mathbb{R}$ form the same composite function independent of the order in which they are taken.

Solution. The mapping procedure is illustrated in Fig. 6.17. For convenience the number lines will only display the integer values for x.

The unbroken lines show $g \circ f{:}x \longmapsto x + 1$

The dotted lines show $f \circ g{:}x \longmapsto x + 1$

It is worth noticing that the composition of two functions is always a function when the composition exists. On some finite sets the composite function may not exist (see Example 17).

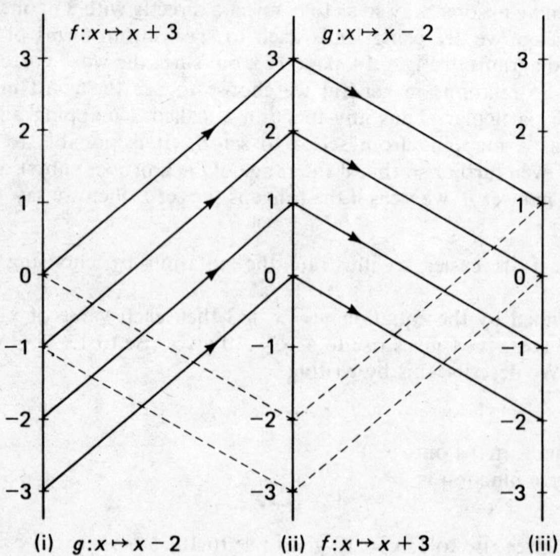

Fig. 6.17

The result we have just obtained implies that if $g = f^{-1}$ then the composite function will leave the original elements as they were. For this example, $f^{-1}: x \longmapsto x - 3$.

Example 16. Find the invariant elements in the composite function $g \circ f$ when the mappings are

$$f: x \longmapsto \frac{4}{x};\ g: x \longmapsto 9x$$

Solution. An invariant element is one whose value remains unchanged by the mapping, e.g. if we start with $x = 5$ and end with the same value 5, then $x = 5$ is an invariant element.

Consider what becomes of $x = 5$ in $g \circ f$

$$x = 5 \text{ is mapped onto } \frac{4}{5} \text{ by } f$$

$$x = \frac{4}{5} \text{ is mapped onto } \frac{36}{5} \text{ by } g$$

$$\therefore\quad x = 5 \text{ is mapped onto } \frac{36}{5} \text{ by } g \circ f$$

In general the mapping $g \circ f$ is given by $g \circ f: x \longmapsto \dfrac{36}{x}$ and we are searching for values of x which make $x = \dfrac{36}{x}$ or make $x^2 = 36$.

$$\therefore\quad x = \pm 6$$ *Answer*

Example 17. Find $g \circ f$ when $f = \{(0, 2), (4, 6), (8, 10)\}$ and $g = \{(2, 9), (6, 11), (10, 13)\}$ and show that $f \circ g$ is not defined.

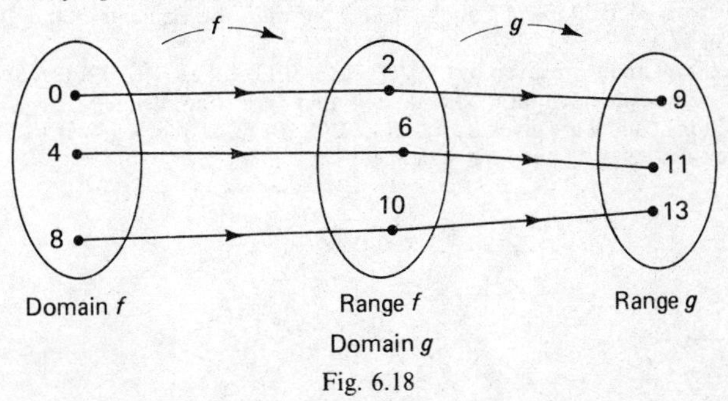

Fig. 6.18

Solution. The mapping diagram of Fig. 6.18 shows the connection between f and g, and clearly displays that using the domain of f we have

$$g \circ f = \{(0, 9), (4, 11), (8, 13)\}$$

But when we attempt to find $f \circ g$, g is only defined to map 2 to 9 and since 9 is not in the domain of f it follows that $f \circ g$ cannot provide a map for the element 2. *Answer*

Exercise 6.7

1. Obtain the inverse relation in each of the following cases, and state whether this relation is a function:

   (i) $\{(0, 1), (0, 2), (1, 3), (2, 5), (6, 7)\}$

(ii) $\{(1, 1), (2, 1), (3, 4), (5, 6)\}$
(iii) $\{(1, 1), (2, 2), (3, 3)\}$
(iv) $\{(1, 1), (1, 2), (2, 1)\}$

2. Obtain the inverse relations of each of the following relations defined in $\mathbb{R} \times \mathbb{R}$:
   (i) $\{(x, y) \mid y = 2x + 1, x \in \mathbb{R}\}$
   (ii) $\{(x, y) \mid y = x^2 - 2, x \in \mathbb{R}\}$
   (iii) $\{(x, y) \mid y = \frac{9}{x}, x \neq 0, x \in \mathbb{R}\}$
   (iv) $\{(x, y) \mid y = 5 + \frac{6}{x}, x \neq 0, x \in \mathbb{R}\}$
3. Obtain the invariant elements of each of the relations of Question 2.
4. Examine Question 6, Exercise 6.6, and state which of the inverse relations for the sketches is a function.
5. Figure 6.19 shows three sketches defined on domains in the real numbers whose graphs are sections of a circle centre 0, radius 1. Sketch the inverse relations and state whether they are functions.

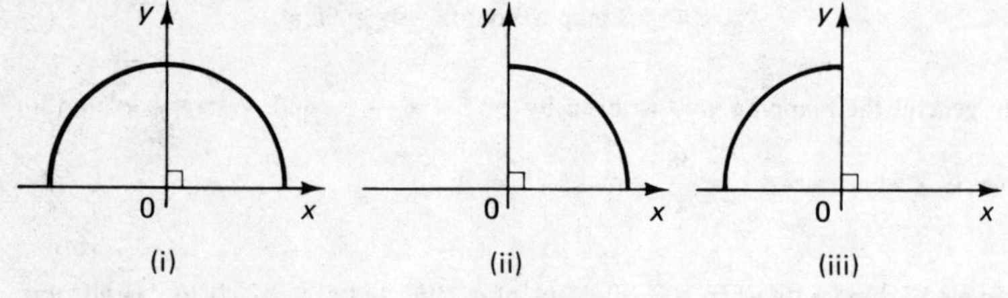

Fig. 6.19

6. Examine worked Example 17 above. Can $f \circ g$ map the elements 6 and 10 in the domain of g?
7. The functions f and g are given by $f = \{(1, 2), (2, 3), (3, 4)\}$; $g = \{(2, 12), (3, 13), (4, 14)\}$. Find the composite function $g \circ f$ and show that $f \circ g$ is not defined.
8. The functions f and g are given by $f = \{(1, 9), (2, 10), (3, 11)\}$; $g = \{(4, 5), (6, 7), (7, 8)\}$. Show that neither $g \circ f$ nor $f \circ g$ is defined.

7

LINEAR PROGRAMMING

In this chapter we shall be concerned with the inequalities which arise from the position of straight lines drawn in the cartesian plane. We have already seen how a straight line given by the equation $y = x$ creates two half planes which represent the inequalities $y < x$ and $y > x$. Now suppose we increased the number of lines in the plane; what type of problem would we be able to solve?

Diagrams like Fig. 6.13 will be our main interest and much of the ease of our understanding will depend on how well we can deal formally with the drawing of straight lines given their respective equations.

Equations for Straight Lines

There are three forms for the equation to a straight line, and they are

$$\text{(i) } y = mx + c$$

$$\text{(ii) } Ax + By = C$$

$$\text{(iii) } \frac{x}{a} + \frac{y}{b} = 1, a \neq 0, b \neq 0$$

Each form can be obtained from the other two by suitable manipulation, as suggested in the next example.

Example 1. Express the equation $y = 2x + 5$ in the forms

$$Ax + By = C \text{ and } \frac{x}{a} + \frac{y}{b} = 1$$

Solution. We first notice that both x and y are on the same side of the equation in forms (ii) and (iii) so we start by writing $y = 2x + 5$ as $-2x + y = 5$, which is now in form (ii) with $A = -2$, $B = 1$, $C = 5$. To gain form (iii) we notice that the right-hand side of form (iii) is 1, so we rewrite $y - 2x = 5$ as

$$\frac{-2x}{5} + \frac{y}{5} = 1$$

which is now rewritten as

$$\frac{x}{-\left(\frac{5}{2}\right)} + \frac{y}{5} = 1$$

which is now in form (iii) with

$$a = \frac{-5}{2}, b = 5$$ *Answer*

The equation $\frac{x}{a} + \frac{y}{b} = 1$ is called the **intercept form** of the equation to the line and is particularly useful for a rapid drawing of the straight-line graph it defines because we can see immediately that if $x = 0$ then $y = b$, i.e. $(0, b)$ is a point on this line. Similarly, if $y = 0$ then $x = a$, i.e. $(a, 0)$ is a point on this line.

Now since there is one and only one straight line passing through these two points we have obtained the line we require.

Example 2. Draw the line $\frac{x}{4} + \frac{y}{6} = 1$ and obtain the inequalities which are represented by each half plane which this line creates.

Solution. Put $x = 0$ and note that $y = 6$. The point (0, 6) lies on this line; this is the point Q in Fig. 7.1.

Put $y = 0$ and note that $x = 4$. The point (4, 0) lies on this line; this is the point P. All we need to do is to insert these points on the cartesian plane, join the points and extend in either direction to gain the line $\frac{x}{4} + \frac{y}{6} = 1$, as shown in Fig. 7.1.

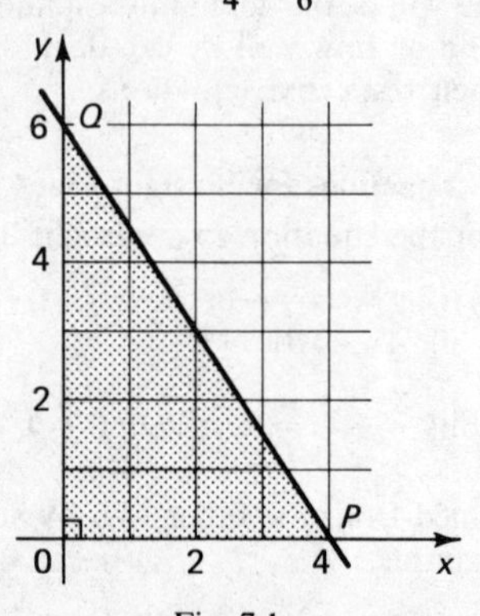

Fig. 7.1

The distance OP is called the intercept on the x axis, and the distance OQ is called the intercept on the y axis, hence the reference to this equation being in the intercept form. (Note that one or both of these intercepts may be negative.)

In order to obtain the half plane inequalities we now rephrase the equation to read

$$\frac{y}{6} = 1 - \frac{x}{4} \text{ and then } y = 6(1 - \frac{x}{4})$$

We now see from our experience in the last chapter that the half plane below the line represents the inequality relation $y < 6(1 - \frac{x}{4})$ and the half plane above the line represents the inequality relation

$$y > 6(1 - \frac{x}{4}).$$ *Answer*

In many examples our attention will be restricted by various equations and inequalities to those points which lie inside regions like the shaded region of Fig. 7.1. For example, the following constraints restrict us to the interior of this particular shaded region of $\triangle\, OPQ$:

(i) $0 < x$, to the right of the y axis,

and (ii) $0 < y$, above the x axis,

and (iii) $y < 6(1 - \frac{x}{4})$ below the line PQ.

Consider some more examples of such constraints.

Example 3. Examine the sketches in Fig. 7.2 and list the constraints necessary to confine our attention to the shaded regions in each sketch.

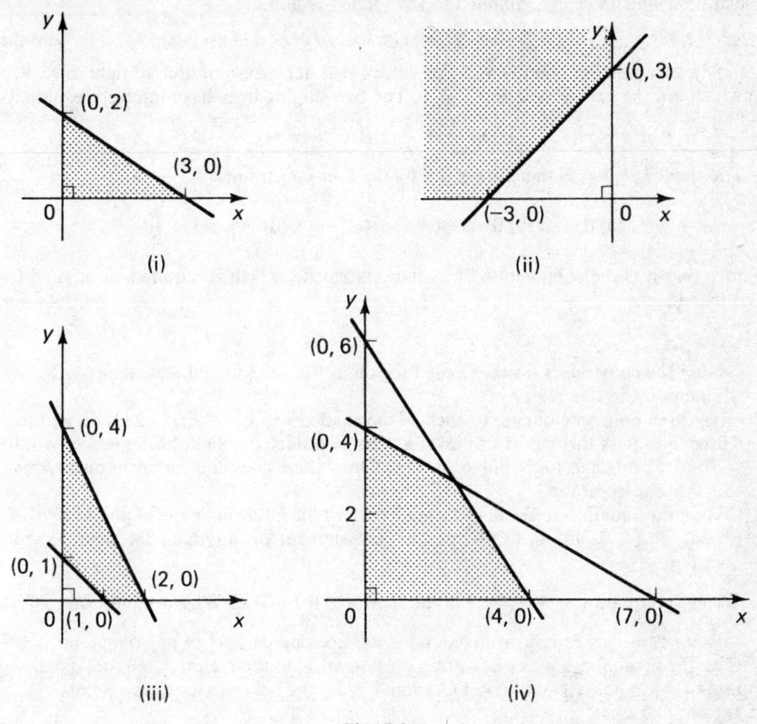

Fig. 7.2

First of all we must obtain the equations to the boundary lines. The axes are of course well known (?) now. Thus the y axis is given by $x = 0$ and the x axis is given by $y = 0$. Now consider each separate sketch in Fig. 7.2:

(i) The intercepts indicate that the line is $\frac{x}{3} + \frac{y}{2} = 1$ which we rewrite as

$$\frac{y}{2} = 1 - \frac{x}{3}$$

$$y = 2(1 - \frac{x}{3})$$

The half plane below the line is therefore represented by $y < 2(1 - \frac{x}{3})$. Hence, the constraints which keep us confined to the shaded region are

$$\text{(a) } 0 < x, \text{ (b) } 0 < y, \text{ (c) } y < 2(1 - \frac{x}{3}) \qquad \textit{Answer}$$

(ii) Again we are considering a region which is contained by three boundary lines and therefore three inequalities. The intercept form of the given line is $\frac{x}{-3} + \frac{y}{3} = 1$, which may be rearranged to $y = 3 + x$.

Since the shaded half plane is above the line it represents $y > 3 + x$. Therefore the constraints which keep us confined to the shaded region are

$$\text{(a) } 0 < y, \text{ (b) } x < 0, \text{ (c) } 3 + x < y$$ *Answer*

(iv) In this case the boundary of the shaded region consists of four straight lines, two of which are the axes $y = 0$ and $x = 0$. The two sloping lines have intercept equations $\frac{x}{7} + \frac{y}{4} = 1$ or $y = 4(1 - \frac{x}{7})$, and $\frac{x}{4} + \frac{y}{6} = 1$ or $y = 6(1 - \frac{x}{4})$.

The shaded region is therefore given by the four constraints:

$$\text{(a) } 0 < y, \text{ (b) } 0 < x, \text{ (c) } y < 4(1 - \frac{x}{7}), \text{ (d) } y < 6(1 - \frac{x}{4})$$

Notice that it is the conjunction of all four statements which is required, i.e. (a) and (b) and (c) and (d). *Answer*

Exercise 7.1

1. Obtain the constraints for the shaded region in Fig. 7.2 (iii) and state the coordinates of one point in this region.
2. Give three points belonging to each of the shaded regions of Fig. 7.2 (i), (ii) and (iv).
3. Three lines pass through the following pairs of points, one line through each pair: (i) (4, 0), (0, 6), (ii) (5, 0), (0, 7), (iii) (6, 0), (0, 8). Draw these lines and comment on anything they have in common.
4. Obtain the equation to the lines through each of the following pairs of points: (i) (8, 0), (0, 16), (ii) (4, 0), (0, 8), (iii) (1, 0), (0, 2). Comment on anything the lines have in common.
5. Write the equation $\frac{x}{4} + \frac{y}{20} = 1$ in the form $Ax + By = C$. What will happen to C as this line is moved further from (0, 0) but still keeping parallel to its original position?
6. Give the inequalities which constrain our attention to the region in the first quadrant and between lines (i) and (ii) in Question 4. State the coordinates of two points in this region.

Optimisation

Returning to our intercept form of the equation to a straight line let us refer to Fig. 7.1. where we see the equation to the line PQ as $\frac{x}{4} + \frac{y}{6} = 1$.

If we double the intercepts the new line will be parallel to PQ and have an equation $\frac{x}{8} + \frac{y}{12} = 1$. If we treble the intercepts our new line, still parallel to PQ, will have the equation $\frac{x}{12} + \frac{y}{18} = 1$. From these examples we see that if we multiply the original intercepts by any non zero number we will obtain a new line parallel to PQ. In other words, the line $\frac{x}{4k} + \frac{y}{6k} = 1, k \neq 0$ is always parallel to $\frac{x}{4} + \frac{y}{6} = 1$.

Another way to regard this is:

(i) If k is greater than 1 and increasing, then the original line is moving further away from the origin (0, 0) but remaining parallel to its original direction.

(ii) If $0 < k < 1$ and decreasing then the original line is moving closer to the origin (0, 0) but remaining parallel to its original direction.

Suppose we rewrite the equation to the line $\frac{x}{4k} + \frac{y}{6k} = 1$ as

$$6x + 4y = 24k$$

Now we see that changing the value of k merely moves the line parallel to itself either towards or away from the origin (0, 0).

Again, rewriting this equation as $6x + 4y = C$ shows that as the line moves parallel to itself in the first quadrant and further away from the origin (0, 0) so the value of C increases, i.e. to increase C we increase the distance of the line from (0, 0). The same conclusion applies to any line given by $Ax + By = C$—that is, keeping the values of A and B unchanged and thereby keeping the line parallel to itself, we increase C by moving the line further away from the origin (0, 0) and across the first quadrant.

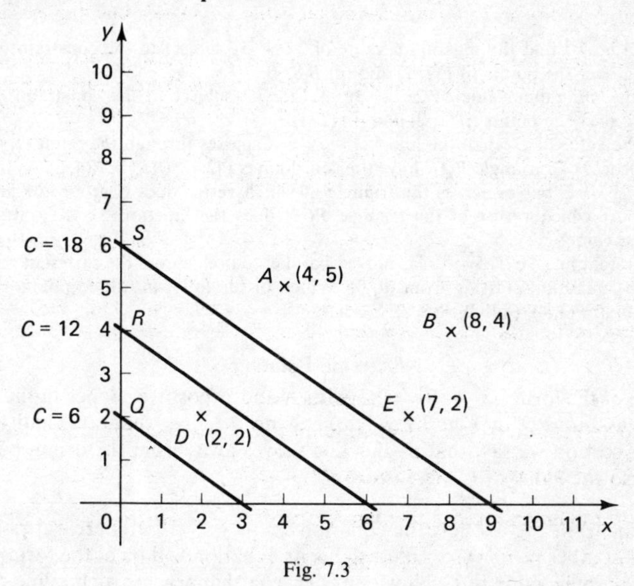

Fig. 7.3

Consider Fig. 7.3, where the line $Ax + By = C$ is displayed in the form $2x + 3y = C$ with different values of C which may be measured by examining the intercept on the y axis.

Thus, $2x + 3y = C$ intersects the y axis at Q where $x = 0$, and at which point $3y = C$ and inspection of the figure shows $C = 6$. Next, $2x + 3y = C$ intersects the y axis at R, at which point $3y = C$, and inspection of the figure shows $C = 12$.

With similar reasoning $C = 18$ when the line passes through S.

Now suppose we have stop points A, B over which the line is not allowed to pass. Clearly, as we move the line parallel to itself until we reach A the maximum permitted value of C will have been reached and obtained by substituting $x = 4$, $y = 5$ in the equation

$$2x + 3y = C$$

When our line is moved to pass through $A(4, 5)$ the value of C is 23, and we say that we have maximised $2x + 3y$ subject to the constraint that we cannot pass over or beyond the point $A(4, 5)$.

The dual process is to consider minimising $2x + 3y$ subject to some constraint which stops the line getting any closer to the origin—for example, not being allowed to pass over the point $D(2, 2)$ means that with $x = 2$, $y = 2$ we get $2x + 3y = C = 10$, and we say that we have minimised $2x + 3y$ subject to the constraint that we cannot pass over the point $D(2, 2)$ towards the origin.

The process of maximising or minimising what is called an objective function like $2x + 3y$ subject to given constraints is called optimisation and the method we are discussing here is called linear programming.

The maximum and minimum values are often referred to as the optimum values.

Exercise 7.2

1. Using Fig. 7.3 find the maximum value of $2x + 3y$ subject to the constraint of not passing over the points (i) $E(7, 2)$ and (ii) $B(8, 4)$.
2. Find the minimum value of $2x + 3y$ in Fig. 7.3 subject to the constraint of not passing over the points (i) (1, 2) and (ii) (2, 1).
3. Find the value of C when the line $5x + 7y = C$ passes through the point (2, 4).
4. The vertices of a triangle PQR have the coordinates $P(2, 4)$, $Q(3, 7)$, $R(4, 5)$. As the line $5x + 7y = C$ passes across the triangle at which vertex does C attain the greatest value? At which vertex of the triangle PQR does the function $3x + y$ attain its greatest value?
5. The straight line $4x + 5y = C$ is moved parallel to itself across the cartesian plane in the first quadrant starting from (0, 0). Which of the following three points does it reach first: $P(3, 4)$, $Q(4, 3)$, $R(5, 2)$, $S(6, 1)$?

Extreme Points

Any line of the form $Ax + By = C$, with A and B positive, slopes in the same manner as line PQ in Fig. 7.1 when crossing the first quadrant, and in the previous section we saw that as this line moves away from the origin parallel to itself so the value of C increases.

The line (i) in Fig. 7.4 has the equation $5x + y = 5$ in its present position and in any other position parallel to this its equation will be of the form $5x + y = C$ for some value of C. Now consider what happens as such a line passes across a line segment like PQ in Fig. 7.4 (line segment means the section of the straight line joining the points P and Q and going no further).

From the diagram we note that as the line $5x + y = C$ moves across the segment PQ so the value of C will be a minimum at Q and a maximum at P. That is, if $5x + y = C$ is constrained to lie between Q and P then the optimum values of C occur at each end of the line segment PQ. In other words, it is useless to consider any intermediate point of PQ when searching for an optimum value for the objective function $5x + y$.

Line (ii) in Fig. 7.4 is given by $1.5x + 4y = 6$ in its present position and so in any other parallel position it will have an equation $1.5x + 4y = C$ and, just

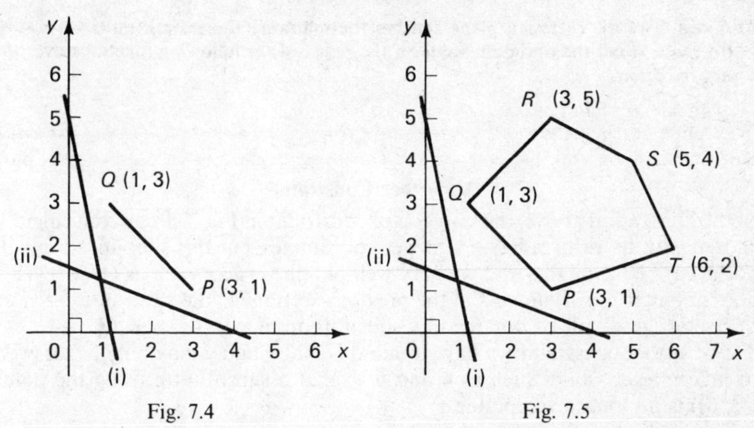

Fig. 7.4

Fig. 7.5

as with line (i), if constrained to lie between P and Q then the optimum values of the objective function $1.5x + 4y$ will be obtained at the ends of the line segment PQ.

By inspection of the diagram, $1.5x + 4y$ has a minimum at P (8.5) and a maximum at Q (13.5), which is the reverse situation to that for $5x + y$. Now suppose our lines (i) and (ii) are constrained to cross the polygon region given by $PQRST$ in Fig. 7.5. Consider line (i) in general position $5x + y = C$: we have just argued that the optimum values of C are at P and Q. But the same argument may be used to cover the passage of $5x + y = C$ across the line QR, and RS and ST and PT.

Hence the optimum values of the objective function $5x + y$ for the whole region of the polygon $PQRST$ will lie at one or other of the vertices or extreme points P, Q, R, S, T and so all we need to do is evaluate $5x + y$ at each vertex and select the required maximum and minimum of $5x + y$ subject to the constraint of having to obtain the solution on the polygon.

The values of C are given as $P(16)$, $Q(8)$, $R(20)$, $S(29)$, $T(32)$. The optimum values of $5x + y$ are therefore 8 at (1, 3) and 32 at (6, 2).

Exercise 7.3

1. The extreme points of a region are $P(-5, 1)$, $Q(-2, 4)$, $R(2, 2)$, $S(3, -3)$ $T(-3, -5)$. Find the optimum values of the following objective functions on this region:

   (i) $x + y$ (iii) $-x + 3y$ (v) $3x - y$
   (ii) $2x + y$ (iv) $-x - 2y$

2. The points of the intersection of the following four lines are $(-2\frac{1}{2}, -4\frac{1}{2})$, $(4, 2\frac{1}{2})$, $(-3, 3)$ and $(-3, 5)$. Identify P, Q, R and S.

   PQ: $5x + 14y = 55$ RS: $15x + y = -42$
   QR: $x + 14y = 39$ SP: $19x + y = -52$

   Obtain the optimum values of each of the following objective functions on $PQRS$: (i) $2x + 2y$; (ii) $2x - 2y$.

3. A region of the cartesian plane satisfies the following inequalities: (a) $2 \leqslant y \leqslant 5$; (b) $0 \leqslant x \leqslant y$. Find the optimum values of the following objective functions over the region enclosed by the above constraints:

   (i) $7x + 3y$ (iii) x
   (ii) $-7x + 3y$ (iv) y

4. A region of the cartesian plane satisfies the following constraints: (a) $0 \leqslant x \leqslant 3$; (b) $y \leqslant x$. Find the optimum values if they exist of the following functions over the above region:

   (i) x (iii) $y - 2x$
   (ii) y (iv) $2y - x$

A Further Constraint

In the illustration above, the polygon of constraint in Fig. 7.5 was conveniently chosen with its vertices having integer coordinates, but this simplification will not always be available and we may well obtain vertices such as $(4\frac{3}{4}, 6\frac{2}{3})$. This does not alter the arguments of the previous section for the value of $5x + y$ at this vertex may still be obtained by substitution of $x = 4\frac{3}{4}$, $y = 6\frac{2}{3}$.

The further constraint we have in mind is that x and y may only be allowed to take integer values such as 4 and 6 so that a substitution using the point $(4\frac{3}{4}, 6\frac{2}{3})$ is no longer acceptable.

This restriction in terms of positive integers only will mean that we shall have to search for possible solutions near to $(4\frac{3}{4}, 6\frac{2}{3})$ and find an optimum by trial and error, testing points such as (4, 6), (5, 6), (4, 7) as long as they are in the allowable region. The next example will illustrate the idea.

Example 4. A shopkeeper has a maximum of £72 to spend on an order for two brands X, Y of soap powder and only a maximum space of 6 m^3 in which to store the order. The manufacturers use different size packets to contain the same volume of powder and this accounts for the different size of packing box used. With the details as given in the following table find the order which gives the maximum profit:

Brand	*Volume of box*	*Cost per box*	*Profit per box*
X	0.5 m^3	£9	£0.5
Y	0.6 m^3	£4	£0.5

Solution. We wish to find the number of boxes bought of each brand. Let x and y be the number of boxes of brand X and Y to be bought, and notice that x and y must be whole numbers.

Volume: The total volume of the order will be $0.5x + 0.6y$ m^3 and this number must be less than or equal to 6 m^3. The solution set is therefore contained in

$$R_1 = \{(x, y) \mid 0.5x + 0.6y \leqslant 6\}$$

Cost: The total cost of the order will be £$(9x + 4y)$ and this must be less than or equal to £72. The solution set is therefore contained in

$$R_2 = \{(x, y) \mid 9x + 4y \leqslant 72\}$$

Profit: The total profit to be made is £$(0.5x + 0.5y)$. The problem is to find the maximum value of the objective function

$$0.5x + 0.5y = C \text{ on } R_1 \cap R_2.$$

A boundary line for R_1 is $0.5x + 0.6y = 6$, which we rewrite in the intercept form

$$\frac{x}{12} + \frac{y}{10} = 1$$

and now draw on Fig. 7.6.

A boundary line for R_2 is $9x + 4y = 72$, which we rewrite in the intercept form

$$\frac{x}{8} + \frac{y}{18} = 1$$

and now draw on Fig. 7.6.

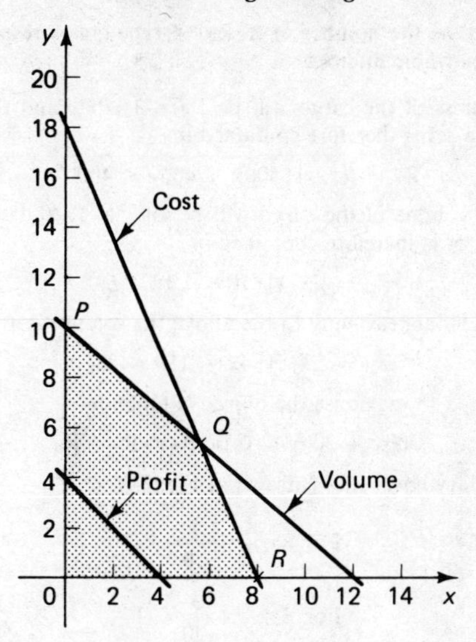

Fig. 7.6

The solution is now constrained to the region in and on the polygon $OPQR$. The profit line is $0.5x + 0.5y = C$ which we may rewrite as $x + y = 2C$, and now draw on Fig. 7.6 in the position which corresponds to $2C = 4$ (i.e. put $x = 0$ and see $y = 2C = 4$).

Consider finding the maximum of $x + y$. We know that without the restriction to whole numbers the maximum would have been found at one of the extreme points P, Q and R. The coordinates of P and R being integers, we find the value of $x + y$ at these points immediately. They are $P(10)$, $R(8)$.

From Fig. 7.6 it is clear that the acceptable points with whole number coordinates nearest to P and Q are (1, 9), (2, 8), (3, 7), (4, 6), (5, 5), (6, 4). At each of these points we have $x + y = 10$, which means that the maximum profit is £5 on each of the orders given by (0, 10), (1, 9), (2, 8), (3, 7), (4, 6), (5, 5), (6, 4).

The corresponding cost of the order will be £40, £45, £50, £55, £60, £65 and £70 respectively.

Reservations. The results indicate that the most sensible choice would be to buy 10 boxes of brand Y and gain a profit of £5 on £40, i.e. $12\frac{1}{2}\%$ of outlay. *Answer*

Example 5. A simplified transport problem reads as follows:

The available cargo space in a ship is 60 m^3 and this is to be filled with two different types of crates of machine tools whose total mass must not exceed 4000 kg. Experience has shown that the cost of packaging should not be greater than £88. The problem data is listed in the following table and we are expected to optimise the possible trade profit on the cargo.

Crates	*X*	*Y*	
Mass	500 kg	400 kg	$\leqslant$ 4000 kg
Volume	10 m^3	3 m^3	$\leqslant$ 60 m^3
Cost of packaging	£4	£11	$\leqslant$ £88
Trade profit	£500	£300	

Solution. Let x and y be the number of X and Y type crates respectively, noting that both x and y must be whole numbers.

Mass: The total mass of the cargo will be $500x + 400y$ and this must not exceed 4000 kg. The solution set is therefore contained in

$$R_1 = \{(x, y) \mid 500x + 400y \leqslant 4000\}$$

Volume: The total volume of the cargo will be $10x + 3y$ and this must not exceed 60 m^3. The solution set is therefore contained in

$$R_2 = \{(x, y) \mid 10x + 3y \leqslant 60\}$$

Packaging: With similar reasoning to the above the solution set is contained in

$$R_3 = \{(x, y) \mid 4x + 11y \leqslant 88\}$$

Profit: We shall need to maximise the objective function

$$500x + 300y = C \text{ on } R_1 \cap R_2 \cap R_3$$

The following boundary lines will be drawn in Fig. 7.7:

$$\text{For } R_1: \frac{x}{8} + \frac{y}{10} = 1$$

$$\text{For } R_2: \frac{x}{6} + \frac{y}{20} = 1$$

$$\text{For } R_3: \frac{x}{22} + \frac{y}{8} = 1$$

The profit line is $500x + 300y = C$, which may be rewritten as

$$\frac{x}{3} + \frac{y}{5} = \frac{C}{1500}$$

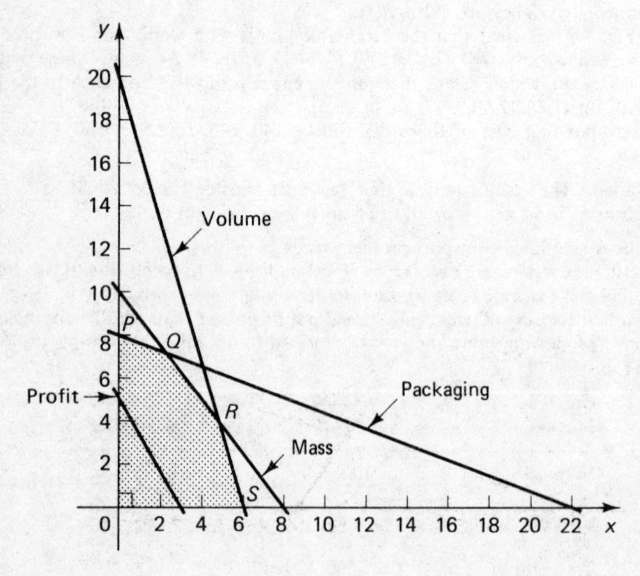

Fig. 7.7

For the profit line we start by drawing $\frac{x}{3} + \frac{y}{5} = \frac{C}{1500}$ in the position, which corresponds to $C = 1500$.

Without the restriction to whole numbers we know that the maximum for $500x + 300y$ would be found at one of the points $P(0, 8)$, $Q(2\frac{10}{39}, 7\frac{7}{39})$, $R(4.8, 4)$ and $S(6, 0)$. As the line is moved parallel to itself and further from 0 we see that the point (4, 5) yields a profit of $500 \times 4 + 300 \times 5 = £3500$, while the point (5, 3) yields a profit of $500 + 5 + 300 \times 3 = £3400$.

Reservations. Notice that the result at the point (5, 3) gives a profit of £3400 and a packaging cost of £53 whereas our optimum result of £3500 takes £71 for packaging, so that the profit gap is not as large as it would appear at first sight and may be even smaller if we take profit as a percentage of the required outlay in sending an extra crate.

Also notice that we have to ignore the finer details of fitting crates of different dimensions into the 60 m^3 cargo space available. *Answer*

Example 6. Find the optimum values for the objective function $f = x + y$ subject to the following constraints:

$$\text{(i) } 6x + 11y \leqslant 132$$
$$\text{(ii) } 10x + 7y \geqslant 140$$
$$\text{(iii) } \frac{2}{7} \leqslant \frac{y}{x} \leqslant \frac{3}{2}$$

Obtain results for (a) x, y real numbers, (b) x, y whole numbers.

Solution. The boundary line for (i) is $6x + 11y = 132$ and clearly the intercepts are given by the points (0, 12), (22, 0), when we rewrite the equation as

$$\frac{x}{22} + \frac{y}{12} = 1$$

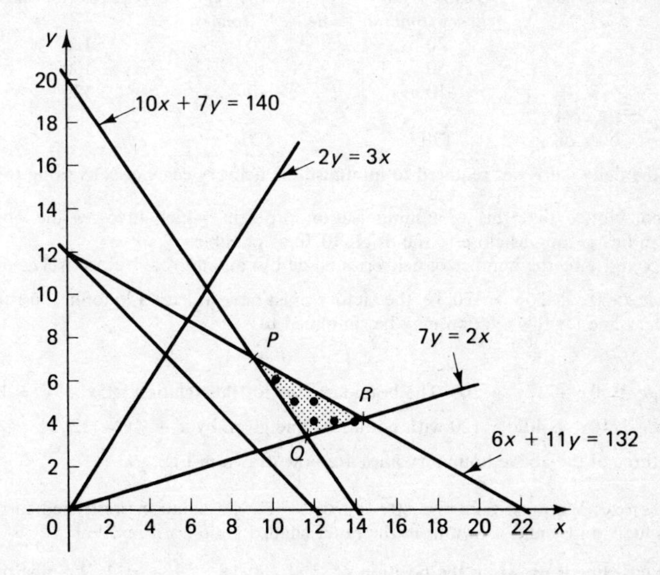

Fig. 7.8

For (ii) the boundary line is $10x + 7y = 140$ and the intercepts are given by the points (0, 20), (14, 0).

Of the two constraints in (iii), $\frac{2}{7} \leqslant \frac{y}{x}$ gives $2x \leqslant 7y$ with boundary line $7y = 2x$ and $\frac{y}{x} \leqslant \frac{3}{2}$ gives $2y \leqslant 3x$ with boundary line $2y = 3x$. The solution region will occur in between these lines. The region of the solution is the triangle PQR in Fig. 7.8, for note the direction of the inequality in (ii).

(a) The graph of $x + y = C$ is drawn for $C = 12$ and it is clear from the slope of this line that the minimum C will occur at Q and the maximum C at R, so we must now solve the boundary line equations to find Q and R.

$$\begin{aligned} \text{For } Q\text{: } 10x + 7y &= 140 \\ 2x - 7y &= 0 \\ 12x &= 140 \\ x &= 11\tfrac{2}{3} \\ y &= 3\tfrac{1}{3} \\ \text{At } Q\text{: } x + y &= 15 \end{aligned} \qquad \begin{aligned} \text{For } R\text{: } 6x + 11y &= 132 \\ 2x - 7y &= 0 \\ 6x - 21y &= 0 \\ 32y &= 132 \\ y &= 4\tfrac{1}{8} \\ x &= 14\tfrac{7}{16} \\ \text{At } R\text{: } x + y &= 18\tfrac{9}{16} \end{aligned}$$

(b) An inspection of $\triangle PQR$ shows that the only possible solutions are (12, 4), (12, 5), (11, 5), (13, 4), (14, 4), (10, 6) and the optimum value of $x + y$ is clearly 18 at (14, 4).

Reservations. The line $2y = 3x$ for the condition $\frac{y}{x} \leqslant \frac{3}{2}$ was superfluous.

Answer

Example 7. A factory is converting its heating and furnacing requirements from oil to three different grades 1, 2 and 3 of coal. The coal may be delivered by rail or road and the extra cost of possible deliveries works out at £30 by rail and £20 by road per delivery.

Coal grade	*Rail Quantity per delivery (tonnes)*	*Road Quantity per delivery (tonnes)*	*Quantity daily required (tonnes)*
1	20	20	120
2	50	10	100
3	10	40	120
extra cost over oil	£30	£20	

Find the daily deliveries required to minimise the delivery costs of converting to coal.

Solution. Notice that this is a minimisation problem, which involves the objective function line getting as close to the origin (0, 0) as possible.

Let x and y be the number of deliveries made by rail and road respectively.

Grade 1: $20x + 20y \geqslant 120$; i.e. the factory must have at least 120 tonnes to run. The boundary line for this relation may be simplified to

$$x + y = 6$$

Grade 2: $50x + 10y \geqslant 100$. The boundary line for this relation is $5x + y = 10$.

Grade 3: $10x + 40y \geqslant 120$ with boundary line given by $x + 4y = 12$.

All three of the above boundary lines are now drawn in Fig. 7.9.

Extra cost: We must minimise $30x + 20y = C$ with solutions restricted to integer values for x and y and drawn from the outer shaded region of Fig. 7.9.

The cost line is drawn in the position of $C = 180$, i.e. $\frac{x}{6} + \frac{y}{9} = 1$. We now need to

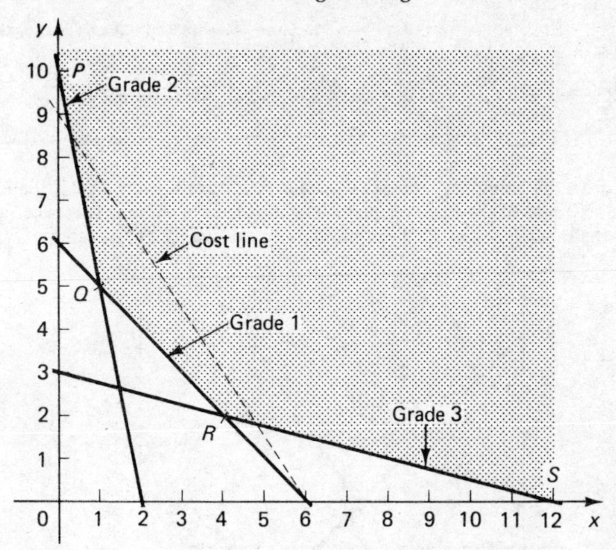

Fig. 7.9

move this line parallel to itself as close as possible to the origin. Fortunately, the extreme points P, Q, R and S all have integer coordinates so we know that the minimum of $30x + 20y$ will occur at Q, the point (1, 5). The minimum extra daily cost is £130.

Reservations. The minimum of the objective function has been produced by 1 load by rail and 5 by road each day. While this gives exact needs for grades 1 and 2 coal, grade 3 coal will be over-delivered by as much as 90 tonnes a day and this cannot be ignored so the daily order will have to be revised every now and again according to storage space available for the grade 3 coal heap, and a new mathematical model for the problem will be required. *Answer*

Example 8. For three horses A, B and C entered in a race the bookmakers are offering the following odds:

A: 3–2 against winning
B: 4–1 against winning
C: evens

A punter intends betting on all three horses. She lays a £10 bet on C but is undecided what to bet on A and B.

Show that it is not possible for her to guarantee winning overall by backing all three horses A, B and C, even assuming that one of these horses wins.

Solution. Let the punter place £x on A and £y on B as well as the £10 on C. Suppose:

A wins: She gets $\frac{3x}{2}$ but loses $10 + y$. To win overall she requires

$$\frac{3x}{2} > 10 + y$$

$$\text{i.e.} \quad y < \frac{3x}{2} - 10$$

B wins: She gets $4y$ but loses $10 + x$. To win overall she requires

$$4y > 10 + x$$

$$\text{i.e.} \quad y > \frac{1}{4}(10 + x)$$

C wins: She gets £10 but loses $x + y$. To win overall she requires

$$10 > x + y$$
$$\text{i.e.} \quad y < 10 - x$$

We now draw the boundary lines of these constraints as shown in Fig. 7.10 and observe that to satisfy each of the three inequalities and be sure of winning it is necessary for us to be on the shaded side of all three lines and this is clearly not possible. *Answer*

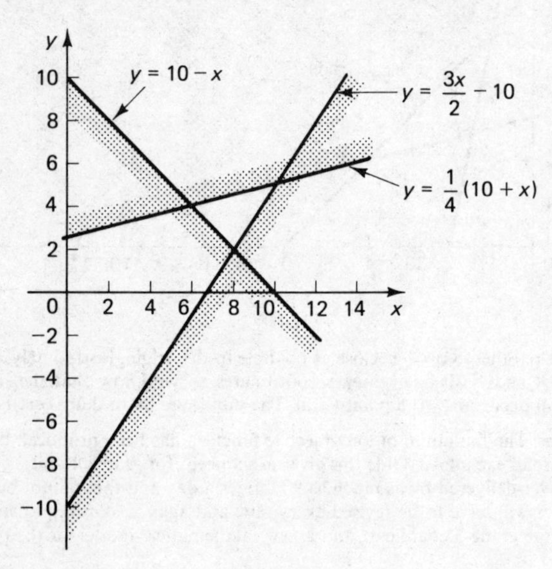

Fig. 7.10

Exercise 7.4

1. Find the maximum profit available in Example 4 and the corresponding orders if the profits per box had been:

   (i) £0.4 for X, £0.6 for Y
   (ii) £0.55 for X, £0.22 for Y
   (iii) £0.4 for X, £0.3 for Y

2. Find the maximum trade profit in Example 5 and the corresponding cargoes if the trade profit per crate had been:

   (i) £1000 for X, £800 for Y
   (ii) £600 for X, £600 for Y
   (iii) £400 for X, £500 for Y

3. Find the optimum of the following objective functions subject to the same constraints as for Fig. 7.8 with x, y positive integers:

   (i) $3x + y$
   (ii) $x + 3y$
   (iii) $x - y$

4. Using the same constraints as in Fig. 7.9 find the minimum values of the following objective functions:

   (i) $x + 6y$ (iii) $9x + 8y$
   (ii) $x + 3y$ (iv) $12x + y$

5. A farmer has 25 hectares of land to prepare for two different crops X and Y but only has 100 units of labour available for the work. The cost of sowing and the expected profit is shown as follows:

	X	Y
Cost of sowing per hectare	£150	£90; £2700 available
Labour unit required per hectare	4	5 ; 100 units available
Profit expected on reaping per hectare	£400	£480

   Find the crop distribution for the maximum profit with complete hectare solutions only.

6. A clothing manufacturer has 600 m² of wool material and 400 m² of nylon lining material available for making coats and dresses. A coat takes 2 m² of each material but a dress takes only 1 m² of lining for 3 m² of wool material due to pleating. The profit offered by three different wholesalers A, B, C is given in the table below. Find his maximum possible profit if he can only sell to one wholesaler. (Let x and y be the number of dresses and coats respectively.)

	X dress	Y coat
A	£3	£3
B	£2.5	£3.5
C	£2.75	£3.25

7. For three horses A, B and C entered in a race the bookmakers are offering the following odds: A 5–2 against winning, B 5–1 against winning, C evens. Assuming that one of these horses wins the race, show that with a fixed bet of £10 on C a punter can arrange bets on A and B to guarantee an overall financial gain.

Non-linear Programming

In non-linear programming the boundaries of the solution sets or regions may no longer be straight lines but apart from the increased difficulty of drawing curved boundaries the principles of optimisation remain the same.

We shall restrict our work to the use of simple curves like $4y = x^2$ (a parabola) and $y = \frac{8}{x}$ (a rectangular hyperbola). First consider Fig. 7.11 in which the shaded region is bounded by the curve given by $4y = x^2$ and the straight line.

$$\frac{x}{4} + \frac{y}{2} = 1 \text{ (both drawn with } x, y \in \mathbb{R})$$

Procedure. To obtain the graph of $4y = x^2$ we merely suggest values of x and find the corresponding values of y which fit the equation $4y = x^2$.

The table of values shows our suggestions of $x = 0, \pm 2, \pm 4, \pm 6$, as being the simplest to make, for plotting $4y = x^2$.

	x	-6	-4	-2	0	2	4	6
	x^2	36	16	4	0	4	16	36
$y =$	$\frac{x^2}{4}$	9	4	1	0	1	4	9

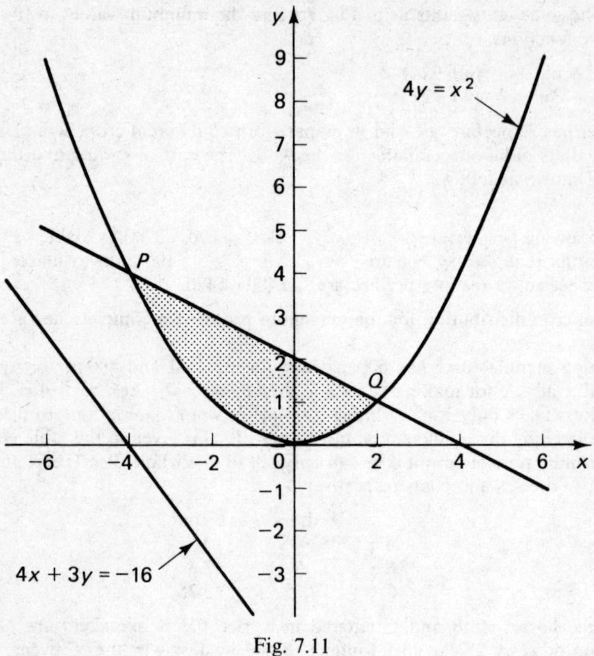

Fig. 7.11

The drawing of the line given by $x + 2y = 4$ may be done by joining the intercepts (0, 2) and (4, 0).

Now rearranging these two equations to $y = \frac{x^2}{4}$ and $y = \frac{1}{2}(4 - x)$, we see that just as

$$y < \tfrac{1}{2}(4 - x) \text{ is the region below the straight line so}$$

$$y < \frac{x^2}{4} \text{ is the region below the curve (parabola).}$$

Similarly $y > \frac{x^2}{4}$ is the region above the curve, i.e. inside the bowl so to speak.

Consequently, the shaded region in Fig. 7.11 is the region for which

$$\frac{x^2}{4} \leqslant y \leqslant \tfrac{1}{2}(4 - x)$$

if we include the boundary.

The points P and Q on Fig. 7.11 are the points of intersection of the straight line with the curve and where

$$\frac{x^2}{4} = \frac{1}{2}(4 - x)$$

i.e. $$x^2 = 2(4 - x) = 8 - 2x$$

Alternatively $x^2 + 2x - 8 = 0 = (x + 4)(x - 2)$
which implies that x is either 2 or -4.

P is the point $(-4, 4)$ and Q is the point $(2, 1)$

The principle of optimising an objective function like $4x + 3y$ is still the same—namely, to place the line $4x + 3y = C$ at or near the extreme points of the region. On Fig. 7.11 this line has been drawn for $C = -16$. As we move this line across the figure from left to right the value of C increases until its maximum value is gained by passing the line through $Q(2, 1)$ to give the maximum value of 11 for the objective function $4x + 3y$.

Example 9. Find the optimum values of $2x + 3y$ subject to the following constraints:

(i) $y^2 \leqslant 9x$

(ii) $\dfrac{4}{x} \leqslant y$

(iii) $x^2 + y^2 \leqslant 64$ subject to $x, y \in \{\text{positive real numbers}\}$

Solution. The boundary curve for the region (i) is $y^2 = 9x$ and for (ii) is $y = \dfrac{4}{x}$.

We compile a table of values for these curves as follows by suggesting easy values for x of 0, 1, 2, 4, 6, 8, 9.

	x	0	1	2	4	6	8	9
	$9x$	0	9	18	36	54	72	81
$y =$	$\sqrt{9x}$	0	3	4.2	6	7.3	8.5	9
$y =$	$\frac{4}{x}$	—	4	2.0	1.0	0.7	0.5	

The boundary for $x^2 + y^2 \leqslant 64$ is the circle $x^2 + y^2 = 8^2$; centre (0, 0) and radius 8, of which only the first quadrant is required. With all these boundary curves drawn on Fig. 7.12 the solution set is represented by the shaded region on the figure.

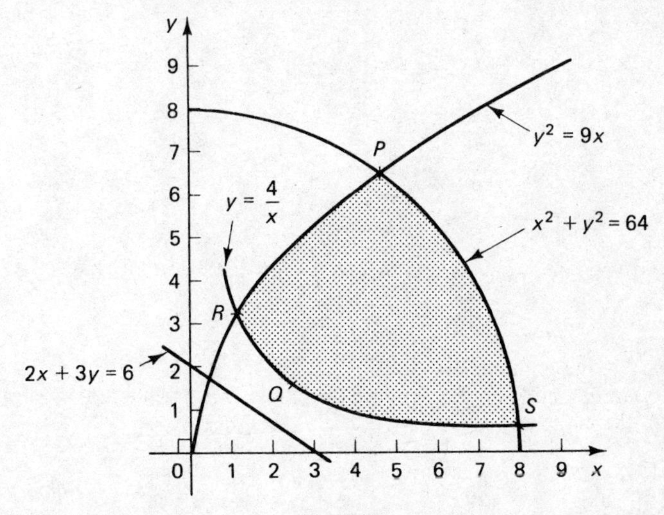

Fig. 7.12

In the now familiar method we slide the line of the objective function $2x + 3y = C$ (drawn for $C = 6$ on the diagram) parallel to itself as far as we can across the solution set. The furthest we can go is the point P, which from the grap is approximately $x = 4.5$ and $y = 6.5$ (by calculation it is really 4.7 and 6.5 to 1 d.p.). Hence the maximum of $2x + 3y$ is 28.5.

The minimum will be found approximately at Q when the line is a tangent to the curve for $y = \frac{4}{x}$. Drawing the line in this position we will see that the intercept on the y axis is given approximately by the point $(0, 3\frac{1}{4})$ so that the minimum of $2x + 3y$ is $9\frac{3}{4}$.

Similarly, an objective function given by $x + y = C$ (say) would have given a maximum for C when this line is a tangent to the circle $x^2 + y^2 = 64$ a result requested in the next exercise. *Answer*

Exercise 7.5

1. Using the same constraints as given by Fig. 7.11, obtain the optimum values of the following objective functions:

   (i) $4x + 3y$ (iv) $4x + y$
   (ii) x (v) $x + y$
   (iii) y

2. Using the same constraints as given by Fig. 7.12 obtain the optimum values of the following objective functions:

   (i) x (iii) $x + y$
   (ii) y (iv) $-x + y$

3. Plot the graphs of $y = x^2$ and $y^2 = x$ for $0 \leqslant x, y \leqslant 1$ and, subject to the constraints $x^2 \leqslant y \leqslant \sqrt{x}$, find the optimum values of the objective functions:

   (i) x (iii) $x + y$
   (ii) y (iv) $-x + y$

8

VECTORS

Quotation of a single number, together with a unit, completely specifies some physical quantities like temperature (10°C), time of day (14.30) or year (1983), mass (5 kg), length (3 m), area (10 m^2), volume (7 m^3) and speed (12 km/h). Such quantities are called **scalars** and the quoted numbers are called their magnitude.

Some quantities are only completely specified when we quote both a magnitude and a direction. The position of one point relative to another is quoted as a displacement (10 km North or 5 km SE), speed in a definite direction becomes a velocity (12 km/h in a direction A to B). Similarly, acceleration needs to specify both magnitude and direction because it is a measure of the rate of change of velocity. This type of quantity is called a **vector**. Force is another example of a vector or vector quantity, since we talk of exerting a force in a certain direction. Weight is a particular example of a force, being the force attracting a mass (scalar number) towards the earth's centre (direction).

When we say that a mass of 5 kg has a 'weight of 5g newtons' we are using an abbreviation for the statement 'a mass of 5 kg is attracted towards the earth's centre with a force of 5g newtons'. Similarly, a speed of 12 km/h in a direction subsequently found to be due South may now be referred to as a velocity of 12 km/h, South.

To describe a scalar as a constant means that its magnitude is constant. Thus, a constant temperature of 10°C for five minutes means just that; it is never anything else for those five minutes. Describing a vector as constant means that not only is its magnitude constant but its direction is also constant. The distinction between a constant scalar and a constant vector is clearly noted in realising that a body is able to travel round a circle with constant speed but never with a constant velocity.

In this text we are only concerned with a geometric illustration of vectors. A straight-line segment of some given length drawn to scale together with an arrowhead to imply direction is an adequate geometric model of a vector for our purposes. Our present simple aim can be illustrated by the following example.

Consider a boat being rowed across a stream. Our mathematical model of this task examines the problem in the following manner:

(i) If there was no current flowing in the stream then the boat would move in the direction in which the boat was pointed, which is the same direction in which the rower makes the effort. Let us say that this effort is worth 8 m a minute all the time he is rowing, i.e. he doesn't spurt, tire or fall in, and that he makes this effort directly across the stream perpendicular to the banks from P to Q as shown in Fig. 8.1.

(ii) If there is a current of, say, 6 m a minute, which we assume is the same right across the stream, then he will clearly be carried downstream.

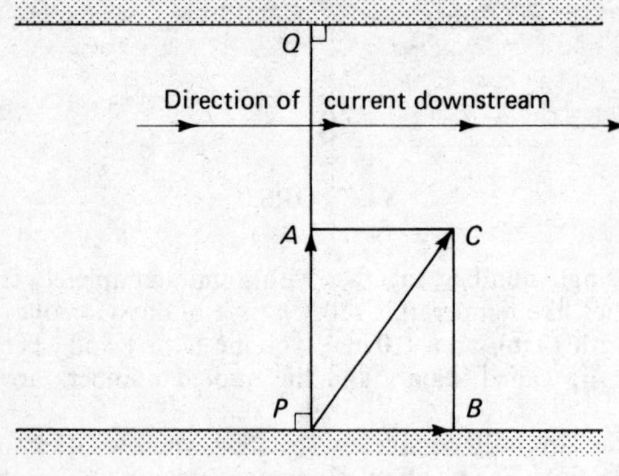

Fig. 8.1

Our model of this problem is to suggest that the boat will move in a direction which results from combining a vector of 8 m/minute directly across stream with another vector of 6 m/minute downstream. This result will be a speed of ? m/minute in a direction ?—that is, another vector which will be appropriately called the resultant of the other two vectors.

Practical laboratory experience with similar vector situations would show us that this resultant vector is obtained in Fig. 8.1 by drawing PA to scale to represent the effort of the rower to gain 8 m/minute directly across stream without the current (i.e. in still water) and drawing PB to represent the current of 6 m/minute downstream. We complete the rectangle $PACB$ (a rectangle because the boat is rowed at right-angles to the current) and PC will represent the resultant vector for the passage of the boat from P to the opposite bank of the stream. That is, PC represents the resultant velocity of the boat.

Finally, using our scale we measure PC and the angle CPB and declare that in our model the boat will move with a resultant velocity of 10 m/minute in the direction of 53° to the banks of the stream as given by angle CPB.

By this example we have shown how, using directed line segments to represent vectors, we can solve problems of this type.

This illustration could have been worded to have the boat start from the point Q, in which case we would have drawn a rectangle $QACB$, using a side through Q to represent the current flow of 6 m/minute. This suggests that any line like PB may be used to represent the current vector—that is, it may be drawn anywhere—and in this respect we call it a 'free vector' because it has an unspecified point of application.

In some cases, especially where forces are concerned, the vector may be restricted to slide along a line of action just like a crane winching a load up and down, so we call this a 'sliding vector'.

Finally, there is the type of vector which is applied or 'bound' to a particular point, like the force in the anchor rope for a suspension bridge.

These three types of vectors are described more formally as follows:

(a) A **bound vector** is one whose point of application is specified. It is said to be *localised* at that point. For example, a force is a bound vector because its point of application must be specified before it can be used.

(b) A **sliding vector** is one whose line of action is specified: the vector is said to be localised in that line of action—for example, a car travelling at a speed of 30 km/h in a straight line.

(c) A **free vector** is one whose line of action is not specified and like the current in the above example we represent it by any parallel line segments of equal length.

Thus we have seen how our geometric vectors will be used to obtain a pictorial solution to a particular vector problem. What we need to do now is establish the rules for dealing with free vectors in general and in what follows we shall use the word 'vector' to mean free vector unless stated otherwise.

In scalars, when we know that two rectangles have equal areas of 36 m^2 say, we do not assume that both rectangles are 18 m $\times$ 2 m because we know that it is possible to speak of two scalar quantities being equal without them being derived from the same figure. One rectangle could be 9 m $\times$ 4 m and the other 12 m $\times$ 3 m.

For two free vectors to be equal they must certainly be of the same magnitude and have the same direction, but they do not need to be in the same straight line. Thus in Fig. 8.2 the directed line segments AB, CD, EF are all

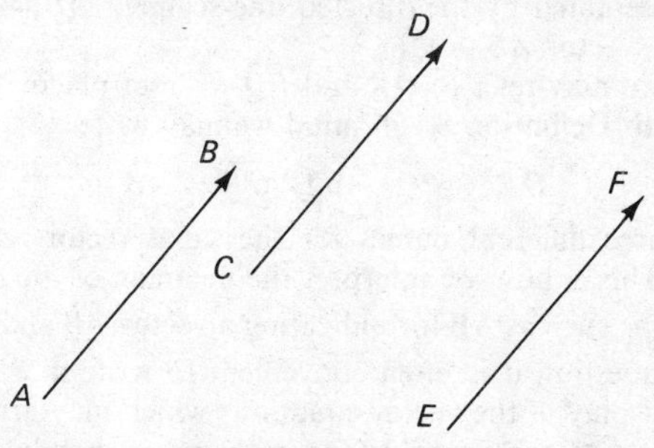

Fig. 8.2

equal in length and parallel to each other and are therefore three of the many possible representations of the same vector or, alternatively, they represent three equal vectors. Hence two trains travelling in the same direction with the same speed on two parallel tracks may have their velocity represented by the same geometric vector even if one is 50 m behind the other.

Definition 8.1. Two vectors are said to be equal if they have the same magnitude and direction. A zero vector will be written **0** and considered to have neither magnitude nor direction.

Let us examine these ideas in some pictorial detail. In Fig. 8.3 (i) we have a line segment with end points A and B complete with arrowhead and an angle of 30° to indicate a direction. What may this represent? To a suitable scale it might be a velocity vector of 20 km/h in a direction E 30°N, or a displacement of 20 km in direction E 30°N not necessarily meaning that we start at A and get mapped onto B but possibly meaning that somewhere something has moved 20 km from one point to another point in a direction E 30°N and our postage stamp diagram in (i) is the best model we have for representing the vector quantity concerned.

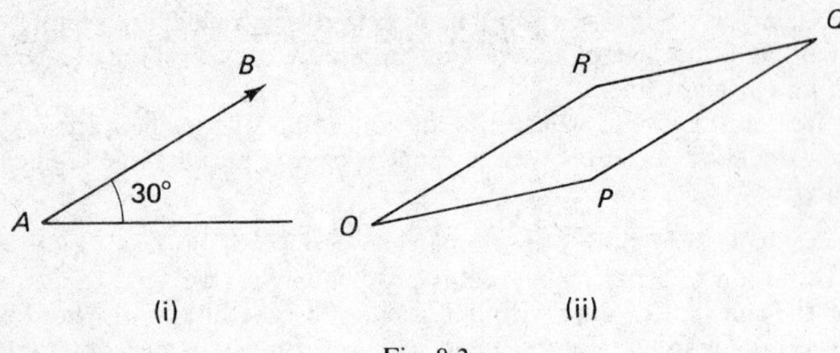

Fig. 8.3

In Fig. 8.3 (ii), we have a parallelogram $OPQR$ in which not only is $OR = PQ = AB$ but OR and PQ have been arranged parallel to AB. All three of these line segments could represent the magnitude of the same vectoring and by inserting arrowheads at R and Q any one of these three directed line segments may be used to represent the same vector in magnitude and direction.

Notation

The vector represented by the directed line segment AB is written $\overrightarrow{AB}$ and referred to as the 'vector AB'.

In Fig. 8.3 (ii) we may refer to $\overrightarrow{OR}$ and $\overrightarrow{PQ}$ without placing the arrowheads on the figure. With Definition 8.1 in mind we may write

$$\overrightarrow{OR} = \overrightarrow{PQ} \quad \text{and} \quad \overrightarrow{OR} = \overrightarrow{AB}$$

We now have three different names for the same vector which is possibly somewhere else. This is how we interpret the meaning of the = sign. Printers also use a bold type such as **AB** for indicating a vector $\overrightarrow{AB}$ and once a vector is established in a question it is often convenient to write $\mathbf{a} = \overrightarrow{AB}$ in order to obtain a neater display of the vector equations which may arise, but it is very difficult for students to copy a bold type in their own handwriting.

Sometimes it is necessary to refer to the magnitude of $\overrightarrow{AB}$ without mentioning its direction and for this we use the notation of $|\overrightarrow{AB}|$, calling this the 'modulus of vector AB' or just 'mod AB'. Similarly, $|\mathbf{a}|$ is referred to as 'mod **a**'. Thus $|\overrightarrow{AB}|$ is the positive number which is the measure of the length of the directed line segment AB. Hence $|\overrightarrow{AB}|$ and $|\mathbf{a}|$ are scalars and statements like $|\overrightarrow{AB}| = 10.46$ (2 d.p.) or $|\mathbf{a}| = 5$ are typical. Finally, if we return to our illustration in Fig. 8.1 we see that we had $|\overrightarrow{PA}| = 8$, $|\overrightarrow{PB}| = 6$ and by measurement or calculation $|\overrightarrow{PC}| = 10$. The direction of $\overrightarrow{PC}$ is then determined by a separate measurement or calculation.

The Triangle Law

Because vectors have both magnitude and direction it is difficult to see how they will combine. It is, however, found in practice that when the directed line segments representing vector quantities such as velocity, force, etc., are combined by the 'triangle law of addition' then the resultants thus obtained are meaningful. Indeed, the triangle law is often used to define what we mean by a vector by saying 'if the composition of any two quantities having magnitude and direction follows the triangle law then those quantities are called vectors'.

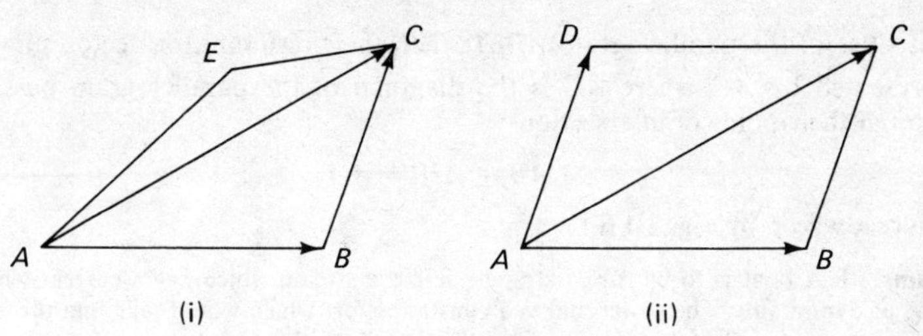

Fig. 8.4

To see what is meant by the 'sum of two vectors' consider the special case of displacement vectors. Let the displacement diagram of Fig. 8.4 (i) represent a two-stage change in position, first from A to B followed by a change from B to C. In the language of vectors this is written as $\vec{AB} + \vec{BC}$ and the + sign is seen to indicate the idea of 'followed by and end on'. If we consider $\vec{AB}$ as 'A mapped onto B', and $\vec{BC}$ as 'B mapped onto C' then the final composite result is 'A mapped onto C' and it is appropriate to represent this result by writing

$$\vec{AB} + \vec{BC} = \vec{AC}$$

and this is called the result of the vector addition of $\vec{AB}$ and $\vec{BC}$—that is, $\vec{AC}$ is the single displacement which would give the same final result as the other two displacements combined. Thus $\vec{AC}$ is the resultant displacement of the two displacements $\vec{AB}$ and $\vec{BC}$. Although we write the usual '+' sign, $\vec{AB} + \vec{BC}$ means a summation carried out according to the triangle law.

Returning to Fig. 8.4, we see that we may deduce that $\vec{AE} + \vec{EC} = \vec{AC}$ where E is any other point as shown. We may go on to say that

$$\vec{AX} + \vec{XC} = \vec{AC}$$

for any point X without indicating its position on the diagram. Such a variety of choice includes the use of D in the parallelogram $ABCD$ of Fig. 8.4 (ii) and so

$$\vec{AD} + \vec{DC} = \vec{AC}$$

which is the result of special interest because $\vec{AD} = \vec{BC}$ and $\vec{DC} = \vec{AB}$. On substituting these names for the same vector we get

$$\vec{BC} + \vec{AB} = \vec{AC}$$

which shows that vector addition is a commutative operation (page 17) because

$$\vec{AB} + \vec{BC} = \vec{AC} = \vec{BC} + \vec{AB}$$

The triangle law is really half a parellelogram law because we can see that $\vec{AB} + \vec{BC} = \vec{AB} + \vec{AD} = \vec{AC}$ by using $\vec{BC} = \vec{AD}$ in Fig 8.4 (ii) (i.e. we have taken two adjacent sides AB and AD of a parallelogram instead of two sides AB and BC of a triangle).

The Parallelogram Law

If two vectors are represented by $\vec{AB}$ and $\vec{AD}$ where AB and AD are adjacent sides of the parallelogram $ABCD$ then their resultant, or vector sum, is represented by $\vec{AC}$ where AC is the diagonal of the parallelogram passing through their point of intersection.

$$\therefore\ \vec{AB} + \vec{AD} = \vec{AC}$$

as already seen in Fig. 8.4 (ii).

Example 1. A boat is to be rowed straight across a stream which has a current which flows at 6 m/minute. The rower makes a constant effort which would take him through still water at the rate of 8 m/minute. Find the velocity with which he must row in order to get directly to the other side of the stream.

Solution. We have used the same type of problem as for Fig. 8.1 in the hope that the significance of the slight changes of conditions will be better appreciated. The boat starts from P and this time must arrive at Q in spite of the current. Using Fig. 8.5 to represent

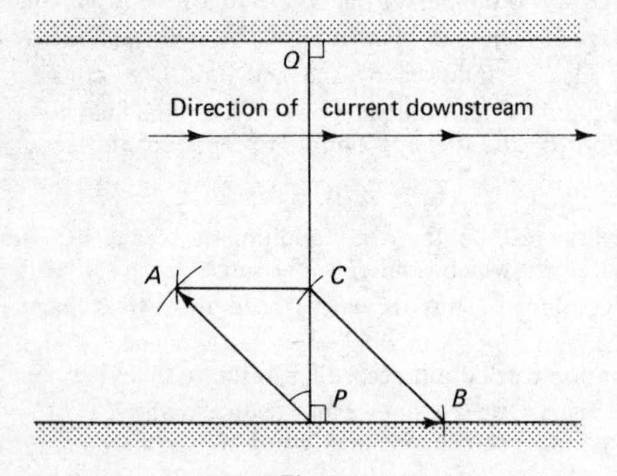

Fig. 8.5

the problem, it is clear that he will need to row partly into the current in order to stop being carried downstream. We suggest that his effort of 8 m/minute is directed along PA and represented by $\vec{PA}$ (we do not know which direction this is yet). As before we let $\vec{PB}$ represent the velocity of the current. Now $\vec{PA}$ must be so arranged that the resultant of $\vec{PA}$ and $\vec{PB}$, by the parallelogram law, must point in the direction of PQ and this fact determines how we draw the parallelogram $PACB$ with diagonal PC lying along PQ.

From the completed triangles and parallelogram we see that $\vec{PA} + \vec{AC} = \vec{PC}$, having noted that $\vec{AC} = \vec{PB}$. Now we concentrate on the magnitude of the vectors with

$$|\vec{PA}| = 8,\ |\vec{AC}| = |\vec{PB}| = 6$$

and from Pythagoras's rule for right-angled triangles we have

$$8^2 = 6^2 + PC^2$$
$$\therefore\ PC = \sqrt{28} = 5.3 \text{ (1dp).}$$

By measurement we find that angle $CPA = 48.6°$ (1 d.p.).

The boat will travel from P to Q with a speed of 5.3 m/minute if the rower points the boat in a direction of 48.6° upstream to PQ. *Answer*

Exercise 8.1

1. Using Fig. 8.1, suppose the stream had been running through parallel banks 50 m

apart. Find: (i) the time the boat takes to get across the stream; (ii) how far downstream the boat would have landed.

2. Repeat the problem of Fig. 8.1 with the same velocity of current but with the rower making an effort of only 6 m/minute directly across the stream. Find answers to the same questions (i) and (ii) as in Question 1 above.
3. Using the example of Fig. 8.5, find the time the boat takes to get across the stream from P to Q if the stream is 50 m wide.
4. Suppose in the problem of Fig. 8.5 that the rower had been able to row at 10 m/minute in still water, what would have been: (i) the direction of $\overrightarrow{PA}$ in order that his resultant velocity was in the direction of PQ; (ii) $|\overrightarrow{PC}|$; (iii) the time taken in crossing from P to Q when $PQ = 60$ m?
5. Two vectors $\overrightarrow{AB}$ and $\overrightarrow{AD}$ are at right angles. Find $\overrightarrow{AB} + \overrightarrow{AD}$, giving the direction of the resultant by quoting its angle of inclination to AB, when $|\overrightarrow{AD}| = 5$, $|\overrightarrow{AB}| = 7$.
6. Using Fig. 8.4 (ii) with $AB = 7$, $AD = 5$ and angle $DAB = 60°$, find $\overrightarrow{AC}$.

Vector Subtraction

The result $\overrightarrow{AB} + \overrightarrow{BC} = \overrightarrow{AC}$ illustrated in Fig. 8.6 is true for any position of the point C. Suppose therefore that the point C is moved closer to the point A until they eventually coincide and make $\overrightarrow{AC} = \overrightarrow{AA} = \mathbf{0}$.

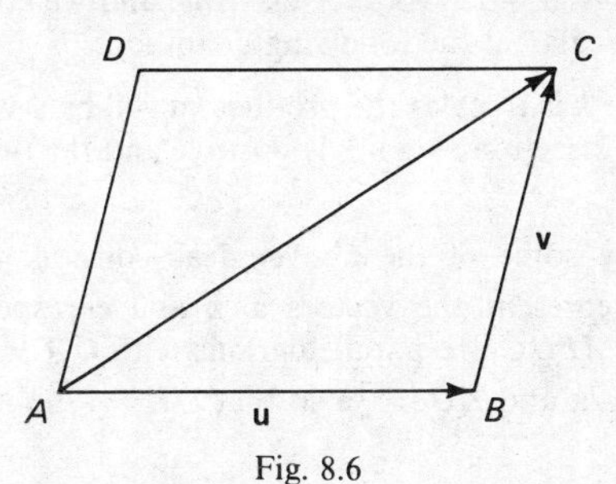

Fig. 8.6

The result $\overrightarrow{AB} + \overrightarrow{BC} = \overrightarrow{AC}$ has now become $\overrightarrow{AB} + \overrightarrow{BA} = \overrightarrow{AA} = \mathbf{0}$, and this suggests that it would be convenient to say that $\overrightarrow{BA} = -\overrightarrow{AB}$ and allow the following equalities to be written

$$\overrightarrow{AB} + \overrightarrow{BA} = \overrightarrow{AB} + (-\overrightarrow{AB}) = \overrightarrow{AB} - \overrightarrow{AB} = \mathbf{0}.$$

To use a different notation let $\overrightarrow{AB}$ represent the vector $\mathbf{u}$ and let $\overrightarrow{BC}$ represent the vector $\mathbf{v}$.

The result $\overrightarrow{AB} + \overrightarrow{BC} = \overrightarrow{AC}$ now reads as $\mathbf{u} + \mathbf{v} = \overrightarrow{AC}$, so that when the points A and C coincide we have $\mathbf{u} + \mathbf{v} = \mathbf{0}$, and again we are led to suggest that $\mathbf{u} + \mathbf{v} = \mathbf{0}$ implies that $\mathbf{v} = -\mathbf{u}$, which in turn leads to the following definition:

Definition 8.2. When the vector $\mathbf{u}$ is represented by $\overrightarrow{AB}$ we define $-\mathbf{u}$ to be the vector represented by $\overrightarrow{BA}$.

Thus $$\mathbf{u} + (-\mathbf{u}) = \mathbf{u} - \mathbf{u} = \mathbf{0}$$

We may also refer to $\overrightarrow{BA}$ or $(-\overrightarrow{AB})$ as the inverse of $\overrightarrow{AB}$ with respect to vector addition.

Returning to Fig. 8.6, let us consider obtaining a representation of $\mathbf{u} - \mathbf{v}$ when points A and C do not coincide. Since $ABCD$ is a parallelogram we may write $\mathbf{v} = \overrightarrow{BC} = \overrightarrow{AD}$ and consequently $-\mathbf{v} = \overrightarrow{CB} = \overrightarrow{DA}$.

$$\therefore \mathbf{u} - \mathbf{v} = \mathbf{u} + (-\mathbf{v}) = \overrightarrow{AB} + \overrightarrow{DA} = \overrightarrow{DA} + \overrightarrow{AB} = \overrightarrow{DB}$$

Thus

$$\overrightarrow{AC} = \mathbf{u} + \mathbf{v} \text{ and } \overrightarrow{DB} = \mathbf{u} - \mathbf{v}$$

Multiplication of a Vector by a Scalar

From Fig. 8.6 we can see that if $\overrightarrow{AB} = \overrightarrow{BC}$ then not only are the vectors $\mathbf{u}$ and $\mathbf{v}$ equal to each other but ABC is a straight line which is twice the length of AB. Hence $\overrightarrow{AC}$ represents a vector having the same direction as $\mathbf{u}$ and with a magnitude which is twice that of $\mathbf{u}$, i.e. $\overrightarrow{AC} = 2\mathbf{u}$.

$$\therefore \mathbf{u} + \mathbf{u} = 2\mathbf{u}$$

As we shall see on page 158, we may use this last result to gain

$$\mathbf{u} + \mathbf{u} + \mathbf{u} = 2\mathbf{u} + \mathbf{u} = 3\mathbf{u}$$

and then $\mathbf{u} + \mathbf{u} + \mathbf{u} + \mathbf{u} + \ldots$ to n terms $= n\mathbf{u}$, and we complete this idea of multiplication by a scalar in the following definition:

Definition 8.3. Where k is a scalar the product $k\mathbf{u}$ will be a vector of magnitude $|k|\,|\mathbf{u}|$ with the same direction as $\mathbf{u}$ if k is positive and the opposite direction to $\mathbf{u}$ if k is negative.

In order to apply some of the above ideas consider Fig. 8.7, in which $\overrightarrow{OA}$, $\overrightarrow{OB}$ and $\overrightarrow{OC}$ represent the vectors $\mathbf{a}$, $\mathbf{b}$ and $\mathbf{c}$ respectively and where $KOBL$, OACB and $APQC$ are parallelograms with $OA = AP = 2KO$. Thus $\overrightarrow{OP} = \overrightarrow{OA} + \overrightarrow{AP} = 2\mathbf{a}$ and $\overrightarrow{KO} = \frac{1}{2}\mathbf{a}$ so that $\overrightarrow{OK} = -\frac{1}{2}\mathbf{a}$ and $\overrightarrow{PO} = -2\mathbf{a}$.

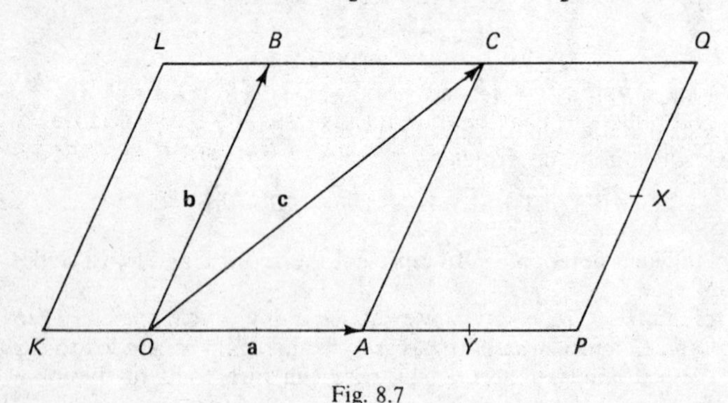

Fig. 8.7

From the parallelogram law in parallelogram $OACB$ we have $\overrightarrow{OA} + \overrightarrow{OB} = \overrightarrow{OC}$ and so $\mathbf{a} + \mathbf{b} = \mathbf{c}$.

Using the triangle law in $\triangle OBA$, we have $\overrightarrow{AO} + \overrightarrow{OB} = \overrightarrow{AB}$ and so $-\mathbf{a} + \mathbf{b} = \overrightarrow{AB}$, or alternatively $\overrightarrow{AB} = \mathbf{b} - \mathbf{a}$.

Similarly in $\triangle\, OPQ$, since $\overrightarrow{PQ} = \mathbf{b}$, we have $\overrightarrow{OQ} = \overrightarrow{OP} + \overrightarrow{PQ} = 2\mathbf{a} + \mathbf{b}$

while in $\triangle KOL$, since $\vec{KL} = \mathbf{b}$, we have $\vec{OL} = \vec{OK} + \vec{KL} = -\frac{1}{2}\mathbf{a} + \mathbf{b}$, or alternatively $\vec{OL} = \mathbf{b} - \frac{1}{2}\mathbf{a}$.

Example 2. Using Fig. 8.7, obtain expressions for $\vec{PB}$, $\vec{PC}$, $\vec{PL}$.

Solution. (i) In $\triangle POB$ we have $\vec{PB} = \vec{PO} + \vec{OB}$

$$\therefore \quad \vec{PB} = -2\mathbf{a} + \mathbf{b} = \mathbf{b} - 2\mathbf{a}$$

(ii) In $\triangle PAC$ we have $\vec{PC} = \vec{PA} + \vec{AC}$

$$\therefore \quad \vec{PC} = -\mathbf{a} + \mathbf{b} = \mathbf{b} - \mathbf{a}$$

Notice that $\vec{AB} = -\mathbf{a} + \mathbf{b} = \mathbf{b} - \mathbf{a} = \vec{PC}$, which proves that $|\vec{AB}| = |\vec{PC}|$ and AB is parallel to PC.

(iii) In $\triangle PKL$ we have $\vec{PL} = \vec{PK} + \vec{KL}$

$$\therefore \quad \vec{PL} = -2\tfrac{1}{2}\mathbf{a} + \mathbf{b} = \mathbf{b} - 2\tfrac{1}{2}\mathbf{a}$$

Example 3. Using Fig. 8.7 with $PX = XQ$ and $AY = YP$, find the following vector sums in terms of **a** and **b**:

$$\text{(i) } \vec{YC} + \vec{LA}$$
$$\text{(ii) } \vec{LY} + \vec{AX}$$

Solution. (i) In $\triangle YAC$ we have $\vec{YC} = \vec{YA} + \vec{AC} = -\frac{1}{2}\mathbf{a} + \mathbf{b}$.

In $\triangle LKA$ we have $\vec{LA} = \vec{LK} + \vec{KA} = -\mathbf{b} + 1\frac{1}{2}\mathbf{a}$.

$$\therefore \quad \vec{YC} + \vec{LA} = (-\tfrac{1}{2}\mathbf{a} + \mathbf{b}) + (-\mathbf{b} + 1\tfrac{1}{2}\mathbf{a})$$
$$= \mathbf{a}$$

Alternatively, since $\vec{YC} = \vec{OL}$ it follows that $\vec{YC} + \vec{LA} = \vec{OL} + \vec{LA} = \vec{OA} = \mathbf{a}$ by the triangle law.

(ii) In $\triangle LKY$ we have $\vec{LY} = \vec{LK} + \vec{KY} = -\mathbf{b} + 2\mathbf{a}$ $(KO = AY = YP)$

In $\triangle APX$, $\vec{PX} = \frac{1}{2}\vec{PQ} = \frac{1}{2}\mathbf{b}$ and so

$$\vec{AX} = \vec{AP} + \vec{PX} = \mathbf{a} + \tfrac{1}{2}\mathbf{b}$$
$$\therefore \quad \vec{LY} + \vec{AX} = (-\mathbf{b} + 2\mathbf{a}) + (\mathbf{a} + \tfrac{1}{2}\mathbf{b})$$
$$= -\mathbf{b} + (2\mathbf{a} + \mathbf{a}) + \tfrac{1}{2}\mathbf{b}$$
$$= -\mathbf{b} + 3\mathbf{a} + \tfrac{1}{2}\mathbf{b}$$
$$= 3\mathbf{a} - \mathbf{b} + \tfrac{1}{2}\mathbf{b} = 3\mathbf{a} - \tfrac{1}{2}\mathbf{b}$$

The justification for these rearrangements rests with the associative law discussed on page 158. We anticipate this law in the spirit of showing why we might need one.

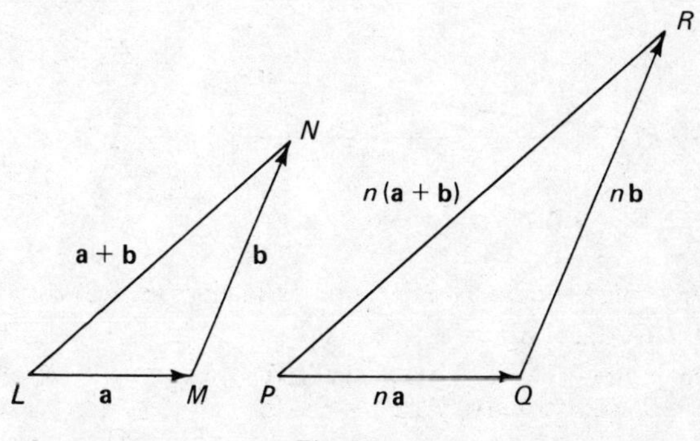

Fig. 8.8

Example 4. Show that the distributive law holds for scalar multiplication over vector addition by proving that $n(\mathbf{a} + \mathbf{b}) = n\mathbf{a} + n\mathbf{b}$.

Solution. For proof we shall consider two similar triangles LMN and PQR. Similar triangles are equiangular and their corresponding sides are in the same ratio. Thus each side of the triangle PQR is obtained from the triangle LMN by multiplying the lengths of its sides by a positive number n.

The arrangement of Fig. 8.8 shows PR parallel to LN, PQ parallel to LM, QR parallel to MN. With $\vec{PQ} = n\mathbf{a}$, $\vec{QR} = n\mathbf{b}$ and $\vec{PR} = n(\mathbf{a} + \mathbf{b})$ the vector sum $\vec{PQ} + \vec{QR} = \vec{PR}$ yields

$$n\mathbf{a} + n\mathbf{b} = n(\mathbf{a} + \mathbf{b})$$ *Answer*

Exercise 8.2

1. $ABCD$ is a square of side 5 units with $\vec{AB} = \mathbf{a}$ and $\vec{AD} = \mathbf{b}$ and E the midpoint of BC. Find the magnitude and direction of each of the following vectors: (i) $\mathbf{a}$, (ii) $\mathbf{b}$, (iii) $\mathbf{a} + \mathbf{b}$, (iv) $\mathbf{a} - \mathbf{b}$, (v) $\vec{AB} + \vec{BE}$, (vi) $\vec{AD} + \vec{DE}$.
2. $ABCD$ is a rectangle with $\vec{AB} = \mathbf{a}$ and $\vec{BC} = \mathbf{b}$. Given that $|\vec{AB}| = 12$ and $|\vec{BC}| = 5$ calculate the magnitude and direction of (i) $\mathbf{a} + \mathbf{b}$, (ii) $\mathbf{a} - \mathbf{b}$.
3. The point O is fixed in space and the point P moves so that $|\vec{OP}|$ is always equal to 10 units. What is the path of P if: (i) P remains in the same plane as O; (ii) P moves anywhere in space?
4. Solve the vector equation $\mathbf{a} + \mathbf{x} = \mathbf{a}$.
5. Solve the vector equation $\mathbf{a} - \mathbf{x} = \mathbf{a}$.
6. Solve the vector equation $\mathbf{a} - \mathbf{x} = \mathbf{b}$.
7. Solve the vector equation $\mathbf{b} - \mathbf{x} = \mathbf{a} - \mathbf{b}$.
8. Using Fig. 8.7, find the following in terms of $\mathbf{a}$, $\mathbf{b}$: (i) $\vec{YB}$, (ii) $\vec{XC}$, (iii) $\vec{YX}$, (iv) $\vec{XL} + \vec{YB}$.

The Associative Law

This law has been discussed elsewhere (pages 17, 89) but not of course for the binary operation of vector addition, which although we know is commutative has yet to be proved associative, i.e. we have yet to interpret $\mathbf{a} + \mathbf{b} + \mathbf{c}$, apart from our intuitive manipulation on pages 156–7.

We wish to prove that $(\mathbf{a} + \mathbf{b}) + \mathbf{c} = \mathbf{a} + (\mathbf{b} + \mathbf{c})$ so that we may write the result as $\mathbf{a} + \mathbf{b} + \mathbf{c}$ without ambiguity. We prove the result by using the diagram of Fig. 8.9 in which vectors $\vec{OK}$, $\vec{KL}$, $\vec{LM}$ and $\vec{MN}$ represent $\mathbf{a}$, $\mathbf{b}$, $\mathbf{c}$ and $\mathbf{d}$ respectively.

In $(\mathbf{a} + \mathbf{b}) + \mathbf{c}$ the vector $\mathbf{a}$ is associated with $\mathbf{b}$ by the bracket and we

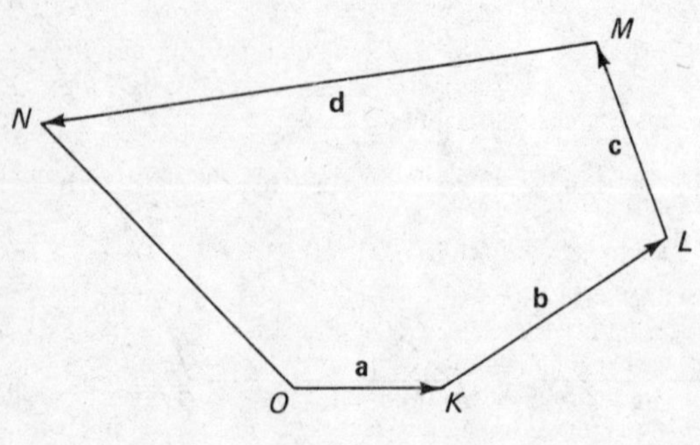

Fig. 8.9

simplify this part of the vector sum first. Noting that $\mathbf{a} + \mathbf{b} = \overrightarrow{OL}$, we have

$$(\mathbf{a} + \mathbf{b}) + \mathbf{c} = \overrightarrow{OL} + \mathbf{c}$$
$$= \overrightarrow{OM} \text{ (triangle law)}$$

In $\mathbf{a} + (\mathbf{b} + \mathbf{c})$ we must start by associating $\mathbf{b}$ with $\mathbf{c}$ and, noting that $\mathbf{b} + \mathbf{c} = \overrightarrow{KM}$, we have

$$\mathbf{a} + (\mathbf{b} + \mathbf{c}) = \mathbf{a} + \overrightarrow{KM}$$
$$= \overrightarrow{OM} \text{ (triangle law)}$$

Therefore $(\mathbf{a} + \mathbf{b}) + \mathbf{c} = \mathbf{a} + (\mathbf{b} + \mathbf{c})$, which may be written as $\mathbf{a} + \mathbf{b} + \mathbf{c}$ because it does not matter which two terms we start with to reduce $\mathbf{a} + \mathbf{b} + \mathbf{c}$ to a final form since we shall always obtain $\mathbf{a} + \mathbf{b} + \mathbf{c} = \overrightarrow{OM}$.

Recall that a comparison with ordinary addition is

$$10 + (9 + 6) = 25 = (10 + 9) + 6$$

Continuing to use Fig. 8.9, we see that it is now possible to extend the vector addition to as many vectors as we please although only four are available at this moment.

$$\therefore \quad \mathbf{a} + \mathbf{b} + \mathbf{c} + \mathbf{d} = (\mathbf{a} + \mathbf{b} + \mathbf{c}) + \mathbf{d}$$
$$= \overrightarrow{OM} + \mathbf{d}$$
$$= \overrightarrow{ON}$$

If this result is written out in full we have

$$\overrightarrow{OK} + \overrightarrow{KL} + \overrightarrow{LM} + \overrightarrow{MN} = \overrightarrow{ON}$$

and the order of lettering on the left-hand side suggests how we might see a simplification without the aid of a diagram—for example

$$\overrightarrow{OA} + \overrightarrow{AQ} + \overrightarrow{QC} + \overrightarrow{CN} + \overrightarrow{NE} + \overrightarrow{EK} = \overrightarrow{OK}$$

Vector Geometry

We should naturally expect that geometric vectors could be used in a new style of proof of some standard results in geometry involving parallelism and equalities. If, for example, we find in some problem that $\overrightarrow{PQ} = \mathbf{a}$ and $\overrightarrow{NM} = -5\mathbf{a}$ then we shall know that $\overrightarrow{PQ}$ is parallel to NM and one fifth of its length. Again, if we obtain a result $\overrightarrow{XY} = \overrightarrow{XA}$ then we shall know that the points Y and A are the same point. As a continuation of the type of exercise we used with Figs. 8.7 and 8.9 consider the regular hexagon of Fig. 8.10 in which we hope to express all the required directed line segments in terms of just two basic vectors $\mathbf{a}$ and $\mathbf{b}$.

Example 5. $OPQRST$ is a regular hexagon with $\overrightarrow{OT} = \mathbf{a}$ and $\overrightarrow{OP} = \mathbf{b}$. Find the following vectors in terms of $\mathbf{a}$ and $\mathbf{b}$:

(i) $\overrightarrow{TP}$, (ii) $\overrightarrow{TQ}$, (iii) $\overrightarrow{PQ}$, (iv) $\overrightarrow{QS}$, (v) $\overrightarrow{PS}$

Solution. Before starting the solution it is necessary to recall that the sides of a regular hexagon are equal in length to the radius OC of the circle centre C which passes through all the vertices of the hexagon. Thus $|\mathbf{a}| = |\mathbf{b}| = r$ the radius of the circle; TCQ is a diameter parallel to OP; and opposite sides of the regular hexagon are parallel. We find the substitute expressions for the required vectors by tracing the line segments of the figure without removing pen from paper.

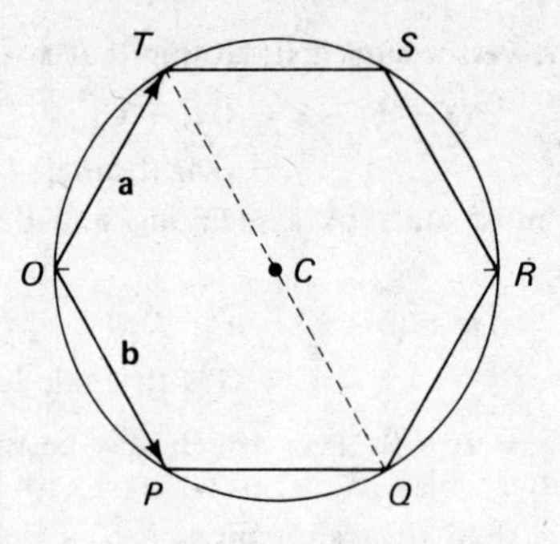

Fig. 8.10

(i) $\overrightarrow{TP} = \overrightarrow{TO} + \overrightarrow{OP} = -\mathbf{a} + \mathbf{b} = \mathbf{b} - \mathbf{a}$. Note that $\overrightarrow{PT} = \mathbf{a} - \mathbf{b}$.

(ii) $\overrightarrow{TQ} = 2\overrightarrow{TC} = 2\mathbf{b}$.

(iii) $\overrightarrow{PQ} = \overrightarrow{PT} + \overrightarrow{TQ} = \mathbf{a} - \mathbf{b} + 2\mathbf{b} = \mathbf{a} + \mathbf{b}$. Note that $\overrightarrow{PQ} = \overrightarrow{TS}$.

(iv) $\overrightarrow{QS} = \overrightarrow{QP} + \overrightarrow{PT} + \overrightarrow{TS} = \overrightarrow{PT} = \mathbf{a} - \mathbf{b}$, i.e. QS is parallel to PT.

(v) $\overrightarrow{PS} = \overrightarrow{PO} + \overrightarrow{OT} + \overrightarrow{TS} = -\mathbf{b} + \mathbf{a} + \overrightarrow{TS} = -\mathbf{b} + \mathbf{a} + \mathbf{a} + \mathbf{b}$

$$\therefore \quad \overrightarrow{PS} = 2\mathbf{a}$$ *Answer*

Example 6. In $\triangle\, OAB$ a point C is chosen so that $n\,\overrightarrow{AC} = m\,\overrightarrow{CB}$ where m, n are positive numbers. Prove that $n\,\overrightarrow{OA} + m\,\overrightarrow{OB} = (m + n)\,\overrightarrow{OC}$.

Solution. Consider $\triangle\, OAB$ drawn in Fig. 8.11 and the given result $n\,\overrightarrow{AC} = m\,\overrightarrow{CB}$, which may be considered as $n\,\overrightarrow{AC} - m\,\overrightarrow{CB} = \mathbf{0}$ or $n\,\overrightarrow{AC} + m\,\overrightarrow{BC} = \mathbf{0}$ or $n\,\overrightarrow{CA} + m\,\overrightarrow{CB}$

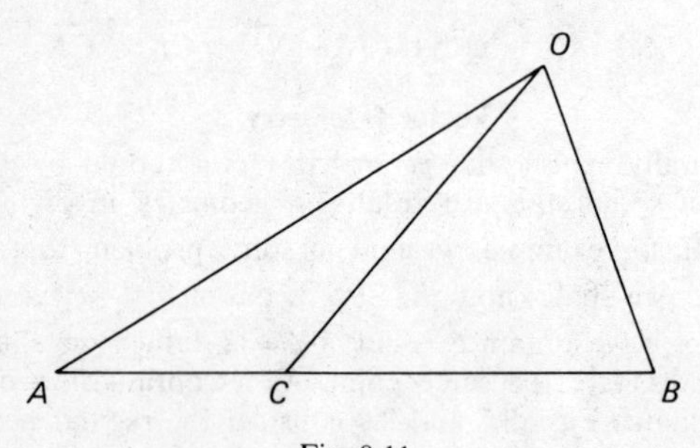

Fig. 8.11

$= \mathbf{0}$. (It is usually convenient to consider a vector result in more than one form.) Taking $n\,\overrightarrow{OA}$ and $m\,\overrightarrow{OB}$ separately we have:

$$n\,\overrightarrow{OA} = n(\overrightarrow{OC} + \overrightarrow{CA}) = n\,\overrightarrow{OC} + n\,\overrightarrow{CA} \text{ (distributive law, page 158)}$$

$$m\,\overrightarrow{OB} = m(\overrightarrow{OC} + \overrightarrow{CB}) = m\,\overrightarrow{OC} + m\,\overrightarrow{CB} \text{ (distributive law, page 158)}$$

$$\therefore \quad n\,\overrightarrow{OA} + m\,\overrightarrow{OB} = n\,\overrightarrow{OC} + m\,\overrightarrow{OC} + n\,\overrightarrow{CA} + m\,\overrightarrow{CB}$$

$$= (n + m)\,\overrightarrow{OC} + \mathbf{0}$$

$$\therefore \quad n\,\overrightarrow{OA} + m\,\overrightarrow{OB} = (n + m)\,\overrightarrow{OC}$$ *Answer*

Notice that if $m = n$ the result becomes $m(\overrightarrow{OA} + \overrightarrow{OB}) = 2m\,\overrightarrow{OC}$, i.e.

$\vec{OA} + \vec{OB} = 2OC$, which is a convenient device for checking if the point C is the midpoint of AB. Again, we might start with C chosen on AB so that $AC/CB = \frac{2}{3}$, i.e. $3AC = 2CB$, which relates to the result just proved by putting $m = 2$ and $n = 3$

$$\therefore \quad 3\vec{OA} + 2\vec{OB} = 5\vec{OC}$$

Such results are fairly common in vector geometry.

Example 7. Prove that the diagonals of a parallelogram bisect each other.

Solution. Recall that a parallelogram has opposite sides of equal length. (This means that squares and rectangles are also parallelograms.) Let Fig. 8.12 represent the problem.

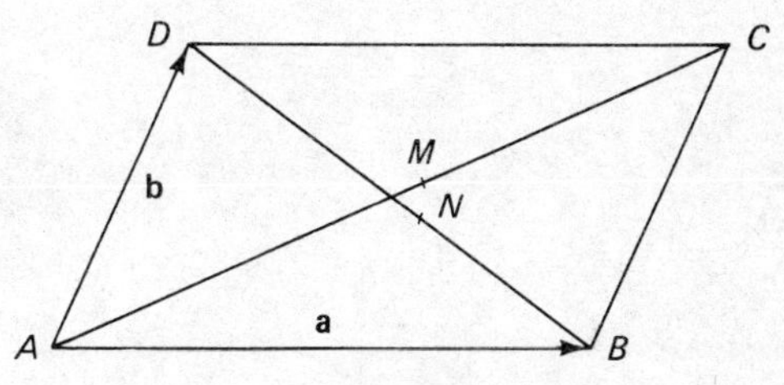

Fig. 8.12

Let $\vec{AB} = \mathbf{a}$, and $\vec{AD} = \mathbf{b}$. We shall suggest that M and N are the midpoints of the diagonals and then deduce that $\vec{AN} = \vec{AM}$, thereby proving that M and N are the same point, which can only be the point of intersection of the diagonals AC, BD. Let M be the midpoint of AC and let N be the midpoint of DB. From the parallelogram law we know that $\vec{AC} = \mathbf{a} + \mathbf{b}$

$$\therefore \quad \vec{AM} = \tfrac{1}{2}\vec{AC} = \tfrac{1}{2}(\mathbf{a} + \mathbf{b})$$

From the result proved in Example 6 above we know that $2\vec{AN} = \mathbf{a} + \mathbf{b}$ (i.e. put $m = n = 1$)

$$\therefore \quad \vec{AN} = \tfrac{1}{2}(\mathbf{a} + \mathbf{b})$$

We have now proved that $\vec{AN} = \vec{AM}$ and since A is the common point of these two equal vectors it follows that M and N must be the same point, and the only point common to both diagonals is where they intersect. Therefore the diagonals intersect at their midpoints. *Answer*

When a system of coplanar forces (vectors) acts at a point it is convenient to know which single force (vector) acting at that point will produce the same result as the original system.

Example 8. Consider the set of vectors given by taking the sides cyclically of the unclosed polygon, $HKLMNO$, with the vectors being abbreviated to **a**, **b**, **c**, **d** and **e** as indicated in Fig. 8.13. Find the resultant of the five vectors and interpret the result in terms of forces acting at a point.

Solution. The resultant of any set of vectors is the name given to their vector sum. Let the resultant $\mathbf{r} = (\mathbf{a} + \mathbf{b}) + \mathbf{c} + \mathbf{d} + \mathbf{e}$, which we now simplify by using the associative law.

$$\begin{aligned}\therefore \quad \mathbf{r} &= (\vec{OM} + \mathbf{c}) + \mathbf{d} + \mathbf{e} \\ &= (\vec{OL} + \mathbf{d}) + \mathbf{e} \\ &= \vec{OK} + \mathbf{e} \\ &= \vec{OH}\end{aligned}$$

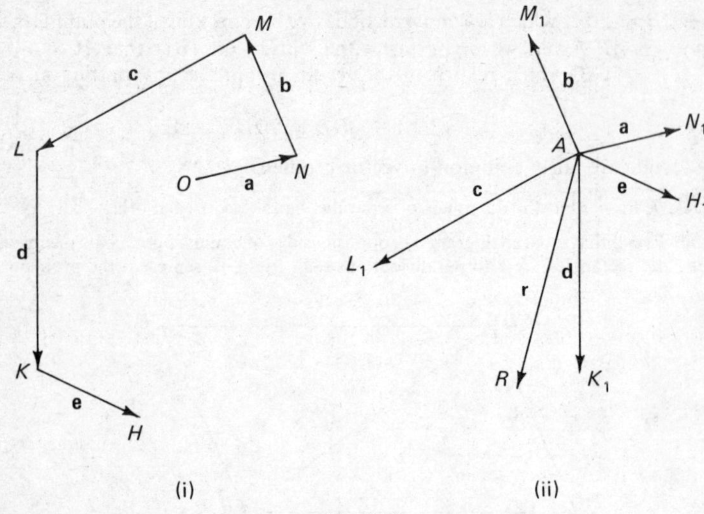

Fig. 8.13

This result is easier to see by writing $\mathbf{r} = \overrightarrow{ON} + \overrightarrow{NM} + \overrightarrow{ML} + \overrightarrow{LK} + \overrightarrow{KH} = \overrightarrow{OH}$. We note that the resultant $\overrightarrow{OH}$ closes the polygon and reverses the cyclic order of the directions of the sides of the polygon.

Now turn to Fig. 8.13 (ii) and consider vectors equal to those in (i) transferred to act at the one point A (i.e. $\overrightarrow{ON} = \overrightarrow{AN_1}$, $\overrightarrow{NM} = \overrightarrow{AM_1}$, $\overrightarrow{ML} = \overrightarrow{AL_1}$, etc.). Knowing that $\mathbf{r} = \overrightarrow{OH}$ and that all the vectors in (ii) are localised at A we can therefore replace this set of five vectors at A by the resultant vector $\overrightarrow{AR}$ (say) equal to $\overrightarrow{OH}$.

It follows, therefore, that if we wish to find the resultant of a given set of forces acting at a point A then we need only draw a polygon of vectors like (i) in which each side represents one force in magnitude and direction with the resultant represented by the closing side and acting at the point A.

Exercise 8.3

1. Using Fig. 8.10, express the following vectors in terms of **a** and **b**: (i) $\overrightarrow{PR}$, (ii) $\overrightarrow{OP} + \overrightarrow{PS} + \overrightarrow{ST}$, (iii) $\overrightarrow{PO} + \overrightarrow{OQ} + \overrightarrow{QT}$.
2. Using Fig. 8.11, find the position of C in each of the following cases when $m\,\overrightarrow{CB} = n\,\overrightarrow{AC}$: (i) $m = 3, n = 4$; (ii) $m = 4, n = 3$; and in (iii) find the resultant of $2\overrightarrow{OA} + 5\overrightarrow{OB}$.
3. In a triangle ABC the mid-point of AB is M and the mid-point of AC is N. Write $\overrightarrow{BA} = \mathbf{a}$ and $\overrightarrow{AC} = \mathbf{b}$ and find (i) $\overrightarrow{BC}$, (ii) $\overrightarrow{MN}$ in terms of **a** and **b**. What conclusion can you deduce from these results?
4. In the square $ABCD$, the diagonals intersect at M and N is the mid-point of MB. Write $\overrightarrow{AB} = \mathbf{b}$ and $\overrightarrow{AD} = \mathbf{a}$ and find the following vectors in terms of **a** and **b**: (i) $\overrightarrow{AC}$, (ii) $\overrightarrow{AM}$, (iii) $\overrightarrow{BD}$, (iv) $\overrightarrow{BM}$, (v) $\overrightarrow{BN}$, (vi) $\overrightarrow{DN}$, (vii) $4\,\overrightarrow{AN}$.
5. In Fig. 8.13 what would have been the resultant **r** if H had coincided with O?

   Four coplanar forces acting at a point A are represented geometrically by the sides of a polygon $ONMLK$ taken cyclically. The figure $ONML$ is a square and K is its centre. Find the resultant of the four forces at A if $|\overrightarrow{ON}| = 100$ units.
6. $OABC$ is a quadrilateral with O outside $\triangle\, ABC$ and K is the mid-point of AB. M is the mid-point of BC, and N the mid-point of AC. (AM and BN are called medians of the triangle ABC.)

Write $\overrightarrow{OA} = \mathbf{a}$, $\overrightarrow{OB} = \mathbf{b}$, $\overrightarrow{OC} = \mathbf{c}$ and express the following vectors in terms of **a**, **b** and **c**: (i) $\overrightarrow{ON}$; (ii) $\overrightarrow{OM}$; (iii) $3\overrightarrow{OG}$ where G lies on AM so that $2MG = GA$; (iv) $3\overrightarrow{OH}$ where H lies on BN so that $2\,NH = HB$; (v) $3\overrightarrow{OL}$ where L lies on CK so that $2KL = LC$.

7. $OABC$ is a quadrilateral with H, L, K, M the mid-points of OA, AB, BC, CO respectively. Write $\overrightarrow{OA} = 2\mathbf{a}$, $\overrightarrow{OB} = 2\mathbf{b}$, $\overrightarrow{OC} = 2\mathbf{c}$, and prove that $HLKM$ is a parallelogram by obtaining the following vectors in terms of **a**, **b**, **c**: (i) $\overrightarrow{HM}$; (ii) $\overrightarrow{LO}$; (iii) $\overrightarrow{OK}$, prove $\overrightarrow{LK} = \overrightarrow{HM}$; (iv) prove $\overrightarrow{HL} = \overrightarrow{MK}$.

Unit Vectors

Definition 8.4. A unit vector is a vector with magnitude 1.

Notation. Unit vectors will be written in the form $\hat{\mathbf{i}}, \hat{\mathbf{j}}, \hat{\mathbf{k}}, \hat{\mathbf{a}}, \hat{\mathbf{u}}$, etc. Thus $|\hat{\mathbf{a}}| = 1$, $|\hat{\mathbf{i}}| = 1$ and $|m\hat{\mathbf{a}}| = |m|$, $|n\hat{\mathbf{j}}| = |n|$, $|-4\hat{\mathbf{a}}| = 4$, $|-\frac{2}{3}\hat{\mathbf{i}}| = \frac{2}{3}$.

To Obtain a Unit Vector in Any Direction $\overrightarrow{AB}$

Suppose that $\hat{\mathbf{i}}$ is the unit vector in the direction of the non-zero vector $\overrightarrow{AB}$. This means that we may write $\overrightarrow{AB} = m\hat{\mathbf{i}}$, where m is a positive scalar.

$$\therefore \quad |\overrightarrow{AB}| = |m\hat{\mathbf{i}}| = m$$

$$\therefore \quad \overrightarrow{AB} = |\overrightarrow{AB}|\hat{\mathbf{i}} \text{ and so } \hat{\mathbf{i}} = \frac{\overrightarrow{AB}}{|\overrightarrow{AB}|}$$

Similarly, the unit vector $\hat{\mathbf{p}}$ (say) in the direction $\overrightarrow{PQ}$ is given by

$$\hat{\mathbf{p}} = \frac{\overrightarrow{PQ}}{|\overrightarrow{PQ}|}$$

Perpendicular Unit Vectors

The most useful unit vectors are those which lie along the x and y axes of coordinates because such an arrangement as seen in Fig. 8.14 enables us to

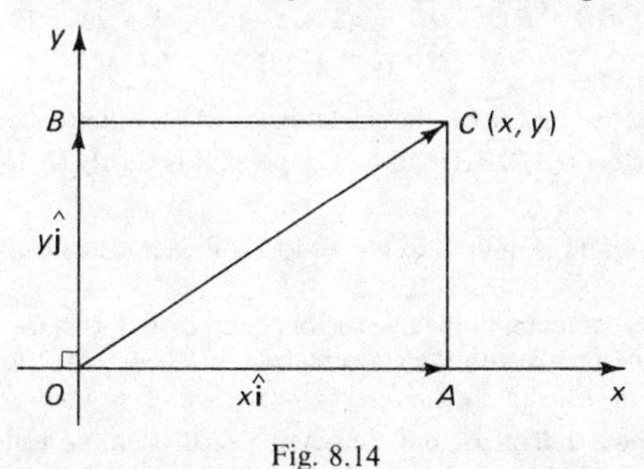

Fig. 8.14

express all our vectors in terms of these unit vectors in the direction of the axes. Suppose that we can represent our vector by $\overrightarrow{OC}$, i.e. we make the origin of coordinates as the initial point of our vector. This will mean that the position of the point C will be given by its x, y coordinates so that

$$OA = x \text{ and } AC = y = OB$$

We now choose unit vectors $\hat{\mathbf{i}}$ and $\hat{\mathbf{j}}$ in the direction of the x and y axes respectively so as to enable us to write

$$\overrightarrow{OC} = \overrightarrow{OA} + \overrightarrow{AC}$$
$$\overrightarrow{OC} = x\hat{\mathbf{i}} + y\hat{\mathbf{j}}$$

$x\hat{\mathbf{i}}$ and $y\hat{\mathbf{j}}$ are referred to as the rectangular component vectors of $\overrightarrow{OC}$ in the x and y directions. Usually we call x the 'x component of $\overrightarrow{OC}$' and y the 'y component of $\overrightarrow{OC}$'.

Hence, if C is the point (9, 4) we say that 9 and 4 are the x and y components of $\overrightarrow{OC}$ and we write $\overrightarrow{OC} = 9\hat{\mathbf{i}} + 4\hat{\mathbf{j}}$. Also notice that if C had been $(-9, 4)$ then the x component vector of $\overrightarrow{OC}$ would be $-9\hat{\mathbf{i}}$ and so we write $\overrightarrow{OC} = -9\hat{\mathbf{i}} + 4\hat{\mathbf{j}}$.

Clearly, by Pythagoras's Theorem we obtain

$$|\overrightarrow{OC}| = \sqrt{x^2 + y^2} \text{ (meaning the positive square root)}$$

We can see the usefulness of the idea of having unit vectors in the x, y direction because with $B(3, 4)$ and $C(10, 6)$ we have

$$\overrightarrow{OB} + \overrightarrow{OC} = (3\hat{\mathbf{i}} + 4\hat{\mathbf{j}}) + (10\hat{\mathbf{i}} + 6\hat{\mathbf{j}}) = 13\hat{\mathbf{i}} + 10\hat{\mathbf{j}}$$

This enables a resultant to be found very easily by writing $\overrightarrow{PQ} = 13\hat{\mathbf{i}} + 10\hat{\mathbf{j}}$ and noting that $|\overrightarrow{PQ}| = \sqrt{(13^2 + 10^2)} = \sqrt{269}$, etc.

Example 9. Given the four points $A(3, 4)$, $B(5, 12)$, $C(-2, 5)$, $D(-4, -8)$ find the vector sum $\overrightarrow{OA} + \overrightarrow{OB} + \overrightarrow{OC} + \overrightarrow{OD}$.

Solution. Without using a diagram we have:

$$\overrightarrow{OA} = 3\hat{\mathbf{i}} + 4\hat{\mathbf{j}}$$
$$\overrightarrow{OB} = 5\hat{\mathbf{i}} + 12\hat{\mathbf{j}}$$
$$\overrightarrow{OC} = -2\hat{\mathbf{i}} + 5\hat{\mathbf{j}}$$
$$\overrightarrow{OD} = -4\hat{\mathbf{i}} - 8\hat{\mathbf{j}}$$
$$\therefore \quad \overrightarrow{OA} + \overrightarrow{OB} + \overrightarrow{OC} + \overrightarrow{OD} = \hat{\mathbf{i}}(3 + 5 - 2 - 4) + \hat{\mathbf{j}}(4 + 12 + 5 - 8)$$
$$\therefore \quad \overrightarrow{OR} = 2\hat{\mathbf{i}} + 13\hat{\mathbf{j}}$$

where we have suggested that $\overrightarrow{OR}$ is the single vector which will produce the same result at O as the four vectors $\overrightarrow{OA}$, $\overrightarrow{OB}$, $\overrightarrow{OC}$, $\overrightarrow{OD}$. The point R is clearly (2, 13).

Finally

$$|\overrightarrow{OR}| = \sqrt{(4 + 169)} = \sqrt{173} = 13.2 \text{ (1 d.p.) if we so require}$$ *Answer*

Example 10. Find the resultant of the five vectors given by $\overrightarrow{OA}$, $\overrightarrow{OB}$, $\overrightarrow{OC}$, $\overrightarrow{OD}$, $\overrightarrow{OE}$, where A, B, C, D, E are the points with the coordinates $(-7, -9)$, $(-8, 10)$, $(11, -4)$, $(5, 6)$, $(-9, -11)$.

Solution. Again without a diagram, but considering each vector as a sum of its x and y components, we have the resultant $\overrightarrow{OR}$ given by

$$\overrightarrow{OR} = \overrightarrow{OA} + \overrightarrow{OB} + \overrightarrow{OC} + \overrightarrow{OD} + \overrightarrow{OE}$$
$$\overrightarrow{OR} = (-7\hat{\mathbf{i}} - 9\hat{\mathbf{j}}) + (-8\hat{\mathbf{i}} + 10\hat{\mathbf{j}}) + (11\hat{\mathbf{i}} - 4\hat{\mathbf{j}}) + (5\hat{\mathbf{i}} + 6\hat{\mathbf{j}}) + (-9\hat{\mathbf{i}} - 11\hat{\mathbf{j}})$$
$$\overrightarrow{OR} = \hat{\mathbf{i}}(-7 - 8 + 11 + 5 - 9) + \hat{\mathbf{j}}(-9 + 10 - 4 + 6 - 11)$$
$$\therefore \quad \overrightarrow{OR} = -8\hat{\mathbf{i}} - 8\hat{\mathbf{j}}$$

Therefore R is the point $(-8, -8)$ in the third quadrant and $\overrightarrow{OR} = \sqrt{(64 + 64)} = 8\sqrt{2}$.

Answer

With this use of unit vectors any vector may be translated to take its initial point to the origin (0, 0) and will be completely specified by quoting the x and y coordinates of its end point. These geometrical vectors are referred to as **position vectors** because they are unambiguously described by the quotation of the coordinates of the end points of the vector. It is for this reason that we may consider a position vector as an ordered pair, since any position vector quoted as (3, 4) (say) has an x vector component of $3\hat{\mathbf{i}}$ and a y vector component of $4\hat{\mathbf{j}}$.

By the same argument an ordered triple like (3, 4, 12) may represent a position vector in three dimensions.

The Dot Product

It is difficult to obtain from first principles an absolute definition of the 'multiplication' of two vectors. However, in practice, two ways of 'multiplying' vectors arise which give useful results. We shall only discuss the easier of the two results where the multiplication of one vector by another vector leads to a scalar result called the scalar product of the two vectors. (This product is also called the dot product of the two vectors.)

The Angle Between Two Vectors

Let two vectors $\mathbf{a}$ and $\mathbf{b}$ be represented by $\overrightarrow{OA}$ and $\overrightarrow{OB}$, as shown in Fig. 8.15. The angle between $\mathbf{a}$ and $\mathbf{b}$ is defined as the angle θ, $0 \leqslant \theta \leqslant \pi$, between $\overrightarrow{OA}$ and $\overrightarrow{OB}$ where both of these vectors are drawn with arrows pointing away from the point of intersection 0. When $\theta = \pi/2$, the vectors are said to be perpendicular and when $\theta = 0$ or π, they are said to be parallel.

Definition 8.5. The scalar product of two vectors $\mathbf{a}$ and $\mathbf{b}$ is defined as the product of their magnitude and the cosine of the angle between them. This product is written $\mathbf{a}.\mathbf{b}$ and hence referred to as the dot product of the two vectors.

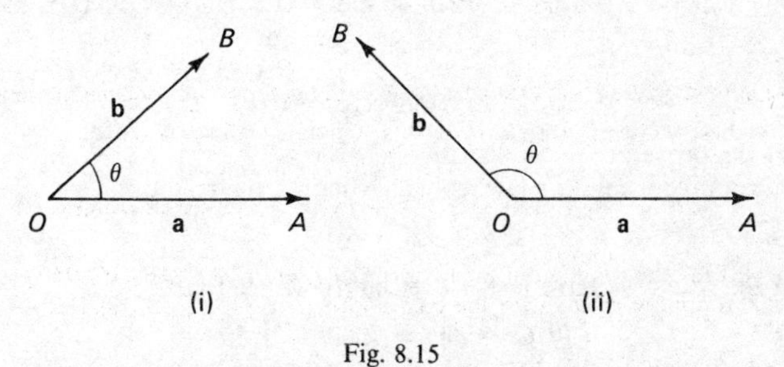

Fig. 8.15

In terms of the diagram of Fig. 8.15 we have

$$\mathbf{a}.\mathbf{b} = |\mathbf{a}|\,|\mathbf{b}| \cos\theta = |\overrightarrow{OA}|\,|\overrightarrow{OB}| \cos\theta$$

We note that $\mathbf{a}.\mathbf{b} = \mathbf{b}.\mathbf{a}$, i.c. thc dot product is commutative.

Before using the dot product in a general manner we look at some particular results:

(i) If $\mathbf{a} = \mathbf{b}$ then $\theta = 0°$ and $\cos\theta = 1$, $\mathbf{a}.\mathbf{b} = |\mathbf{a}|\,|\mathbf{b}|$. Now suppose $\mathbf{a} = \hat{\mathbf{i}}$ or $\hat{\mathbf{j}}$.

$$\therefore \quad \hat{\mathbf{i}}.\hat{\mathbf{i}} = 1, \; \hat{\mathbf{j}}.\hat{\mathbf{j}} = 1$$

(ii) If $\mathbf{a}$ is perpendicular to $\mathbf{b}$ then $\theta = 90°$ and $\cos\theta = 0$. $\therefore$ $\mathbf{a}.\mathbf{b} = 0$, i.e. the dot product of any two perpendicular vectors is 0.

$$\therefore \quad \hat{\mathbf{i}}.\hat{\mathbf{j}} = 1 \times 1 \cos 90° = 0$$

The Distributive Law

Here we are referring to the distribution of 'dot' over '+' and we want to prove that $\mathbf{x}.(\mathbf{b} + \mathbf{c}) = \mathbf{x}.\mathbf{b} + \mathbf{x}.\mathbf{c}$. For this purpose we consider the diagram of Fig. 8.16 with $\overrightarrow{OX} = \mathbf{x}$, $\overrightarrow{OH} = \mathbf{b}$, $\overrightarrow{HL} = \mathbf{c}$. Note that the angle between $\overrightarrow{HL}$ and $\overrightarrow{OX}$ is β.

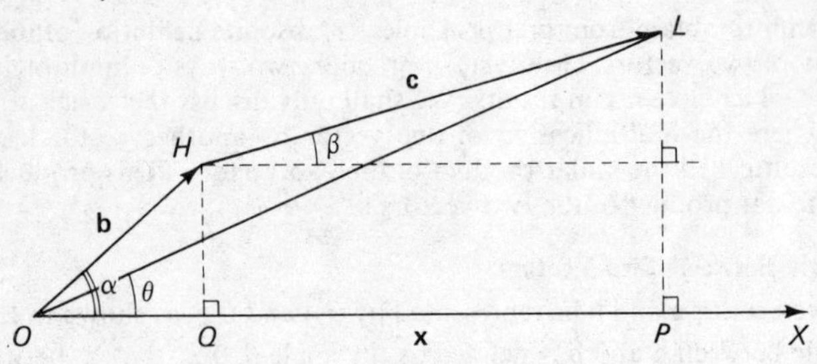

Fig. 8.16

We observe that

$$\mathbf{b} + \mathbf{c} = \overrightarrow{OL}$$

$$\begin{aligned}
\therefore \quad \mathbf{x}.(\mathbf{b} + \mathbf{c}) = \mathbf{x}.\overrightarrow{OL} &= OX.OL\cos\theta = OX.(OL\cos\theta) = OX.OP \\
\mathbf{x}.\mathbf{b} &= OX.OH\cos\alpha = OX.OQ \text{ (i.e. } OH\cos\alpha = OQ) \\
\mathbf{x}.\mathbf{c} &= OX.HL\cos\beta = OX.QP \text{ (i.e. } HL\cos\beta = QP) \\
\therefore \quad \mathbf{x}.\mathbf{b} + \mathbf{x}.\mathbf{c} &= OX.OQ + OX.QP \\
&= OX(OQ + QP) \\
&= OX.OP = OX(OL\cos\theta) \\
\therefore \quad \mathbf{x}.\mathbf{b} + \mathbf{x}.\mathbf{c} &= \mathbf{x}.(\mathbf{b} + \mathbf{c}) \qquad \text{(i)}
\end{aligned}$$

and we have proved that . is distributive over +.

As an example using this law, consider finding the result of

$$7\hat{\mathbf{i}}.(2\hat{\mathbf{i}} + 5\hat{\mathbf{j}})$$

Using the result we have just proved in (i), we see that

$$\begin{aligned}
7\hat{\mathbf{i}}.(2\hat{\mathbf{i}} + 5\hat{\mathbf{j}}) &= 7\hat{\mathbf{i}}.2\hat{\mathbf{i}} + 7\hat{\mathbf{i}}.5\hat{\mathbf{j}} \\
&= 14\hat{\mathbf{i}}.\hat{\mathbf{i}} + 35\hat{\mathbf{i}}.\hat{\mathbf{j}} \\
&= 14 + 0 = 14
\end{aligned}$$

Equally conveniently we see that

$$\begin{aligned}
9\hat{\mathbf{j}}.(2\hat{\mathbf{i}} + 5\hat{\mathbf{j}}) &= 18\hat{\mathbf{j}}.\hat{\mathbf{i}} + 45\hat{\mathbf{j}}.\hat{\mathbf{j}} \\
&= 0 + 45
\end{aligned}$$

Finally, putting these two results together we get

$$(7\hat{\mathbf{i}} + 9\hat{\mathbf{j}}).(2\hat{\mathbf{i}} + 5\hat{\mathbf{j}}) = 7\hat{\mathbf{i}}.(2\hat{\mathbf{i}} + 5\hat{\mathbf{j}}) + 9\hat{\mathbf{j}}.(2\hat{\mathbf{i}} + 5\hat{\mathbf{j}})$$
$$= 14 + 45$$
$$= 59$$

We now see from the right-hand side an easy method for finding the dot product of two vectors without involving any trigonometry because all we have to do is multiply the corresponding components of each vector and add the result, i.e. in

$$(7\hat{\mathbf{i}} + 9\hat{\mathbf{j}}).(2\hat{\mathbf{i}} + 5\hat{\mathbf{j}})$$

7 corresponds to 2 being the x components, and 9 corresponds to 5 being the y components. The dot product is therefore $7 \times 2 + 9 \times 5 = 59$ as we obtained above.

Example 11. Find the dot product of the two vectors $\overrightarrow{OA}$ and $\overrightarrow{OB}$ where A and B are the points (4, 5), (9, 3) respectively. Find also the angle between OA and OB. (In this context the point O is always taken as (0, 0).)

Solution. As already discovered we know that

$$\overrightarrow{OA}.\overrightarrow{OB} = (4\hat{\mathbf{i}} + 5\hat{\mathbf{j}}).(9\hat{\mathbf{i}} + 3\hat{\mathbf{j}})$$
$$= 36\hat{\mathbf{i}}.\hat{\mathbf{i}} + 12\hat{\mathbf{i}}.\hat{\mathbf{j}} + 45\hat{\mathbf{j}}.\hat{\mathbf{i}} + 15\hat{\mathbf{j}}.\hat{\mathbf{j}}$$
$$= 36 + 0 + 0 + 15$$
$$= 51$$
$$\therefore \quad \overrightarrow{OA}.\overrightarrow{OB} = 51$$

From the definition 8.5 we know that $\overrightarrow{OA}.\overrightarrow{OB} = |\overrightarrow{OA}|\ |\overrightarrow{OB}| \cos\theta$, where θ is the angle between the vectors. Now $|\overrightarrow{OA}|^2 = \overrightarrow{OA}.\overrightarrow{OA} = 16 + 25 = 41$ and $|\overrightarrow{OB}|^2 = \overrightarrow{OB}.\overrightarrow{OB} = 81 + 9 = 90$.

$$\therefore \quad \overrightarrow{OA}.\overrightarrow{OB} = \sqrt{41}.\sqrt{90}\cos\theta = \sqrt{3690}\cos\theta = (60.75)\cos\theta \text{ (2 d.p.)}$$

Using the result already obtained for $\overrightarrow{OA}.\overrightarrow{OB}$, we have

$$51 = (60.75)\cos\theta$$
$$\therefore \quad \cos\theta = 0.8396 \text{ and so } \theta = 32.9° \text{ (1 d.p.)}$$ *Answer*

Example 12. Three points A, B, C in the cartesian plane have coordinates $(-6, -7)$, $(9, 3)$ and $(-2, -3)$ respectively. Find the vector sum $\overrightarrow{OA} + \overrightarrow{OB} + \overrightarrow{OC}$ and find the angle between this resultant and $\overrightarrow{OA}$.

Solution. We have $\overrightarrow{OA} + \overrightarrow{OB} + \overrightarrow{OC} = (-6\hat{\mathbf{i}} - 7\hat{\mathbf{j}}) + (9\hat{\mathbf{i}} + 3\hat{\mathbf{j}}) + (-2\hat{\mathbf{i}} - 3\hat{\mathbf{j}}) = \overrightarrow{OR}$ (say).

$\therefore \quad \overrightarrow{OR} = \hat{\mathbf{i}} - 7\hat{\mathbf{j}}$, a vector which points into the fourth quadrant (page 120).

We note that

$$|\overrightarrow{OR}| = \sqrt{(1 + 49)} = \sqrt{50}$$

We also note that

$$|\overrightarrow{OA}| = \sqrt{(36 + 49)} = \sqrt{85}$$

Let the angle that $\overrightarrow{OR}$ makes with $\overrightarrow{OA}$ be θ.

$$\therefore \quad \overrightarrow{OR}.\overrightarrow{OA} = |\overrightarrow{OR}|\,|\overrightarrow{OA}|\cos\theta = \sqrt{50}.\sqrt{85}\cos\theta$$

But $\overrightarrow{OR}.\overrightarrow{OA} = (\hat{\mathbf{i}} - 7\hat{\mathbf{j}}).(-6\hat{\mathbf{i}} - 7\hat{\mathbf{j}}) = -6 + 49 = 43$

$$\therefore \quad 43 = \sqrt{50}\,\sqrt{85}\cos\theta \text{ and } \cos\theta = \frac{43}{\sqrt{4250}} = 0.6596$$

$$\therefore \quad \theta = 48.7° \text{ (1 d.p.)}$$ *Answer*

Exercise 8.4

1. Find the modulus of the vector $\overrightarrow{OA}$ and the angle between $\overrightarrow{OA}$ and the x-axis when $O = (0, 0)$ and A is the point: (i) (1, 1), (ii) (2, 2), (iii) (4, 3), (iv) (11, 60).
2. Find the resultant $\overrightarrow{OR}$ of the five vectors formed by $\overrightarrow{OA}$, $\overrightarrow{OB}$, $\overrightarrow{OC}$, $\overrightarrow{OD}$, $\overrightarrow{OE}$, where A, B, C, D, E are the points (1, 2), (−1, 2), (2, −7), (−3, 4), (−6, 7) and find the angle between $\overrightarrow{OR}$ and $\overrightarrow{OC}$.
3. The vectors **a**, **b** and **c** are given by $\mathbf{a} = 5\hat{\imath} - 4\hat{\jmath}$, $\mathbf{b} = -4\hat{\imath} + 5\hat{\jmath}$, $\mathbf{c} = -5\hat{\imath} + 5\hat{\jmath}$. Find: (i) $\mathbf{a}.\mathbf{b}$; (ii) $\mathbf{b}.\mathbf{c}$; (iii) $\mathbf{a}.\mathbf{c}$; (iv) the angle between **a** and **b**. Interpret $\mathbf{a}\,(\mathbf{b}.\mathbf{c})$.
4. The points A and B are given by (3, 9), (4, 17). Find the components of the unit vector in the direction $\overrightarrow{AB}$, relative to the x, y axes.
5. In three dimensions the coordinates of two points A and B are (2, 5, 9) and (−3, 6, 1) respectively. Using the same principle as for two component vectors, find the dot product of $\overrightarrow{OA}$ and $\overrightarrow{OB}$.

More Dot Products

The distributive law for vectors discussed on page 166 enabled us to write the expansion $\mathbf{e}.(\mathbf{b} - \mathbf{c}) = \mathbf{e}.\mathbf{b} - \mathbf{e}.\mathbf{c}$ for any three vectors **e**, **b** and **c**.

Suppose we put $\mathbf{e} = \mathbf{b} - \mathbf{c}$. The expansion will now read

$$\begin{aligned} (\mathbf{b} - \mathbf{c}).(\mathbf{b} - \mathbf{c}) &= (\mathbf{b} - \mathbf{c}).\mathbf{b} - (\mathbf{b} - \mathbf{c}).\mathbf{c} \\ &= \mathbf{b}.\mathbf{b} - \mathbf{c}.\mathbf{b} - \mathbf{b}.\mathbf{c} + \mathbf{c}.\mathbf{c} \\ \therefore \quad (\mathbf{b} - \mathbf{c}).(\mathbf{b} - \mathbf{c}) &= b^2 + c^2 - 2\mathbf{b}.\mathbf{c} \text{ since } \mathbf{b}.\mathbf{c} = \mathbf{c}.\mathbf{b} \end{aligned}$$

The next illustration uses this result in the context of both the triangle and parallelogram law for vectors.

Consider Fig. 8.17 as a diagram which not only illustrates a triangle of vectors ABC but also a parallelogram of vectors $ABKC$. In mathematics the

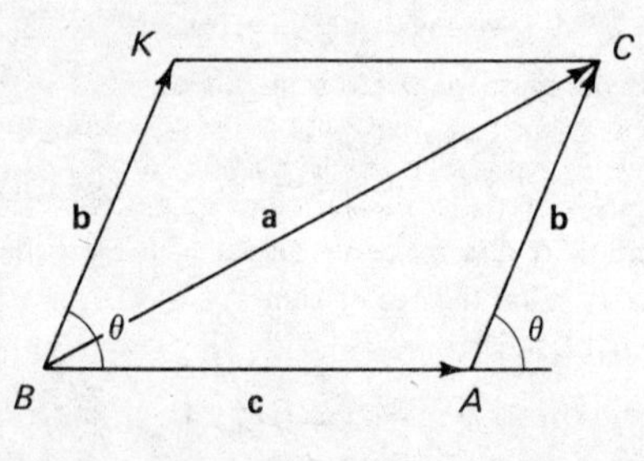

Fig. 8.17

usual notation for lettering the sides of a triangle is a for the side opposite angle A, b for the side opposite angle B and c for the side opposite angle C.

In $\triangle ABC$: Here we see that $\mathbf{a} = \mathbf{b} + \mathbf{c}$ by the triangle law.

$$\begin{aligned} \therefore \quad \mathbf{a}.\mathbf{a} &= (\mathbf{b} + \mathbf{c}).(\mathbf{b} + \mathbf{c}) \\ &= b^2 + c^2 + 2\mathbf{b}.\mathbf{c} \\ \therefore \quad a^2 &= b^2 + c^2 + 2bc \cos \theta \end{aligned}$$

because θ is the angle **b** and **c** as shown in Fig. 8.17 (see page 165).

Since $\theta = 180° - \angle BAC$ we have $\cos \theta = -\cos \angle BAC = -\cos A$ for short.

$$\therefore \quad a^2 = b^2 + c^2 - 2bc \cos A$$

a result which holds for any triangle ABC whatsoever using the a, b, c notation. Obviously there are two other results involving $\cos B$ and $\cos C$ which are requested in the next exercise.

This result is known as the cosine rule for the triangle ABC; it is a relation between the lengths of the sides of the triangle.

In parallelogram $ABKC$: Here we again see θ as the angle between the two vectors **b** and **c**

$$\therefore\ a^2 = b^2 + c^2 + 2bc \cos \theta$$

is the formula for the magnitude of the resultant **a** of two vectors **b**, **c** where θ is the angle between **b** and **c**.

We now have a method for calculating the magnitude of a resultant instead of finding it from a scaled diagram.

Example 13. The angle between two vectors is 35° and their magnitudes are 9 and 6 units. Calculate the magnitude of their resultant.

Solution. Using Fig. 8.17, we imagine $BA = 9$ and $BK = 6$ with $\theta = 35°$. Let **a** be the resultant of the two vectors, just as in Fig. 8.17.

$$\therefore\ a^2 = 9^2 + 6^2 + 2 \times 9 \times 6 \cos 35°$$
$$a^2 = 117 + 108 \cos 35° = 205.5 \text{ (1 d.p.)}$$
$$\therefore\ a = 14.3 \text{ (1 d.p.)}$$

A quotation of the angle between **a** and one of the given vectors will complete the specification of **a**. *Answer*

The concurrency (passing through a common point) of the medians of a triangle (Question 6, Exercise 8.3) is only one of several similar relations for a $\triangle ABC$

Example 14. The concurrency of the altitudes of a triangle at what is called the orthocentre of the triangle can be proved quite simply using dot products, particularly since the dot product for two perpendicular vectors is always zero.

Consider Fig. 8.18, in which two of the altitudes AD and BE are drawn to intersect at H. We wish to prove that CH will be perpendicular to BA, i.e. that $\overrightarrow{CH}.\mathbf{c} = 0$.

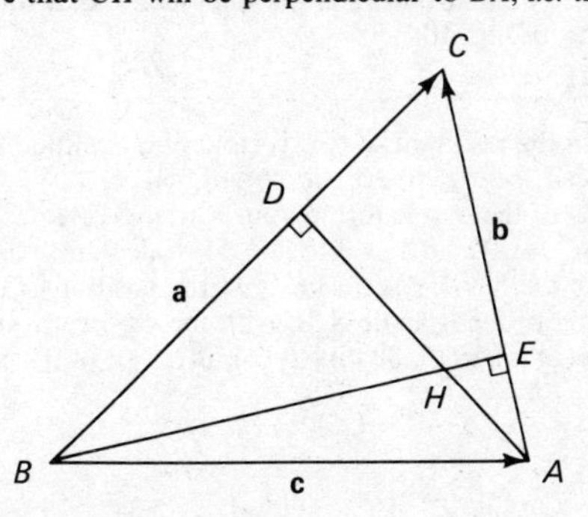

Fig. 8.18

Solution. Notice that the vector arrangement on the diagram gives $\mathbf{a} = \mathbf{c} + \mathbf{b}$.

Since AH is perpendicular to BC we have

$$\overrightarrow{AH}.\mathbf{a} = 0$$

Since BH is perpendicular to AC we have

$$\vec{BH}.\mathbf{b} = 0$$

Rewrite $\qquad \vec{AH}.\mathbf{a} = (\mathbf{b} + \vec{CH}.).\,\mathbf{a} = \mathbf{b}.\mathbf{a} + \vec{CH}.\mathbf{a} = 0$

and $\qquad \vec{BH}.\mathbf{b} = (\mathbf{a} + \vec{CH}).\mathbf{b} = \mathbf{a}.\mathbf{b} + \vec{CH}.\mathbf{b} = 0$

using the distributive law.

$$\therefore \vec{CH}.\mathbf{a} = \vec{CH}.\mathbf{b}$$

$$\text{i.e. } \vec{CH}.(\mathbf{a} - \mathbf{b}) = 0$$

but $\mathbf{a} - \mathbf{b} = \mathbf{c}$

$$\therefore \vec{CH}.\mathbf{c} = 0$$

and so CH is perpendicular to BA. Hence all three altitudes of a triangle are concurrent.

Answer

Example 15. Using Fig. 8.18, suppose that the vertices have coordinates given by $A(1, 3)$, $B(7, 9)$, $C(4, 12)$. Calculate the position of the orthocentre H and comment on the result.

Solution. Suppose the position of H is (x, y), and consider using the fact that $\vec{CH}.\vec{AB} = 0$.

$$\vec{AB} = \vec{AO} + \vec{OB} = \vec{OB} - \vec{OA} = (7\hat{\imath} + 9\hat{\jmath}) - (\hat{\imath} + 3\hat{\jmath}) = 6\hat{\imath} + 6\hat{\jmath}$$

where 0 is the origin $(0, 0)$.

$$\vec{CH} = \vec{CO} + \vec{OH} = \vec{OH} - \vec{OC} = (x\hat{\imath} + y\hat{\jmath}) - (4\hat{\imath} + 12\hat{\jmath}) = (x - 4)\hat{\imath} + (y - 12)\hat{\jmath}$$

$\therefore \vec{CH}.AB = 0$ leads to $6(x - 4) + 6(y - 12) = 0$

i.e. $x + y = 16 \qquad$ (i)

Now we must use $\vec{AH}.\vec{BC} = 0$, where

$$\vec{BC} = \vec{BO} + \vec{OC} = \vec{OC} - \vec{OB} = (4\hat{\imath} + 12\hat{\jmath}) - (7\hat{\imath} + 9\hat{\jmath}) = -3\hat{\imath} + 3\hat{\jmath}$$

$$\vec{AH} = \vec{AO} + \vec{OH} = \vec{OH} - \vec{OA} = (x\hat{\imath} + y\hat{\jmath}) - (\hat{\imath} + 3\hat{\jmath}) = (x - 1)\hat{\imath} + (y - 3)\hat{\jmath}$$

$\therefore \vec{AH}.\vec{BC} = 0$ leads to $-3(x - 1) + 3(y - 3) = 0$

i.e. $-x + y = 2 \qquad$ (ii)

The solution to equations (i) and (ii) is $x = 7$, $y = 9$ and H is the point $(7, 9)$.

It is of some interest to note that we have carried out the vector procedure of finding H automatically, and now we find that H is the same point as B. In other words angle ABC was a right-angle to begin with.

Answer

Exercise 8.5

1. Find the magnitude of the resultant of two vectors of magnitude 10 and 8 units each when the angle between them is: (i) 30°, (ii) 45°, (iii) 60°.
2. Quote the three forms of the cosine formula for a triangle ABC.
3. In $\triangle ABC$, angle $ABC = 100°$, $BA = 9$, $BC = 11$. Calculate AC.
4. In $\triangle ABC$, $AB = 60$, $BC = 61$, $CA = 11$. Find the position of the orthocentre.
5. Find the orthocentre H of the triangle ABC with the vertices having the coordinates given by: (i) $A(0, 0)$, $B(14, 0)$, $C(10, 8)$; (ii) $A(6, 5)$, $B(2, 2)$, $C(8, -1)$.

9

MATRICES

We shall introduce matrices through the study of reflections and rotations of points in the cartesian plane. Thereafter we shall manipulate the matrices as the elements in a new algebra.

Reflections in the Axes

The simplest reflections take place in the x and y axes, because one of the coordinates of the point always remains unchanged. For example, in Fig. 9.1

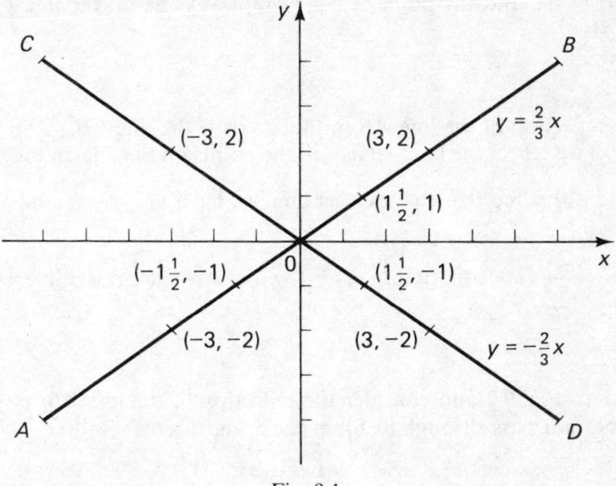

Fig. 9.1

the images of the points $(1\frac{1}{2}, 1)$, $(3, 2)$, $(-3, 2)$ after reflection in the x axis are $(1\frac{1}{2}, -1)$, $(3, -2)$, $(-3, -2)$ respectively. The reflection of the general point (x, y) in the x axis mirror produces the image point $(x, -y)$.

This reflection in the x axis mirror of points (x, y) of the cartesian plane may be written formally as a one-to-one mapping $f:(x, y) \longmapsto (x, -y)$ and referred to as a reflection in the line $y = 0$.

Notation

In general, the original point before the reflection or rotation takes place will be represented by (x_0, y_0) and will become (x_1, y_1) as a result of the transformation. The point (x_1, y_1) is called the image of (x_0, y_0) in the transformation.

A reflection of points using the y axis for a mirror in Fig. 9.1 maps the points $(3, 2)$, $(1\frac{1}{2}, -1)$, $(3, -2)$ onto the points $(-3, 2)$, $(-1\frac{1}{2}, -1)$, $(-3, -2)$ respectively, and the reflection of the general point (x, y) in the y axis mirror produces the image point $(-x, y)$.

This reflection in the line $x = 0$ may be written formally as the mapping

$$f:(x, y) \longmapsto (-x, y)$$

Overall it is not the reflection of the individual points which is important but what happens to any relations between the points of the original set. That is, if the original points lie on a straight line will their images lie on a straight line? Or to put it another way, what happens when we reflect the whole line?

Fig. 9.1 shows the line AB, $y = \frac{2x}{3}$ so that any original point on this line is (x_0, y_0) or $(x_0, \frac{2x_0}{3})$. After reflection in the x axis this point becomes, or is called, (x_1, y_1) where $x_1 = x_0$, $y_1 = \frac{-2x_0}{3}$.

$\therefore \quad y_1 = \frac{-2x_1}{3}$ and the point (x_1, y_1) will always lie on the line CD where $y = \frac{-2x}{3}$.

Thus the reflection of the line AB in the x axis is the line CD.
Similarly, if we reflect the line AB (i.e. all the points on the line) in the y axis, we again obtain the line DC because this time we have $x_1 = -x_0$ and $y_1 = \frac{2x_0}{3}$.

$\therefore \quad y_1 = \frac{-2x_1}{3}$ and the point (x_1, y_1) will always lie on the line DC where $y = \frac{-2x}{3}$.

Now refer to Fig. 9.2 and consider the reflection of the line AB, $y = 2x + 4$ (which does not pass through (0, 0)) in the x and y axes.

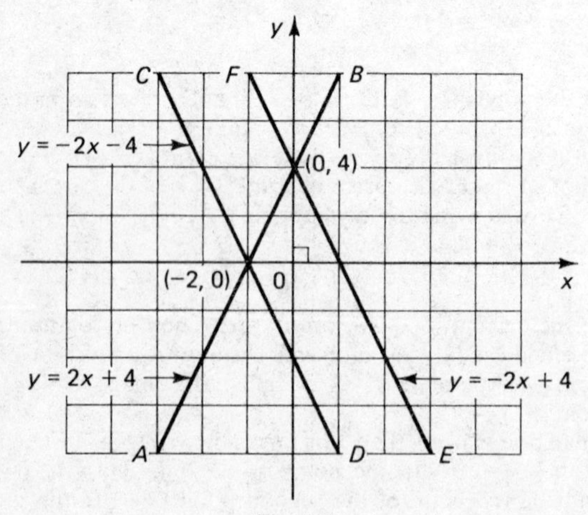

Fig. 9.2

(i) Reflection in the x axis ($y = 0$).
The image of (x_0, y_0) is (x_1, y_1), where $x_1 = x_0$ and $y_1 = -y_0$. but

$y_0 = 2x_0 + 4$ on the original line, AB.

$\therefore \quad y_1 = -y_0 = -(2x_0 + 4) = -2x_1 - 4$ on the new line.

Thus, the point (x_1, y_1) is always on the line $y = -2x - 4$ and so, the line CD is the reflection of the line AB in the x axis as shown in Fig. 9.2.

(ii) Reflection in the y axis ($x = 0$).
The image of (x_0, y_0) is (x_1, y_1), where $x_1 = -x_0$ and $y_1 = y_0$, but

$y_0 = 2x_0 + 4$ on the original line, AB.

$\therefore \quad y_1 = y_0 = 2x_0 + 4 = -2x_1 + 4$ on the new line.

Thus, the point (x_1, y_1) is always on the line $y = -2x + 4$ and so, the line EF is the reflection of the line AB in the y axis as shown in Fig. 9.2.

Exercise 9.1

1. Find the images of the following points after reflection in the line (a) $y = 0$ and (b) $x = 0$: (i) (4, 5), (ii) -4, 5), (iii) (-4, -5), (iv) (4, -5), (v) (5, 4).
2. Find the reflection of each of the following lines in each axis separately: (i) $y = 3x$; (ii) $y = \frac{5x}{4}$; (iii) $y = 4x + 2$, giving the reasons as described in the discussion for $y = 2x + 4$ above; (iv) $y = -2x + 4$, as in (iii).
3. What is the final image of the point (x_0, y_0) when it is transformed by: (i) reflection in $y = 0$ followed by reflection in $y = 0$; (ii) reflection in $x = 0$ followed by reflection in $x = 0$; (iii) reflection in $y = 0$ followed by reflection in $x = 0$.

Reflection in $y = \pm x$

You will recall that we obtained an inverse relation R^{-1} by reflecting the ordered pairs of R in the line $y = x$ (page 127). With a relation defined by $R = \{(x, y) \mid y = f(x), x, y \in \mathbb{R}\}$ the inverse relation R^{-1} was defined as $R^{-1} = \{(y, x) \mid (x, y) \in R\}$ and the interchange of coordinates was seen to be equivalent to a reflection in the line $y = x$, e.g. the points (1, 3), (4, 7), (-2, 1) have images of (3, 1), (7, 4), (1, -2) respectively when reflected in the mirror line $y = x$.

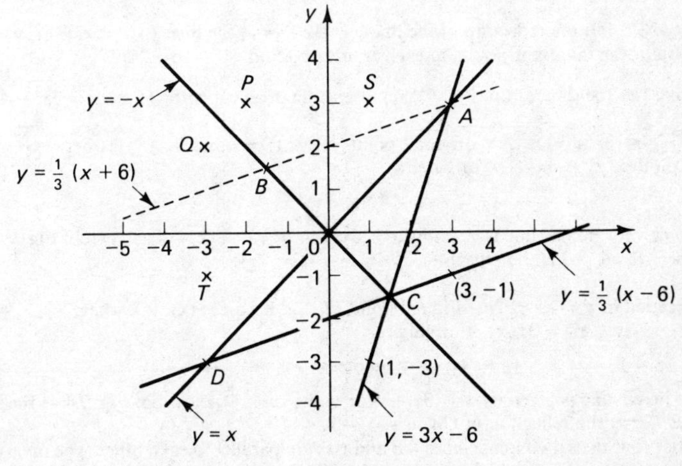

Fig. 9.3

Reflection in $y = x$

Now suppose we consider what happens to all the points on the line $y = 3x - 6$, which is the line *AC* in Fig. 9.3, in the one-to-one mapping $f:(x, y) \longmapsto (y, x)$ which represents a reflection in the line $y = x$. An original point (x_0, y_0) becomes (x_1, y_1), where $x_1 = y_0$ and $y_1 = x_0$. But, $y_0 = 3x_0 - 6$ originally.

$$\therefore y_0 = 3x_0 - 6 \text{ becomes } x_1 = 3y_1 - 6$$

Thus (x_1, y_1) will always lie on the line $3y = x + 6$ or $y = \frac{1}{3}(x + 6)$—the dotted line *AB* in Fig. 9.3.

Reflection in $y = -x$

In Fig. 9.3 the point Q is the reflection of the point $P(-2, 3)$ in the line $y = -x$. By inspection we see that the point Q is $(-3, 2)$. (The coordinates have been interchanged after changing their signs.) Likewise the points S and T are reflections of each other in the line $y = -x$, so that with $T(-3, -1)$ we see the point S as $(1, 3)$. We can therefore write the reflection of points in the line $y = -x$ as the mapping given by

$$f:(x, y) \longmapsto (-y, -x)$$

Now suppose we consider the reflection of the whole line *AC*, $y = 3x - 6$ (i.e. all the points on the line) in the line $y = -x$.

An original point (x_0, y_0) becomes (x_1, y_1), where $x_1 = -y_0$ (or $-x_1 = y_0$) and $y_1 = -x_0$ (or $-y_1 = x_0$). But, $y_0 = 3x_0 - 6$ originally.

$$\therefore y_0 = 3x_0 - 6 \text{ becomes } -x_1 = -3y_1 - 6$$

Thus (x_1, y_1) will always lie on the line $x = 3y + 6$, or $y = \frac{1}{3}(x - 6)$ as shown in Fig. 9.3.

$$\therefore AB \text{ is the reflection of } AC \text{ in } y = x,$$

and *CD* is the reflection of *CA* in $y = -x$, and we notice that *AB* is parallel to *CD*.

Example 1. Find the reflection of the line $y = 3x - 4$ in the lines (i) $y = x$, (ii) $y = -x$, and obtain the invariant points in each transformation.

Solution. Let the diagram of Fig. 9.4 represent the problem with *AC* as $y = 3x - 4$.

Reflection in $y = x$. An original point (x_0, y_0) becomes (x_1, y_1) where $x_1 = y_0$, $y_1 = x_0$. But, $y_0 = 3x_0 - 4$ originally.

$$\therefore y_0 = 3x_0 - 4 \text{ becomes } x_1 = 3y_1 - 4$$

i.e. the new (or image) line is given by $x = 3y - 4$ or $y = \frac{1}{3}(x + 4)$, shown in Fig. 9.4 as the line *AB*, i.e. *AB* is the reflection of *AC* in $y = x$.

Reflection in $y = -x$. The original point (x_0, y_0) becomes (x_1, y_1), where $x_1 = -y_0$, $y_1 = -x_0$. But, $y_0 = 3x_0 - 4$ originally.

$$\therefore y_0 = 3x_0 - 4 \text{ becomes } -x_1 = -3y_1 - 4$$

i.e. the image line is given by $x = 3y + 4$ or $y = \frac{1}{3}(x - 4)$, shown in Fig. 9.4 as the line *CD*, i.e. *CD* is the reflection of *CA* in $y = -x$.

Notice that these two image lines *AB* and *CD* are parallel to each other. The invariant points are those points which are the same after reflection as they were before. These

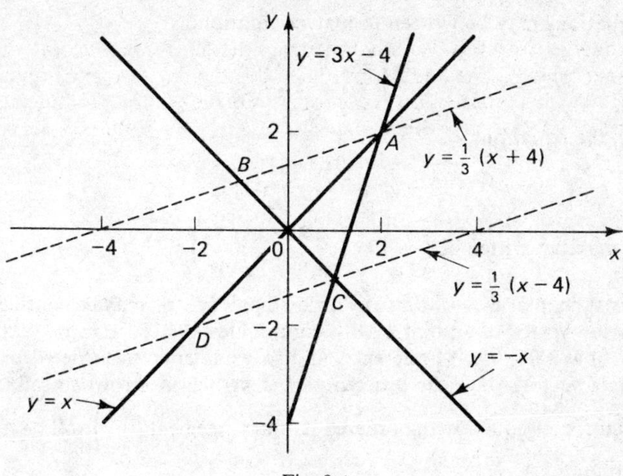

Fig. 9.4

points are evidently where the line AC intersects $y = x$ or $y = -x$, i.e. the point $A(2, 2)$ for reflection in $y = x$ and the point $C(1, -1)$ for the reflection in $y = -x$. *Answer*

Exercise 9.2

1. Find the reflection of each of the following lines in the lines (a) $y = x$ and (b) $y = -x$, and state the invariant points in each case: (i) $y = 5x + 3$, (ii) $y = -3x + 5$.
2. The line $y = 3x - 6$ is reflected in the line $y = x$ followed by reflection in the line $y = -x$. Find the final image line.
3. The line $y = 3x - 6$ is reflected in the line $y = -x$ followed by reflection in the line $y = x$. Find the final image line and give the coordinates of any points which remain invariant in the double reflection.

Matrices for Reflections

The most general linear relation possible between the points (x_0, y_0) and (x_1, y_1) is given by

$$x_1 = ax_0 + by_0 \qquad \text{(i)}$$
$$y_1 = cx_0 + dy_0 \qquad \text{(ii)}$$

where a, b, c and d are constant numbers for any one particular relation.

These equations may be written in what is called a matrix notation as follows:

$$\begin{pmatrix} x_1 \\ y_1 \end{pmatrix} = \begin{pmatrix} a & b \\ c & d \end{pmatrix} \begin{pmatrix} x_0 \\ y_0 \end{pmatrix}$$

and this statement defines the meaning of the matrix $\begin{pmatrix} a & b \\ c & d \end{pmatrix}$ as a set of numbers each having its own significance in the given pair of equations (i) and (ii).

At present we may think of this matrix, or array, of numbers as being attached to these equations only, but later on we shall see how to manipulate these matrices as elements in a matrix algebra. Just for the moment we consider the matrix as a shorthand form for writing pairs of equations. Again, consider

$$x_1 = 4x_0 + 5y_0$$
$$y_1 = 7x_0 - 3y_0$$

These equations may be written in matrix notation as

$$\begin{pmatrix} x_1 \\ y_1 \end{pmatrix} = \begin{pmatrix} 4 & 5 \\ 7 & -3 \end{pmatrix} \begin{pmatrix} x_0 \\ y_0 \end{pmatrix}$$

Similarly, the equations

$$x_1 = 0x_0 - 1y_0$$
$$y_1 = -x_0 + 0y_0$$

may be written as $\begin{pmatrix} x_1 \\ y_1 \end{pmatrix} = \begin{pmatrix} 0 & -1 \\ -1 & 0 \end{pmatrix} \begin{pmatrix} x_0 \\ y_0 \end{pmatrix}$

In the last chapter we saw that any ordered pair (x_0, y_0) may be considered to be a position vector of a point P with coordinates $x = x_0$, $y = y_0$, taking the origin (0, 0) as the point of reference, and so we refer to an ordered pair like (x_0, y_0) as a vector. Since the ordered pair is written in a row we call (x_0, y_0) a row vector; consequently the ordered pair arrangement $\begin{pmatrix} x_0 \\ y_0 \end{pmatrix}$ will be referred to as a column vector.

The array $\begin{pmatrix} a & b \\ c & d \end{pmatrix}$ is called a 2×2 ('2 by 2') matrix since it has two rows and two columns, a suitable gap (sometimes a comma) being inserted to keep the respective elements separated from each other. An array like $\begin{pmatrix} a & b & p \\ c & d & q \end{pmatrix}$ would be called a 2×3 matrix because it has two rows and three columns. With this dimensional description in mind it is clear that we may refer to the array $\begin{pmatrix} x_0 \\ y_0 \end{pmatrix}$ as a 2×1 matrix, i.e. two rows and one column.

This description of a matrix always refers first to the number of rows and second to the number of columns.

The four reflections $f:(x_0, y_0) \longmapsto (x_1, y_1)$ discussed in the previous section will now be expressed in matrix form.

Matrix notation

1. Reflection in x axis, i.e. mirror line $y = 0$.

   Call this $\mathbf{M}_{y=0}$ $\quad x_1 = x_0$, $\quad y_1 = -y_0$ $\qquad \begin{pmatrix} x_1 \\ y_1 \end{pmatrix} = \begin{pmatrix} 1 & 0 \\ 0 & -1 \end{pmatrix} \begin{pmatrix} x_0 \\ y_0 \end{pmatrix}$

2. Reflection in y axis, i.e. mirror line $x = 0$.

   Call this $\mathbf{M}_{x=0}$ $\quad x_1 = -x_0$, $\quad y_1 = y_0$ $\qquad \begin{pmatrix} x_1 \\ y_1 \end{pmatrix} = \begin{pmatrix} -1 & 0 \\ 0 & 1 \end{pmatrix} \begin{pmatrix} x_0 \\ y_0 \end{pmatrix}$

3. Reflection in $y = x$, i.e. mirror line $y = x$.

   Call this $\mathbf{M}_{y=x}$ $\quad x_1 = y_0$, $\quad y_1 = x_0$ $\qquad \begin{pmatrix} x_1 \\ y_1 \end{pmatrix} = \begin{pmatrix} 0 & 1 \\ 1 & 0 \end{pmatrix} \begin{pmatrix} x_0 \\ y_0 \end{pmatrix}$

4. Reflection in $y = -x$, i.e. the mirror line $y = -x$.

   Call this $\mathbf{M}_{y=-x}$ $\quad x_1 = -y_0$, $\quad y_1 = -x_0$ $\qquad \begin{pmatrix} x_1 \\ y_1 \end{pmatrix} = \begin{pmatrix} 0 & -1 \\ -1 & 0 \end{pmatrix} \begin{pmatrix} x_0 \\ y_0 \end{pmatrix}$

Dot Products

The dot product for two vectors was defined in the previous chapter. Here we speak of the dot product of a row vector by a column vector assuming the product exists. Thus the dot product of row vector (9, 6) by column vector $\begin{pmatrix}3\\11\end{pmatrix}$ will still follow the dot rule because they both have two and only two components and we are able to identify which are the corresponding components. The product is written without the dot as follows:

$$(9\ \ 6)\begin{pmatrix}3\\11\end{pmatrix} = (27 + 66) = 93$$

Do note the significance of saying 'multiply by' when referring to (9 6) multiplied by $\begin{pmatrix}3\\11\end{pmatrix}$. Thus $(9\ \ 6)\begin{pmatrix}3\\11\end{pmatrix}$ represents (9 6) multiplied by $\begin{pmatrix}3\\11\end{pmatrix}$, a procedure which may be referred to as 'multiplication of (9 6) on the right'. We get an entirely different result from $\begin{pmatrix}3\\11\end{pmatrix}(9\ \ 6)$ which represents $\begin{pmatrix}3\\11\end{pmatrix}$ multiplied by (9 6) and may be referred to as 'multiplication of (9 6) on the left'.

Similarly $$(3\ \ 11)\begin{pmatrix}9\\6\end{pmatrix} = (27 + 66) = 93$$

and $$(-4\ \ 2)\begin{pmatrix}9\\6\end{pmatrix} = (-36 + 12) = -24.$$

Putting these last two results together, the following results will be intuitively clear. Thus the multiplication of a 2 × 2 matrix by a 2 × 1 matrix yields the following type of results:

$$\begin{pmatrix}3 & 11\\-4 & 2\end{pmatrix}\begin{pmatrix}9\\6\end{pmatrix} = \begin{pmatrix}93\\-24\end{pmatrix} \text{ and } \begin{pmatrix}-4 & 2\\3 & 11\end{pmatrix}\begin{pmatrix}9\\6\end{pmatrix} = \begin{pmatrix}-24\\93\end{pmatrix}$$

noting the difference between these results and

$$\begin{pmatrix}3 & 11\\-4 & 2\end{pmatrix}\begin{pmatrix}6\\9\end{pmatrix} = \begin{pmatrix}117\\-6\end{pmatrix} \text{ and } \begin{pmatrix}-4 & 2\\3 & 11\end{pmatrix}\begin{pmatrix}6\\9\end{pmatrix} = \begin{pmatrix}-6\\117\end{pmatrix}$$

Finally, in general we have $(a\ \ b)\begin{pmatrix}x\\y\end{pmatrix} = (ax + by)$, and so it follows that we may use the dot product to suggest that the reassembly of the left-hand side of (iii)

$$\begin{pmatrix}a & b\\c & d\end{pmatrix}\begin{pmatrix}x\\y\end{pmatrix} = \begin{pmatrix}ax + by\\cx + dy\end{pmatrix} \qquad \text{(iii)}$$

to give the right-hand side of (iii) is a form of multiplication of the 2 × 2 matrix $\begin{pmatrix}a & b\\c & d\end{pmatrix}$ by the 2 × 1 matrix $\begin{pmatrix}x\\y\end{pmatrix}$.

Exercise 9.3

1. Write the following sets of equations in matrix form:

(i) $x_1 = 2x_0 + 3y_0$, $y_1 = 11x_0 + y_0$ (ii) $x_1 = -x_0 + y_0$, $y_1 = x_0$ (iii) $x_1 = 7x_0 - y_0$, $y_1 = \quad y_0$ (iv) $x_1 = -3x_0$, $y_1 = 5y_0$

2. Obtain the original equations from the following matrix forms:

(i) $\begin{pmatrix} x_1 \\ y_1 \end{pmatrix} = \begin{pmatrix} -1 & 1 \\ 1 & 0 \end{pmatrix}\begin{pmatrix} x_0 \\ y_0 \end{pmatrix}$ (ii) $\begin{pmatrix} x_1 \\ y_1 \end{pmatrix} = \begin{pmatrix} 7 & -1 \\ 0 & 1 \end{pmatrix}\begin{pmatrix} x_0 \\ y_0 \end{pmatrix}$

(iii) $\begin{pmatrix} x_1 \\ y_1 \end{pmatrix} = \begin{pmatrix} -3 & 0 \\ 0 & 5 \end{pmatrix}\begin{pmatrix} x_0 \\ y_0 \end{pmatrix}$ (iv) $\begin{pmatrix} x_1 \\ y_1 \end{pmatrix} = \begin{pmatrix} 2 & 3 \\ 11 & 1 \end{pmatrix}\begin{pmatrix} x_0 \\ y_0 \end{pmatrix}$

3. Find the following products:

(i) $\begin{pmatrix} 2 & 4 \\ 6 & 8 \end{pmatrix}\begin{pmatrix} 1 \\ 1 \end{pmatrix}$ (ii) $\begin{pmatrix} 2 & 4 \\ 6 & 8 \end{pmatrix}\begin{pmatrix} 1 \\ -1 \end{pmatrix}$ (iii) $\begin{pmatrix} 2 & 4 \\ 6 & 8 \end{pmatrix}\begin{pmatrix} -1 \\ 1 \end{pmatrix}$

(iv) $\begin{pmatrix} 2 & 4 \\ 6 & 8 \end{pmatrix}\begin{pmatrix} -1 \\ -1 \end{pmatrix}$ (v) $\begin{pmatrix} -2 & 4 \\ 3 & -5 \end{pmatrix}\begin{pmatrix} 2 \\ 3 \end{pmatrix}$ (vi) $\begin{pmatrix} 0 & 1 \\ 1 & 0 \end{pmatrix}\begin{pmatrix} 13 \\ 17 \end{pmatrix}$

(vii) $\begin{pmatrix} 1 & 0 \\ 0 & 1 \end{pmatrix}\begin{pmatrix} 13 \\ 17 \end{pmatrix}$ (viii) $\begin{pmatrix} -2 & 2 \\ 2 & -2 \end{pmatrix}\begin{pmatrix} 13 \\ 17 \end{pmatrix}$ (ix) $\begin{pmatrix} -1 & 1 \\ 1 & -1 \end{pmatrix}\begin{pmatrix} 13 \\ 17 \end{pmatrix}$

Rotations

A rotation of the point $P(x_0, y_0)$ about the origin $O(0, 0)$ means keeping the line OP a constant length and turning the line OP, about O as centre, through a given angle which will be called positive for anticlockwise rotation and negative for clockwise rotation. A rotation of 90° about O will map the point $P(x_0, y_0)$ onto the point $Q(x_1, y_1)$, seen in Fig. 9.5 with $OQ = OP$, and it will

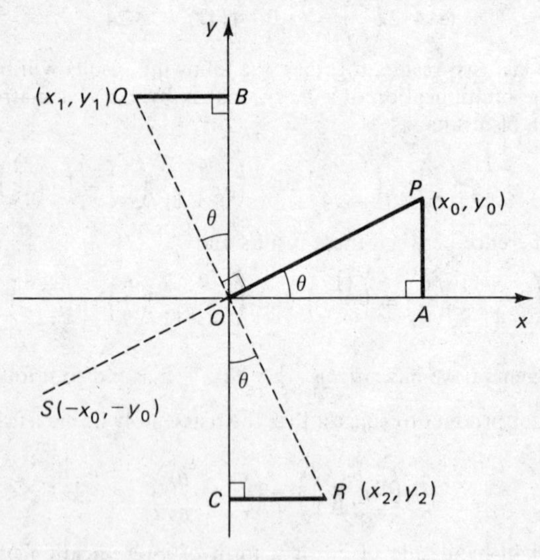

Fig. 9.5

also map $A(x_0, 0)$ onto $B(0, x_0)$—that is, we may consider the whole of the triangle OPA to be rotated through 90° from position OPA to position OQB. A comparison of these two triangles shows that

$$x_1 = -y_0 \quad \text{because} \quad BQ = AP$$

and

$$y_1 = x_0 \quad \text{because} \quad OB = OA$$

This rotation, which we shall represent by R(0, 90°) (i.e. rotation about *O* through 90° anticlockwise), is a mapping

$$f:(x_0, y_0) \longmapsto (x_1, y_1)$$

which may be written

$$f:(x, y) \longmapsto (-y, x)$$

with a matrix representation of

$$\begin{pmatrix} x_1 \\ y_1 \end{pmatrix} = \begin{pmatrix} 0 & -1 \\ 1 & 0 \end{pmatrix}\begin{pmatrix} x_0 \\ y_0 \end{pmatrix}$$

The rotation R(0, −90°) will map $P(x_0, y_0)$ onto the point $R(x_2, y_2)$ in Fig 9.5, and in fact this rotation takes the whole of the triangle *OPA* from position *OPA* onto position *ORC*. A comparison of these two triangles shows that this time

$$x_2 = y_0 \quad \text{because} \quad CR = AP$$
and
$$y_2 = -x_0 \quad \text{because} \quad OC = OA$$

Hence, R(0, −90°) is a mapping

$$f:(x_0, y_0) \longmapsto (x_2, y_2)$$

which may be written

$$f:(x, y) \longmapsto (y, -x)$$

with a matrix representation of

$$\begin{pmatrix} x_2 \\ y_2 \end{pmatrix} = \begin{pmatrix} 0 & 1 \\ -1 & 0 \end{pmatrix}\begin{pmatrix} x_0 \\ y_0 \end{pmatrix}$$

Naturally a double application of either of these rotations will produce R(0, 180°) and map $P(x_0, y_0)$ onto the point $S(x_3, y_3)$, with the obvious result that

$$x_3 = -x_0$$
$$y_3 = \quad -y_0$$

with a matrix representation of

$$\begin{pmatrix} x_3 \\ y_3 \end{pmatrix} = \begin{pmatrix} -1 & 0 \\ 0 & -1 \end{pmatrix}\begin{pmatrix} x_0 \\ y_0 \end{pmatrix}$$

Finally, R(0, 360°) maps *P* back onto itself. In fact, every point in the plane is mapped onto itself in this rotation so that we may as well leave everything unchanged and put

$x_1 = x_0$, $y_1 = \quad y_0$ with a matrix representation of $\begin{pmatrix} x_1 \\ y_1 \end{pmatrix} = \begin{pmatrix} 1 & 0 \\ 0 & 1 \end{pmatrix}\begin{pmatrix} x_0 \\ y_0 \end{pmatrix}$

A list summarising these last four results is as follows:

5. Rotation about (0, 0) through 90° anticlockwise. *Matrix notation*
   Call this R(0, 90°) $x_1 = \quad -y_0$, $y_1 = x_0$ $\qquad \begin{pmatrix} x_1 \\ y_1 \end{pmatrix} = \begin{pmatrix} 0 & -1 \\ 1 & 0 \end{pmatrix}\begin{pmatrix} x_0 \\ y_0 \end{pmatrix}$

6. Rotation about (0, 0) through 90° clockwise.
   Call this R(0, −90°) $x_1 = y_0$, $y_1 = -x_0$
   $$\begin{pmatrix} x_1 \\ y_1 \end{pmatrix} = \begin{pmatrix} 0 & 1 \\ -1 & 0 \end{pmatrix}\begin{pmatrix} x_0 \\ y_0 \end{pmatrix}$$
7. Rotation about (0, 0) through 180°.
   Call this R(0, 180°) $x_1 = -x_0$, $y_1 = -y_0$
   $$\begin{pmatrix} x_1 \\ y_1 \end{pmatrix} = \begin{pmatrix} -1 & 0 \\ 0 & -1 \end{pmatrix}\begin{pmatrix} x_0 \\ y_0 \end{pmatrix}$$
8. Leave everything unchanged:
   Call this I $x_1 = x_0$, $y_1 = y_0$
   $$\begin{pmatrix} x_1 \\ y_1 \end{pmatrix} = \begin{pmatrix} 1 & 0 \\ 0 & 1 \end{pmatrix}\begin{pmatrix} x_0 \\ y_0 \end{pmatrix}$$

We now have eight different mappings due to reflections or rotations, each one being a particular case of the general mapping given by

$$\begin{pmatrix} x_1 \\ y_1 \end{pmatrix} = \begin{pmatrix} a & b \\ c & d \end{pmatrix}\begin{pmatrix} x_0 \\ y_0 \end{pmatrix}$$

For example, the identity mapping (because it leaves everything unchanged) is given by $a = 1$, $b = 0$, $c = 0$, $d = 1$ and so on for all the other mappings. Hence any particular mapping is gained by choosing the appropriate numerical values for a, b, c and d.

Radial Expansion

Any line through (0, 0) is called a radial line and any point which moves along such a line is said to be moving radially. Starting with the point $P(x_0, y_0)$ in Fig. 9.5, we may double the distance OP by mapping P onto the point $(2x_0, 2y_0)$. In fact, all distances from (0, 0) would be doubled in the mapping

$$f:(x, y) \longmapsto (2x, 2y)$$

Similarly, all distances from (0, 0) would be halved by the mapping

$$f:(x, y) \longmapsto \left(\frac{x}{2}, \frac{y}{2}\right)$$

We can represent all radial mappings by

$$f:(x, y) \longmapsto (kx, ky)$$

where k is any positive number (in particular we have just taken $k = 2$ and $\frac{1}{2}$) and the corresponding matrix notation is given in 9 below.

9. Radial expansion about the point $O(0, 0)$.
   Call this E(0, k) $x_1 = kx_0$, $y_1 = ky_0$
   $$\begin{pmatrix} x_1 \\ y_1 \end{pmatrix} = \begin{pmatrix} k & 0 \\ 0 & k \end{pmatrix}\begin{pmatrix} x_0 \\ y_0 \end{pmatrix}$$

Many other descriptive linear transformations of the plane in which straight lines are mapped onto straight lines are available but they are all like the above nine particular cases of the general transformation

$$\begin{pmatrix} x_1 \\ y_1 \end{pmatrix} = \begin{pmatrix} a & b \\ c & d \end{pmatrix}\begin{pmatrix} x_0 \\ y_0 \end{pmatrix}$$

For the radial expansion E(0, k) we choose $a = k$, $b = 0$, $c = 0$, $d = k$.

Example 2. Find the result of the mapping R(0, 90°) on the point (−6, 7) followed by R(0, 90°). Show that R(0, 180°) gives the same final result.

Solution. R(0, 90°) for (−6, 7) is represented by

$$\begin{pmatrix} x \\ y \end{pmatrix} = \begin{pmatrix} 0 & -1 \\ 1 & 0 \end{pmatrix}\begin{pmatrix} -6 \\ 7 \end{pmatrix} = \begin{pmatrix} -7 \\ -6 \end{pmatrix}$$

i.e. $x = -7$, $y = -6$. (Recall that if we write $\begin{pmatrix} x \\ y \end{pmatrix} = \begin{pmatrix} 10 \\ 20 \end{pmatrix}$ then this means that $x = 10$ and $y = 20$.)

R(0, 90°) for (−7, −6) is represented by

$$\begin{pmatrix} x \\ y \end{pmatrix} = \begin{pmatrix} 0 & -1 \\ 1 & 0 \end{pmatrix}\begin{pmatrix} -7 \\ -6 \end{pmatrix} = \begin{pmatrix} 6 \\ -7 \end{pmatrix}$$

Therefore the result of R(0, 90°) followed by R(0, 90°) on the original point (−6, 7) is to map (−6, 7) onto (6, −7).

R(0, 180°) for (−6, 7) is given by

$$\begin{pmatrix} x \\ y \end{pmatrix} = \begin{pmatrix} -1 & 0 \\ 0 & -1 \end{pmatrix}\begin{pmatrix} -6 \\ 7 \end{pmatrix} = \begin{pmatrix} 6 \\ -7 \end{pmatrix}$$

and $x = 6$, $y = -7$ the same final result as before. *Answer*

If we examine the two stages of the last result we see the following connection:

1st stage $\begin{pmatrix} x_1 \\ y_1 \end{pmatrix} = \begin{pmatrix} 0 & -1 \\ 1 & 0 \end{pmatrix}\begin{pmatrix} x_0 \\ y_0 \end{pmatrix}$ 2nd stage $\begin{pmatrix} x_2 \\ y_2 \end{pmatrix} = \begin{pmatrix} 0 & -1 \\ 1 & 0 \end{pmatrix}\begin{pmatrix} x_1 \\ y_1 \end{pmatrix}$

$\therefore \begin{pmatrix} x_2 \\ y_2 \end{pmatrix} = \begin{pmatrix} 0 & -1 \\ 1 & 0 \end{pmatrix}\begin{pmatrix} 0 & -1 \\ 1 & 0 \end{pmatrix}\begin{pmatrix} x_0 \\ y_0 \end{pmatrix}$ i.e. $\begin{pmatrix} 6 \\ -7 \end{pmatrix} = \begin{pmatrix} 0 & -1 \\ 1 & 0 \end{pmatrix}\begin{pmatrix} 0 & -1 \\ 1 & 0 \end{pmatrix}\begin{pmatrix} -6 \\ 7 \end{pmatrix}$

But we achieved this result by R(0, 180°) in one single mapping

$$\begin{pmatrix} 6 \\ -7 \end{pmatrix} = \begin{pmatrix} -1 & 0 \\ 0 & -1 \end{pmatrix}\begin{pmatrix} -6 \\ 7 \end{pmatrix}$$

The comparison of these results suggests that

$$\begin{pmatrix} -1 & 0 \\ 0 & -1 \end{pmatrix} = \begin{pmatrix} 0 & -1 \\ 1 & 0 \end{pmatrix}\begin{pmatrix} 0 & -1 \\ 1 & 0 \end{pmatrix}$$

and so we are compelled to consider the two matrices on the right to be multiplied in some way so as to gain the single matrix on the left.

Example 3. Find the image of the line $y = x + 2$ in the mapping R(0, 90°).

Solution. Any original point of the line $y = x + 2$ may be represented by $(x_0, x_0 + 2)$ because $y_0 = x_0 + 2$.

The equations for R(0, 90°) are

$$x_1 = -y_0 = -(x_0 + 2) = -x_0 - 2$$

and $$y_1 = x_0$$

Therefore any point (x_0, y_0) on the line $y = x + 2$ becomes (x_1, y_1), where $y_1 + x_1 = -2$, i.e. (x_1, y_1) is always on the line $y = -x - 2$. Hence $y = x + 2$ is mapped onto $y = -x - 2$ by R(0, 90°). *Answer*

Example 4. A mapping is given by the pair of equations $x_1 = 2x_0 + 3y_0$ and $y_1 = 5x_0 - 7y_0$. Find the image of (i) point (4, 6) and (ii) line $y = 2x - 1$.

Solution. (i) The image point is given by

$$x_1 = 8 + 18 = 26$$
$$y_1 = 20 - 42 = -22 \quad \text{(i.e. } (26, -22)\text{)}$$ *Answer*

(ii) Any original point on the line $y = 2x - 1$ is given by (x_0, y_0), where $y_0 = 2x_0 - 1$. Substitution of this point into the mapping equations gives

$$x_1 = 2x_0 + 3(2x_0 - 1) = 8x_0 - 3$$
$$y_1 = 5x_0 - 7(2x_0 - 1) = -9x_0 + 7$$

To get the equation relating x_1 to y_1 only, we eliminate x_0 by using the results

$$\frac{x_1 + 3}{8} = x_0 = \frac{y_1 - 7}{-9}$$

$$\therefore \quad -9x_1 - 27 = 8y_1 - 56$$
$$\therefore \quad 8y_1 + 9x_1 = 29$$

i.e. the image (x_1, y_1) is always on the line $8y + 9x = 29$. Hence $y = 2x - 1$ is mapped onto $8y + 9x = 29$ by the given mapping. *Answer*

Exercise 9.4

1. Apply R(0, 180°) twice in succession to the point (2, 5) and show that the point is mapped onto itself.
2. Show that R(0, 90°) followed by R(0, 180°) achieves the same result as R(0, −90°) for the points: (i) (3, 1); (ii) (−2, −7); (iii) (x_0, y_0).
3. Find the image of the point (1, 2) in the mapping R(0, 90°) followed by $M_{y=x}$. Now find the image of any point (x_0, y_0) in the same mapping and show that the mapping $M_{y=0}$ would have produced the same end result.
4. Find the image of the line $y = x + 2$ in the mapping R(0, 180°).
5. Using the mapping of the worked example 4 above find the image of (i) the point (−3, 7) and (ii) the line $y = x - 2$.

Matrix Multiplication

The combination of two mappings into one, as we saw on page 181, has already suggested the possibility of detaching the matrices from the context of the equations which produced them and considering their multiplication independently.

A row vector may be considered as a 1×2 matrix and a column vector may be considered as a 2×1 matrix and we have already defined a multiplication of two such matrices—namely, the dot product on page 177. Consistency will therefore require that we use dot products in multiplying any two matrices of other sizes like 2×2 wherever such multiplication is possible.

Definition 9.1. The scalar or dot product of two ordered sets is defined as the sum of the products of corresponding elements, and is only defined if the sets are equivalent.

For example, suppose we have four ordered sets

$$A = \{2, 6, -3, 7\}, B = \{2, 5\}, C = \{9, -2\}, D = \{-5, 4, 10, -8\}$$

The scalar product of A and D, written $A.D$ in dot product notation, is given by

$$A.D = 2 \times (-5) + 6 \times 4 + (-3) \times 10 + 7 \times (-8)$$
$$= -10 + 24 - 30 - 56 = -72$$

Similarly

$$B.C = 2 \times 9 + 5 \times (-2) = 18 - 10 = 8$$

But clearly we cannot form the dot product $A.B$ or $A.C$ or $B.D$.

We can build up to the multiplication of 2×2 matrices by using what we already know about dot products and observing the progress of the following illustrations.

Example 5

$$(2\ \ 5)\begin{pmatrix}3\\4\end{pmatrix} = (26)$$

i.e. a (1×2) matrix multiplied by a (2×1) matrix gives a 1×1 matrix, a single element.

Example 6. Similarly to Example 5

$$(2\ \ 5)\begin{pmatrix}10\\7\end{pmatrix} = (55)$$

Example 7. Putting together the previous two results we get

$$(2\ \ 5)\begin{pmatrix}3 & 10\\4 & 7\end{pmatrix} = (26\ \ 55)$$

i.e. a (1×2) matrix multiplied by a (2×2) matrix gives a (1×2) matrix which is a (1×2) row vector.

Note that $\begin{pmatrix}3 & 10\\4 & 7\end{pmatrix}(2\ \ 5)$ is meaningless, i.e. it is not possible to multiply a (2×2) matrix by a (1×2) matrix.

Example 8. Extending the result of example 7 we have

$$(2\ \ 5)\begin{pmatrix}3 & 10 & -1\\4 & 7 & 2\end{pmatrix} = (26\ \ 55\ \ 8)$$

i.e. a (1×2) matrix multiplied by a (2×3) matrix gives a (1×3) matrix.

Thus each element in the result is found by obtaining the dot product of the row $(2\ \ 5)$ with the corresponding column in the following matrix.

Now suppose we introduce a second row into the first matrix, and repeat the dot products by taking a row at a time.

Example 9

$$\begin{pmatrix}2 & 5\\6 & 8\end{pmatrix}\begin{pmatrix}3\\4\end{pmatrix} = \begin{pmatrix}26\\50\end{pmatrix}$$

i.e. (2×2) matrix multiplied by (2×1) matrix $=$ (2×1) matrix.

Example 10. Similarly

$$\begin{pmatrix}2 & 5\\6 & 8\end{pmatrix}\begin{pmatrix}10\\7\end{pmatrix} = \begin{pmatrix}55\\116\end{pmatrix}$$

Example 11

$$\begin{pmatrix}2 & 5\\6 & 8\end{pmatrix}\begin{pmatrix}3 & 10\\4 & 7\end{pmatrix} = \begin{pmatrix}26 & 55\\50 & 116\end{pmatrix}$$

i.e. (2×2) matrix multiplied by (2×2) matrix $=$ (2×2) matrix.

Example 12

$$\begin{pmatrix}2 & 5\\6 & 8\end{pmatrix}\begin{pmatrix}3 & 10 & -1\\4 & 7 & 2\end{pmatrix} = \begin{pmatrix}26 & 55 & 8\\50 & 116 & 10\end{pmatrix}$$

i.e. (2×2) matrix multiplied by (2×3) matrix $=$ (2×3) matrix.

Comparing this example with Example 8 we see that the first row in the final result has the same elements as before. Thus the second row (6, 8) of the first matrix contributes to the elements of the second row of the final result.

Example 13. Let $A = \begin{pmatrix} 2 & 5 \\ 6 & 8 \end{pmatrix}$, $B = \begin{pmatrix} 1 & 2 \\ 3 & 4 \end{pmatrix}$. Find the result $AB = C$.

Solution

$$AB = \begin{pmatrix} 2 & 5 \\ 6 & 8 \end{pmatrix}\begin{pmatrix} 1 & 2 \\ 3 & 4 \end{pmatrix} = \begin{pmatrix} 17 = (2 \times 1) + (5 \times 3) & 24 = (2 \times 2) + (5 \times 4) \\ 30 = (6 \times 1) + (8 \times 3) & 44 = (6 \times 2) + (8 \times 4) \end{pmatrix} = \begin{pmatrix} 17 & 24 \\ 30 & 44 \end{pmatrix}.$$

We note that the elements of C are identified as follows:

17 is the dot product of the first row of A with the first column of B. 24 is the dot product of the first row of A with the second column of B, and so on (see Question 1, Exercise 9.5).

When we look at Example 12 we should realise that in general if A and B are two matrices we shall only be able to multiply A by B to get the product AB if:

A has the same number of elements in a row as B has elements in a column or, to put it another way, if A has the same number of columns as B has rows.

As an aid to seeing the order of the final matrix result it is always useful to write the order of a matrix underneath its representative letter thus,

$$\underset{2 \times 2}{A} \; \underset{2 \times 1}{B} = \underset{2 \times 1}{C}; \qquad \underset{2 \times 2}{A} \; \underset{2 \times 3}{B} = \underset{2 \times 3}{C}$$

$\underset{2 \times 3}{A} \; \underset{2 \times 2}{B}$ does not exist, as indicated by the mismatch of the inner numbers of the order statement. Again $\underset{4 \times 9}{A} \, B$ is only possible if B has 9 rows.

We have tried to capture the idea of matrix multiplication intuitively without a formal definition but we do need a definition of equal matrices because we shall be asking you to show that $AB \neq BA$, i.e. matrix multiplication is not commutative.

Definition 9.2. Two matrices A and B are equal if they are of the same order (i.e. each have the same number of rows and each have the same number of columns) and if corresponding elements are equal to each other.

Recall that our statement for two equal vectors (page 151) meant that if $(a, b) = (p, q)$ then $a = p$ and $b = q$.

Thus $A = \begin{pmatrix} 1 & 2 \\ 3 & 4 \end{pmatrix}$ is equal to $B = \begin{pmatrix} a & b \\ c & d \end{pmatrix}$ only if $a = 1, b = 2, c = 3, d = 4$.

Definition 9.3. The unit matrix is a square matrix which has 1 in each diagonal position and 0 everywhere else.

The unit matrix is written

$$I = \begin{pmatrix} 1 & 0 \\ 0 & 1 \end{pmatrix}, \quad I = \begin{pmatrix} 1 & 0 & 0 \\ 0 & 1 & 0 \\ 0 & 0 & 1 \end{pmatrix}, \quad I = \begin{pmatrix} 1 & 0 & 0 & 0 \\ 0 & 1 & 0 & 0 \\ 0 & 0 & 1 & 0 \\ 0 & 0 & 0 & 1 \end{pmatrix}$$

the context of the problem determining the order of I required.

Definition 9.4. The zero matrix has every element equal to 0. We shall write the zero matrix as 0.

$$\therefore \quad 0 = \begin{pmatrix} 0 & 0 \\ 0 & 0 \end{pmatrix}, \quad 0 = \begin{pmatrix} 0 & 0 & 0 \\ 0 & 0 & 0 \\ 0 & 0 & 0 \end{pmatrix}$$

Definition 9.5. When n is a scalar number the matrix nA is obtained by multiplying every element of A by n.

$$\text{If } A = \begin{pmatrix} 2 & 4 \\ 5 & 9 \end{pmatrix} \text{ then } 3A = \begin{pmatrix} 6 & 12 \\ 15 & 27 \end{pmatrix}$$

$$\text{If } B = \begin{pmatrix} 0 & 2 \\ 7 & -3 \end{pmatrix} \text{ then } -5B = \begin{pmatrix} 0 & -10 \\ -35 & +15 \end{pmatrix}$$

Exercise 9.5

Find the following matrix products:

1. (i) $(2 \;\; -7)\begin{pmatrix} 3 \\ 1 \end{pmatrix}$ (iii) $(2 \;\; -7)\begin{pmatrix} 3 & 11 \\ 1 & 2 \end{pmatrix}$

   (ii) $(2 \;\; -7)\begin{pmatrix} 11 \\ 2 \end{pmatrix}$ (iv) $(2 \;\; -7)\begin{pmatrix} 11 & 3 \\ 2 & 1 \end{pmatrix}$

2. (i) $(2 \;\; 3)\begin{pmatrix} 4 \\ 5 \end{pmatrix}$ (iii) $\begin{pmatrix} 2 & 3 \\ -2 & 5 \end{pmatrix}\begin{pmatrix} 4 \\ 5 \end{pmatrix}$

   (ii) $(-2 \;\; 5)\begin{pmatrix} 4 \\ 5 \end{pmatrix}$ (iv) $\begin{pmatrix} -2 & 5 \\ 2 & 3 \end{pmatrix}\begin{pmatrix} 4 \\ 5 \end{pmatrix}$

3. (i) $\begin{pmatrix} 5 & 7 \\ 3 & 2 \end{pmatrix}\begin{pmatrix} 1 \\ 2 \end{pmatrix}$ (iii) $\begin{pmatrix} 5 & 7 \\ 3 & 2 \end{pmatrix}\begin{pmatrix} 1 & 0 \\ 2 & 3 \end{pmatrix}$

   (ii) $\begin{pmatrix} 5 & 7 \\ 3 & 2 \end{pmatrix}\begin{pmatrix} 0 \\ 3 \end{pmatrix}$ (iv) $\begin{pmatrix} 5 & 7 \\ 3 & 2 \end{pmatrix}\begin{pmatrix} 0 & 1 \\ 3 & 2 \end{pmatrix}$

4. (i) $(2 \;\; 3)\begin{pmatrix} -2 \\ 5 \end{pmatrix}$ (iii) $(2 \;\; 3)\begin{pmatrix} 5 \\ -2 \end{pmatrix}$

   (ii) $(3 \;\; 2)\begin{pmatrix} 5 \\ -2 \end{pmatrix}$ (iv) $(3 \;\; 2)\begin{pmatrix} -2 \\ 5 \end{pmatrix}$

5. (i) $\begin{pmatrix} 1 \\ 2 \end{pmatrix}(4 \;\; 5)$ (ii) $\begin{pmatrix} 3 \\ 7 \end{pmatrix}(4 \;\; 5)$ (iii) $\begin{pmatrix} 1 & 3 \\ 2 & 7 \end{pmatrix}(4 \;\; 5)$

6. (i) $\begin{pmatrix} 2 & 7 \\ 3 & 9 \end{pmatrix}\begin{pmatrix} 9 & -7 \\ -3 & 2 \end{pmatrix}$ (ii) $\begin{pmatrix} 9 & -7 \\ -3 & 2 \end{pmatrix}\begin{pmatrix} 2 & 7 \\ 3 & 9 \end{pmatrix}$

7. With $A = \begin{pmatrix} 2 & 3 \\ 4 & 5 \end{pmatrix}$, $B = \begin{pmatrix} 1 & 2 \\ 4 & 7 \end{pmatrix}$, $C = \begin{pmatrix} -1 & 2 \\ -2 & 3 \end{pmatrix}$

   Show that (i) $AB \neq BA$
   (ii) $AC \neq CA$
   (iii) $BC \neq CB$
   (iv) Find (a) $A(BC)$, (b) $(AB)C$, (c) ABC.

8. With $D = \begin{pmatrix} 1 & 2 & 3 \\ -1 & 4 & 7 \end{pmatrix}$ and using the matrices of question 7 find: (i) AD, (ii) BD, (iii) CD, (iv) DA.

9. Using the matrices of question 7 find the following results: (i) $2A$, (ii) $3B$, (iii) $2A + A$, (iv) $B + A$.
10. Solve the matrix equation $X + A = 3A$.
11. Using the matrices of the mappings on pages 176 and 179 obtain the following products and identify the one equivalent mapping: (i) $M_{y=0}.M_{x=0}$; (ii) $M_{x=0}.M_{y=0}$; (iii) $M_{x=0}.R(0, 90°)$; (iv) $R(0, 90°).R(0, -90°)$.

The (2 × 2) Inverse Matrix

For 2 × 2 matrices the identity element with respect to matrix multiplication is the identity matrix

$$I = \begin{pmatrix} 1 & 0 \\ 0 & 1 \end{pmatrix}$$

so that for any 2 × 2 matrix A we have $AI = A = IA$.

$$\therefore \quad \begin{pmatrix} 2 & 9 \\ 7 & 3 \end{pmatrix}\begin{pmatrix} 1 & 0 \\ 0 & 1 \end{pmatrix} = \begin{pmatrix} 2 & 9 \\ 7 & 3 \end{pmatrix} = \begin{pmatrix} 1 & 0 \\ 0 & 1 \end{pmatrix}\begin{pmatrix} 2 & 9 \\ 7 & 3 \end{pmatrix}$$

Also
$$\begin{pmatrix} -4 & 6 \\ -7 & 5 \end{pmatrix}\begin{pmatrix} 1 & 0 \\ 0 & 1 \end{pmatrix} = \begin{pmatrix} -4 & 6 \\ -7 & 5 \end{pmatrix} = \begin{pmatrix} 1 & 0 \\ 0 & 1 \end{pmatrix}\begin{pmatrix} -4 & 6 \\ -7 & 5 \end{pmatrix}$$

Having obtained the identity matrix with respect to matrix multiplication we can now define the inverse matrix.

Definition 9.6. The inverse of the matrix A is a matrix called A^{-1} such that

$$AA^{-1} = I = A^{-1}A$$

A^{-1} is called the 'inverse of A' but we shall see that not all matrices have inverses. Any matrix which does not have an inverse is said to be singular.

It is easy to see that

$$\begin{pmatrix} 8 & 3 \\ 5 & 2 \end{pmatrix}\begin{pmatrix} 2 & -3 \\ -5 & 8 \end{pmatrix} = \begin{pmatrix} 1 & 0 \\ 0 & 1 \end{pmatrix} = \begin{pmatrix} 2 & -3 \\ -5 & 8 \end{pmatrix}\begin{pmatrix} 8 & 3 \\ 5 & 2 \end{pmatrix}$$

Hence, $\begin{pmatrix} 2 & -3 \\ -5 & 8 \end{pmatrix}$ is the multiplicative inverse of $\begin{pmatrix} 8 & 3 \\ 5 & 2 \end{pmatrix}$ and vice versa, so at first sight the method for creating an inverse for 2 × 2 matrices looks obvious, i.e. interchanging elements in the leading diagonal and changing signs in the other diagonal.

Example 14. Find the inverse of the matrix $A = \begin{pmatrix} 9 & 2 \\ 10 & 4 \end{pmatrix}$, with respect to multiplication.

Solution. Following the above illustration we try $\begin{pmatrix} 4 & -2 \\ -10 & 9 \end{pmatrix}$ and find that

$$\begin{pmatrix} 9 & 2 \\ 10 & 4 \end{pmatrix}\begin{pmatrix} 4 & -2 \\ -10 & 9 \end{pmatrix} = \begin{pmatrix} 16 & 0 \\ 0 & 16 \end{pmatrix} = \begin{pmatrix} 4 & -2 \\ -10 & 9 \end{pmatrix}\begin{pmatrix} 9 & 2 \\ 10 & 4 \end{pmatrix}$$

We are obviously moving in the correct direction because our result is $16I$ instead of I, so we suggest

$$A^{-1} = \frac{1}{16}\begin{pmatrix} 4 & -2 \\ -10 & 9 \end{pmatrix}$$

and find that $AA^{-1} = A^{-1}A = I$ as desired. *Answer*

Example 15. Find the inverse of the matrix $A = \begin{pmatrix} 4 & 3 \\ 5 & 6 \end{pmatrix}$ with respect to multiplication.

Solution. We try the interchange of elements in the leading diagonal and the change of sign in the other diagonal to gain

$$B = \begin{pmatrix} 6 & -3 \\ -5 & 4 \end{pmatrix}$$

Experimenting, we find

$$A\,B = \begin{pmatrix} 4 & 3 \\ 5 & 6 \end{pmatrix}\begin{pmatrix} 6 & -3 \\ -5 & 4 \end{pmatrix} = \begin{pmatrix} 9 & 0 \\ 0 & 9 \end{pmatrix} = 9I$$

We now amend the result to gain

$$A^{-1} = \begin{pmatrix} \frac{6}{9} & \frac{-3}{9} \\ \frac{-5}{9} & \frac{4}{9} \end{pmatrix} \text{ and } AA^{-1} = I = A^{-1}A$$

Notice that we may also write

$$A^{-1} = \frac{1}{9}\begin{pmatrix} 6 & -3 \\ -5 & 4 \end{pmatrix}$$

Example 16. Find the inverse of the general 2×2 matrix $A = \begin{pmatrix} a & b \\ c & d \end{pmatrix}$ with respect to multiplication.

Solution. Suggest $B = \begin{pmatrix} d & -b \\ -c & a \end{pmatrix}$ and find

$$AB = \begin{pmatrix} a & b \\ c & d \end{pmatrix}\begin{pmatrix} d & -b \\ -c & a \end{pmatrix} = \begin{pmatrix} ad - bc & 0 \\ 0 & ad - bc \end{pmatrix}$$

$$\therefore \quad AB = (ad - bc)\begin{pmatrix} 1 & 0 \\ 0 & 1 \end{pmatrix}$$

and it follows that

$$A^{-1} = \frac{1}{(ad - bc)}.B$$

$$\text{i.e.} \quad A^{-1} = \frac{1}{(ad - bc)}\begin{pmatrix} d & -b \\ -c & a \end{pmatrix} = \begin{pmatrix} \frac{d}{ad - bc} & \frac{-b}{ad - bc} \\ \frac{-c}{ad - bc} & \frac{a}{ad - bc} \end{pmatrix}$$

always provided that $ad - bc \neq 0$. *Answer*

Definition 9.7. The numerical value of $ad - bc$ is called the determinant of the matrix $A = \begin{pmatrix} a & b \\ c & d \end{pmatrix}$. This is sometimes written det A or $\begin{vmatrix} a & b \\ c & d \end{vmatrix}$.

$$\text{Thus det } A = \begin{vmatrix} a & b \\ c & d \end{vmatrix} \text{ or det } A = ad - bc.$$

If det $A = 0$ then A^{-1} does not exist and we say that the matrix A is singular.

Example 17. Given that matrix $A = \begin{pmatrix} 8 & -2 \\ 5 & 3 \end{pmatrix}$ find (i) det A and (ii) A^{-1}.

Solution. (i) det $A = \begin{vmatrix} 8 & -2 \\ 5 & 3 \end{vmatrix} = (8 \times 3) - (-2 \times 5) = 34.$

(ii) $$A^{-1} = \frac{1}{34}\begin{pmatrix} 3 & 2 \\ -5 & 8 \end{pmatrix} = \begin{pmatrix} \frac{3}{34} & \frac{2}{34} \\ \frac{-5}{34} & \frac{8}{34} \end{pmatrix} = \begin{pmatrix} \frac{3}{34} & \frac{1}{17} \\ \frac{-5}{34} & \frac{4}{17} \end{pmatrix}$$

Answer

Notice once again our insistence on $A^{-1}A = I = AA^{-1}$. The inverse A^{-1} must commute with A. In general, as we saw in Exercise 9.5, matrix multiplication is not commutative, i.e. $AB \neq BA$ in general.

We may also have a product $AC = 0$ without either A or C being the zero matrix (Definition 9.4). For example, $\begin{pmatrix} 4 & 2 \\ 2 & 1 \end{pmatrix}\begin{pmatrix} 1 & -2 \\ -2 & 4 \end{pmatrix} = 0$. This is quite different from ordinary multiplication in algebra where we have $xy = 0$ implying that $x = 0$ or $y = 0$.

Exercise 9.6

$$A = \begin{pmatrix} 4 & 1 \\ 7 & 3 \end{pmatrix}, B = \begin{pmatrix} 2 & 3 \\ 1 & -1 \end{pmatrix}, C = \begin{pmatrix} 4 & 8 \\ 1 & 2 \end{pmatrix}$$

1. Show that (i) $AB \neq BA$, i.e. A and B are not commutative with respect to multiplication and (ii) $(AB)C = A(BC)$, i.e. A, B, C are associative with respect to multiplication.
2. Find A^{-1}, B^{-1} and explain why C^{-1} does not exist.
3. Find the product $A^{-1}B$ and show that $BA^{-1} \neq A^{-1}B$.
4. The 2×2 matrix X satisfies the equation $AX = B$. Put $X = A^{-1}B$ as found in Question 3 and show that the equation $AX = B$ is satisfied.
5. A matrix D has an inverse $D^{-1} = \begin{pmatrix} 3 & 4 \\ 1 & 2 \end{pmatrix}$. Find D.
6. Solve the equations (i) $AX = C$ and (ii) $XA = C$.
7. Find the following products:

(i) $\begin{pmatrix} 1 & -2 \\ -2 & 4 \end{pmatrix}\begin{pmatrix} 4 & 2 \\ 2 & 1 \end{pmatrix}$ (iii) $\begin{pmatrix} 1 & -2 \\ -2 & 4 \end{pmatrix}\begin{pmatrix} 2 & 1 \\ -4 & -2 \end{pmatrix}$

(ii) $\begin{pmatrix} 2 & 1 \\ -4 & -2 \end{pmatrix}\begin{pmatrix} 1 & -2 \\ -2 & 4 \end{pmatrix}$ (iv) $\begin{pmatrix} 1 & -1 \\ -1 & 1 \end{pmatrix}\begin{pmatrix} 1 & 1 \\ 1 & 1 \end{pmatrix}$

Simultaneous Equations

We shall accept without proof that matrices obey the associative law for multiplication—that is, if A, B and C are three matrices then $(AB)C = A(BC)$, assuming that all the products exist, and we may write $(AB)C = A(BC) = ABC$ without ambiguity. Suppose now that we have a matrix equation $AX = B$ from which we wish to find X, knowing that A^{-1} exists. We may therefore form the following products.

$A^{-1}(AX) = A^{-1}(B)$ (i.e. multiplying on the left, see page 177).

$\therefore \quad (A^{-1}A)X = IX = A^{-1}B$ since $AA^{-1} = I$

But $IX = X$, so we have $X = A^{-1}B$ (as suggested in Question 4, Exercise 9.6). Compare this procedure with the solution of an ordinary algebraic equation $ax = b$.

The inverse of a with respect to multiplication is $a^{-1} = \frac{1}{a}$

$$\therefore \quad \frac{1}{a}.(ax) = \frac{1}{a}.b$$

$$\therefore \quad \left(\frac{1a}{a}\right)x = 1x = \frac{1}{a}.b$$

$$\text{i.e. } x = a^{-1}\,b$$

In particular suppose $4x = 7$

$$\therefore \quad (4^{-1})4x = (4^{-1})7, \left(4^{-1} = \frac{1}{4}\right)$$

$$\therefore \quad x = (4^{-1})7$$

This use of the inverse matrix provides an attractive matrix method for solving simultaneous equations.

Example 18. Solve the simultaneous equations $7x + 4y = 3$
$3x + 2y = 1$

Solution. Writing these equations in matrix form we have

$$\begin{pmatrix} 7 & 4 \\ 3 & 2 \end{pmatrix}\begin{pmatrix} x \\ y \end{pmatrix} = \begin{pmatrix} 3 \\ 1 \end{pmatrix}$$

We now find the inverse of $\begin{pmatrix} 7 & 4 \\ 3 & 2 \end{pmatrix}$ which is $\frac{1}{2}\begin{pmatrix} 2 & -4 \\ -3 & 7 \end{pmatrix}$ since det $\begin{pmatrix} 7 & 4 \\ 3 & 2 \end{pmatrix} = 2$.

We multiply the given equation on the left by this inverse matrix.

$$\therefore \quad \frac{1}{2}\begin{pmatrix} 2 & -4 \\ -3 & 7 \end{pmatrix}\begin{pmatrix} 7 & 4 \\ 3 & 2 \end{pmatrix}\begin{pmatrix} x \\ y \end{pmatrix} = \frac{1}{2}\begin{pmatrix} 2 & -4 \\ -3 & 7 \end{pmatrix}\begin{pmatrix} 3 \\ 1 \end{pmatrix}$$

$$\therefore \quad \begin{pmatrix} 1 & 0 \\ 0 & 1 \end{pmatrix}\begin{pmatrix} x \\ y \end{pmatrix} = \begin{pmatrix} x \\ y \end{pmatrix} = \frac{1}{2}\begin{pmatrix} 2 \\ -2 \end{pmatrix} = \begin{pmatrix} 1 \\ -1 \end{pmatrix}$$

$\therefore$ $x = 1$ and $y = -1$ and the equations are solved. *Answer*

The method can be extended to any number of equations and it is this generality that makes it mathematically attractive. For two equations in two unknowns the method is tedious but means well.

The method will not work if the inverse does not exist. There are usually one of two reasons for this.

1. The two equations are really one and the same equation, e.g. $x + y = 10$, $2x + 2y = 20$. In this case there is an infinite number of solutions. (0, 10), (1, 9), (2, 8) . . .

2. The equations may be inconsistent, e.g. $x + y = 0$, $x + y = 1$. In this case there are not any solutions.

Exercise 9.7

Solve the following sets of simultaneous equations using the matrix method in Questions 1–4:

1. $4x + y = 7$
   $5x + 2y = 11$
2. $4x + y = -5$
   $3x + 2y = 0$
3. $5x + 2y = 12$
   $3x - y = 5$
4. $2x - 5y = 11$
   $x + y = -5$
5. A transformation A is given by

$$\begin{aligned} x_1 &= 3x_0 + 2y_0 \\ y_1 &= 2x_0 + 3y_0 \end{aligned} \quad \text{i.e.} \begin{pmatrix} x_1 \\ y_1 \end{pmatrix} = \mathrm{A} \begin{pmatrix} x_0 \\ y_0 \end{pmatrix}$$

   Find the inverse transformation A^{-1}.
   Starting with the point $x = 2$, $y = 5$ find the image of this point in the transformation A. Now find the image of your answer in the transformation A^{-1}.
6. The transformation $B \begin{pmatrix} x_0 \\ y_0 \end{pmatrix} = \begin{pmatrix} x_1 \\ y_1 \end{pmatrix}$ has $B = \begin{pmatrix} 7 & 4 \\ 3 & 2 \end{pmatrix}$. Find B^{-1} and calculate the point (x_0, y_0) whose image in B was (2, 0).

A Group of Matrices

On page 101 we obtained a group based on the operation of 'followed by'. The elements of this group were rotations about different lines. In this chapter we have been able to use matrices to represent both rotations and reflections. For example we have been able to say the matrix $\begin{pmatrix} 1 & 0 \\ 0 & -1 \end{pmatrix}$ represents a reflection of the cartesian plane (i.e. all points in the cartesian plane) in the line $y = 0$ because it may be used in $\begin{pmatrix} x_1 \\ y_1 \end{pmatrix} = \begin{pmatrix} 1 & 0 \\ 0 & -1 \end{pmatrix}\begin{pmatrix} x_0 \\ y_0 \end{pmatrix}$ and these are the equations which *do* the reflections.

We next show that the set of matrices

$$A = \begin{pmatrix} 1 & 0 \\ 0 & -1 \end{pmatrix}, B = \begin{pmatrix} -1 & 0 \\ 0 & 1 \end{pmatrix}, C = \begin{pmatrix} -1 & 0 \\ 0 & -1 \end{pmatrix}, I = \begin{pmatrix} 1 & 0 \\ 0 & 1 \end{pmatrix}$$

together with the binary operation of matrix multiplication will form a group.

We shall need to carry out all possible multiplications of these four matrices so we make up the usual multiplication Table 9.1 (some entries are left open for the reader to fill in).

×	*I*	*A*	*B*	*C*
I	*I*	*A*	*B*	*C*
A	*A*	*I*		*B*
B	*B*	*C*	*I*	
C	*C*			*I*

Table 9.1

In order to check whether we have a group with × on $\{I, A, B, C\}$ turn to Table 5.9. The comparison reveals the isomorphism

$$I \to I, A \to V, B \to H, C \to E$$

or

$$I \to I, A \to H, B \to V, C \to E$$

so we do no further work but declare that these four matrices form a commutative group under the operation of matrix multiplication. We notice from the table that each matrix is its own inverse because I appears everywhere in the leading diagonal, e.g. $AA = I \Rightarrow A^{-1} = A$. We can check this by finding each inverse independently, e.g.

$$\det A = -1 \text{ and } A^{-1} = \frac{1}{-1}\begin{pmatrix} -1 & 0 \\ 0 & +1 \end{pmatrix} = A$$

The form of the matrices suggests a possibility of finding another group from the rotation and reflections listed elsewhere in this chapter.

Unfortunately the set of all 2×2 matrices does not form a group with respect to matrix multiplication because not all matrices have multiplicative inverses, a fact which we have noted already in Exercise 9.6.

Exercise 9.8

1. Given that

$$I = \begin{pmatrix} 1 & 0 \\ 0 & 1 \end{pmatrix}, P = \begin{pmatrix} 0 & 1 \\ -1 & 0 \end{pmatrix}, Q = \begin{pmatrix} -1 & 0 \\ 0 & -1 \end{pmatrix}, R = \begin{pmatrix} 0 & -1 \\ 1 & 0 \end{pmatrix}$$

(i) Compile a multiplication table for $S = \{I, P, Q, R\}$.
(ii) Is the multiplication commutative?
(iii) Which is the identity element?
(iv) State the inverse of each element.
(v) Does $\{I, P, Q, R\}$ form a group under matrix multiplication?
(vi) Is there an isomorphism between $(S, \times)$ and $(T, +)$ where $T = \{0, 1, 2, 3\}$ and $+$ is arithmetic modulo 4. Check with Table 5.14 and establish the correspondence between S and T.
(vii) Name a proper subgroup in $\{I, P, Q, R\}$ under matrix multiplication.

2. Show that the set of matrices

$$I = \begin{pmatrix} 1 & 0 \\ 0 & 1 \end{pmatrix}, A = \begin{pmatrix} 1 & 0 \\ 0 & -1 \end{pmatrix}, R = \begin{pmatrix} 0 & -1 \\ 1 & 0 \end{pmatrix}, K = \begin{pmatrix} 0 & -1 \\ -1 & 0 \end{pmatrix}$$

does not form a group with respect to matrix multiplication.

3. Show that $A = \begin{pmatrix} 1 & 1 \\ 1 & 1 \end{pmatrix}$ could never be in a group under the operation of matrix multiplication.

Addition

For (2×1) matrices the addition operation is that which has been defined as vector addition on page 164. Thus, taking vectors $\mathbf{a} = 5\hat{\mathbf{i}} + 4\hat{\mathbf{j}}$ and $\mathbf{b} = 6\hat{\mathbf{i}} + 3\hat{\mathbf{j}}$ we have $\mathbf{a} + \mathbf{b} = 11\hat{\mathbf{i}} + 7\hat{\mathbf{j}}$ already defined. If these vectors are now written in (2×1) matrix form the equivalent statements to the above will read as

$$\mathbf{a} + \mathbf{b} = \begin{pmatrix} 5 \\ 4 \end{pmatrix} + \begin{pmatrix} 6 \\ 3 \end{pmatrix} = \begin{pmatrix} 11 \\ 7 \end{pmatrix}$$

and the resulting vector is obtained by ordinary addition of the corresponding components.

For consistency we shall continue to use this idea to define the addition of matrices.

Definition 9.8. The sum of two matrices if it exists is equal to the matrix of the sum of the corresponding elements. The sum will only exist if the matrices are of the same order.

The definition means that

$$\begin{pmatrix}4 & 3\\2 & 9\end{pmatrix}+\begin{pmatrix}-1 & 5\\6 & -2\end{pmatrix}=\begin{pmatrix}3 & 8\\8 & 7\end{pmatrix} \text{ or } \begin{pmatrix}2 & 9\\6 & 7\end{pmatrix}+\begin{pmatrix}1 & 0\\-4 & -9\end{pmatrix}=\begin{pmatrix}3 & 9\\2 & -2\end{pmatrix}$$

and in general

$$\begin{pmatrix}a & b\\c & d\end{pmatrix}+\begin{pmatrix}p & q\\r & s\end{pmatrix}=\begin{pmatrix}a+p & b+q\\c+r & d+s\end{pmatrix}$$

Since ordinary addition is both commutative and associative it follows that if A, B, C are three matrices then $A + B = B + A$ and $(A + B) + C = A + (B + C) = A + B + C$, assuming that all the separate sums exist. The subtraction of two matrices will follow the same principle based on ordinary subtraction.

Clearly the zero matrix $0 = \begin{pmatrix}0 & 0\\0 & 0\end{pmatrix}$ is the identity element for (2×2) matrix addition because $0 + A$ for any (2×2) matrix A leaves A unchanged, i.e. $0 + A = A = A + 0$ for all (2×2) matrices A.

Therefore if $A + B = 0$ then B will be called the additive inverse of A.

Example 19. Find the additive inverse of the matrix $A = \begin{pmatrix}-2 & 4\\6 & -8\end{pmatrix}$ and solve the equation $X + A = C$ where $C = \begin{pmatrix}1 & 9\\7 & 3\end{pmatrix}$.

Solution. The additive inverse of A is given by $(-A)$ such that

$$A + (-A) = 0 = (-A) + A$$

Working from the Definition 9.8 of matrix addition it follows that

$$(-A) = \begin{pmatrix}2 & -4\\-6 & 8\end{pmatrix} \text{ because } \begin{pmatrix}-2 & 4\\6 & -8\end{pmatrix}+\begin{pmatrix}2 & -4\\-6 & 8\end{pmatrix} = 0$$

Solving equation $X + A = C$ involves the step

$$X + A + (-A) = C + (-A)$$
$$\therefore \quad X + 0 = C + (-A)$$
$$\therefore \quad X = \begin{pmatrix}1 & 9\\7 & 3\end{pmatrix}+\begin{pmatrix}2 & -4\\-6 & 8\end{pmatrix}=\begin{pmatrix}3 & 5\\1 & 11\end{pmatrix}$$

a result which may be written down without the manipulative fuss now that we have seen how it is properly found. We shall accordingly write $A + (-B)$ as $A - B$. *Answer*

Example 20. Solve the equation $3X + B = A$ when $A = \begin{pmatrix}4 & 9\\7 & 3\end{pmatrix}$ and $B = \begin{pmatrix}2 & -7\\5 & 1\end{pmatrix}$.

Solution. We have $3X + B + (-B) = A + (-B)$.

$$\therefore \quad 3X = A - B = \begin{pmatrix}2 & 16\\2 & 2\end{pmatrix}$$

$$\therefore \quad X = \frac{1}{3}\begin{pmatrix}2 & 16\\2 & 2\end{pmatrix} = \begin{pmatrix}\frac{2}{3} & \frac{16}{3}\\\frac{2}{3} & \frac{2}{3}\end{pmatrix}$$ *Answer*

Exercise 9.9

$$A = \begin{pmatrix}7 & 10\\21 & 15\end{pmatrix}, B = \begin{pmatrix}9 & 7\\3 & 10\end{pmatrix}, C = \begin{pmatrix}-1 & -2\\-4 & -9\end{pmatrix}$$

Using the given matrices, A, B, C answer the following questions:

1. (i) $A + B$, (ii) $A - B$.
2. Find X when $X + B = A$.
3. Find (i) $B + C - A$, (ii) $B - A + C$.
4. Find X when $X + A = B + C$.
5. Find X when $CX = A + B$.
6. Find X when $XC = A + B$.
7. Find X when $CX = A + C$.

10

TESSELLATIONS

Tiling the kitchen floor or the bathroom wall using tiles of the same square shape is the most elementary example of a tessellation, which is the covering of an area without overlapping or gaps between the tiles. Square or rectangular tiles are the most convenient because the regions to be tiled are usually rectangular in shape, although even so we may have to trim the finish tiles to shape the edge of the region.

For our purposes we shall ignore the difficulties of trimming the tiles and just concentrate on the coverage of the inner region with tiles of various shapes. The mathematical purpose of this activity is to first see how we may describe the position of any one tile with reference to an adjacent tile and then to capture intuitively an awareness of the basic requirements for fitting shapes together in order to tessellate a region.

Quadrilaterals

For our starting point we shall consider the fact that it is possible to tessellate a region using a tile with any quadrilateral shape.

We proved in the vector section (Question 7, Exercise 8.3) that the joins of the mid-points of the sides of any quadrilateral will form a parallelogram. What we wish to accept now is that starting with the parallelogram $PQRS$ we may draw a quadrilateral $ABCD$ around it so that each vertex P, Q, R, S is a mid-point of one of the sides of the quadrilateral.

Consider a parallelogram $PQRS$ already drawn as in Fig. 10.1, and suppose that D is a suggested vertex for the quadrilateral $ABCD$. (D must not be too far away from R and S, as experience will show.) Starting from D draw the straight line DSA with S as mid-point. Now that A is found draw the straight line APB with P as mid-point and similarly draw BQC with Q as mid-point. We shall now have the points C, R and D on a straight line with R as the mid-

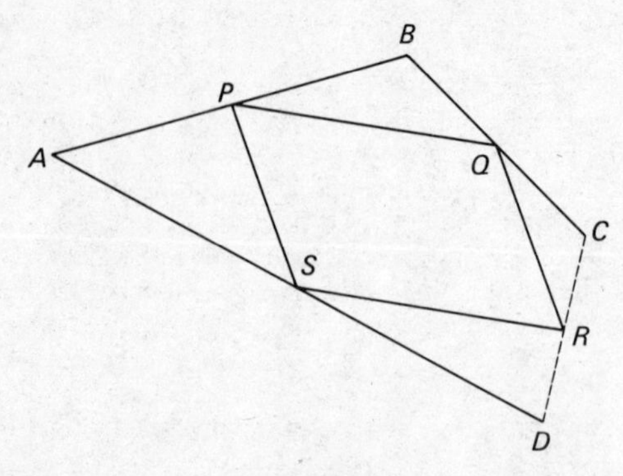

Fig. 10.1

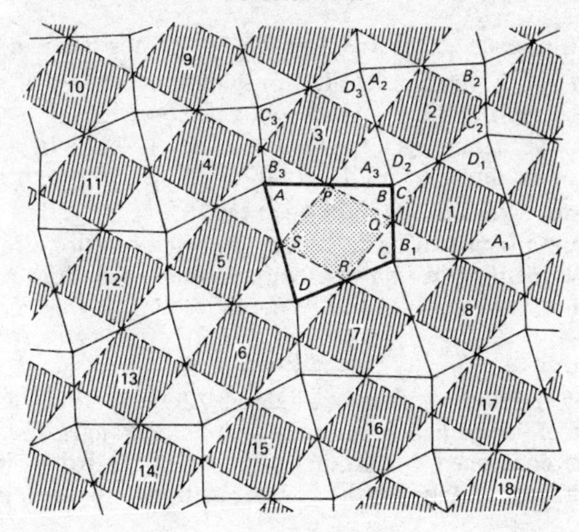

Fig. 10.2

point of CD. Clearly by varying the position of the starting point D we may obtain an infinite number of different quadrilaterals each having P, Q, R and S as mid-points of its sides.

Now consider a region tessellated by parallelograms as in Fig. 10.2. The first quadrilateral tile $ABCD$ is obtained in the manner discussed with Fig. 10.1 and the diagram shows the whole region is tessellated with quadrilaterals congruent to $ABCD$. (Congruent means equal in all respects.) Three adjacent tile positions to $ABCD$ have been numbered 1, 2 and 3. (The reader ought to make a cut-out copy of $ABCD$ to check the following instructions.)

Keeping the tile on the paper rotate $ABCD$ through 180° about Q, which is the mid-point of BC, onto position 1 and note the new position of the vertices on $A_1B_1C_1D_1$.

We get from position 1 to position 2 by rotating $A_1B_1C_1D_1$ through 180° about the mid-point of C_1D_1 and this will bring the vertices onto new positions given by $A_2B_2C_2D_2$. Notice that we could have gone from $ABCD$ straight onto position 2 by a translation (i.e. slide) without rotation. This means that the translation taking A onto A_2 is equivalent to the two successive rotations, 180° about the mid-point of BC followed by 180° about the mid-point of C_1D_1.

Next we get from position 2 to position 3 by a rotation of $A_2B_2C_2D_2$ through 180° about the mid-point of A_2D_2 onto the position $A_3B_3C_3D_3$.

Examining the final picture we know that each side of the quadrilateral fits into the picture because in each rotation the end points of the side are interchanged when pivoted about the mid-point of that side. We also know that there are no gaps or overlaps because each corner is exactly covered by the four different angles of the quadrilateral and the sum of these angles is 360°.

If we maintain this method of rotation about mid-points we will cover (i.e. tessellate) the whole plane.

Examination of the overall picture in Fig. 10.2 shows that we could describe the coverage of half the plane in terms of translations alone by using the addition of multiples of two basic vectors $\overrightarrow{DB}$ and $\overrightarrow{CA}$. Thus, adding the vector $\overrightarrow{DB}$ to each of the positions of A, B, C and D translates $ABCD$ onto position 2, whereas the addition of vector $\vec{CA}$ to each of the positions A, B, C and D translates $ABCD$ onto position 4. The combination $\vec{CA} - \vec{DB}$ will translate $ABCD$ onto position 4 and then finally onto position 12.

Unfortunately we cannot get to positions like 1, 7, 15 by translation of $ABCD$ alone. For example, we need a rotation about Q through 180° together with a translation $-2\vec{DB}$ to get onto position 15.

Thus, combinations of two basic vectors and one rotation are sufficient to enable us to describe the position of any tile in the tessellations relative to the original position of $ABCD$.

Quadrisides

For the type of tessellation shown in Fig. 10.2 the mid-points like Q are necessary in order to gain a rotation which interchanges the ends of the straight-line side concerned, but in fact any join of the two points B and C would be acceptable for the rotation onto position 1 if this join were sym-

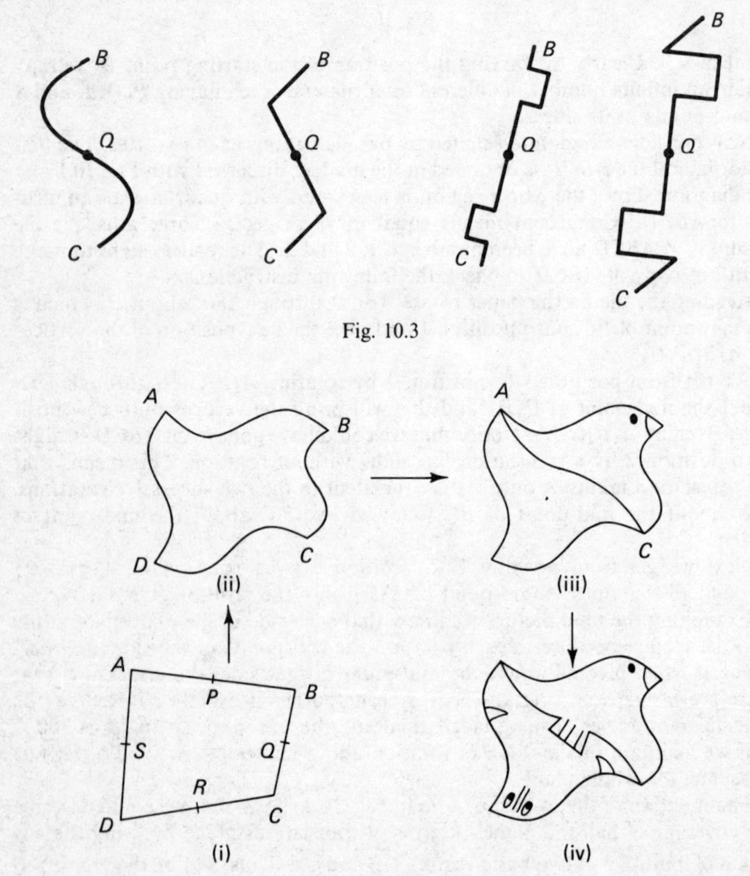

Fig. 10.3

Fig. 10.4

metrical about the mid-point Q. Suggested joins for BC could be any of those in Fig. 10.3, since each join is symmetrical about the mid-point Q.

But what applies for this side BC also applies to the other three sides and the resulting four-vertex figure, although no longer a quadrilateral (i.e. lateral = straight line), may now appropriately be called a quadriside. Again the tessellation may be described in terms of two translations and a rotation.

To pursue a more artistic end result let us consider transforming our original quadrilateral $ABCD$ into a quadriside as shown in Fig. 10.4. We first transform the sides of $ABCD$ into simple curves which are symmetrical about the same mid-point as before. Now add a superfluous line of any shape from A to C, to produce (iii) as well as adding a beak and eye for the beginnings of a bird image. For (iv) we use our imagination (?) to think of what to put into the lower half of the figure—another bird? no—a frog. Of course, the interior design of the tile is irrelevant but it produces the picturesque tessellation of Fig. 10.5 with the position of each tile completely specified with reference to

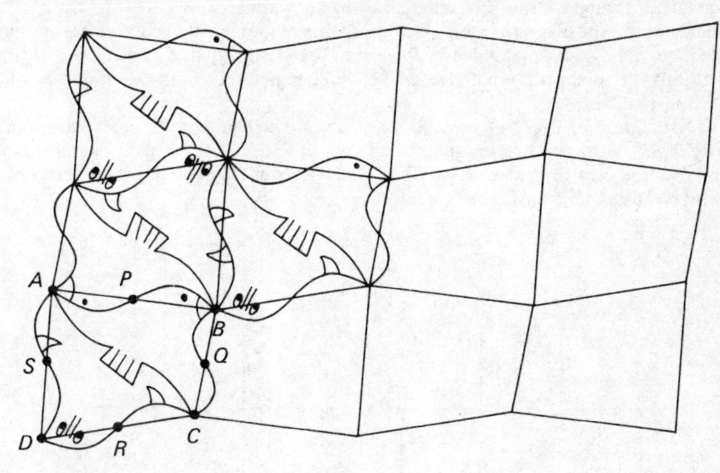

Fig. 10.5

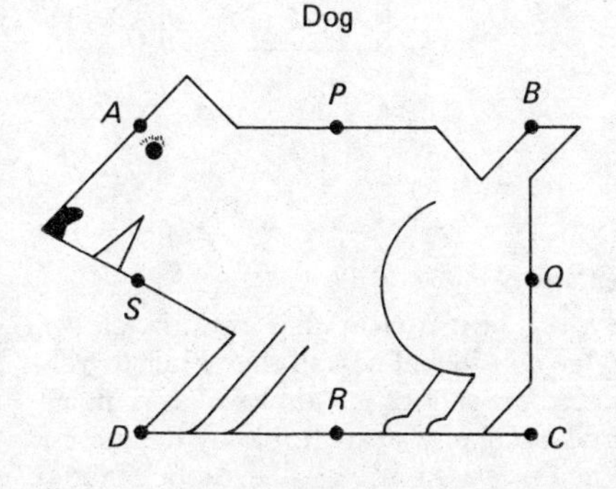

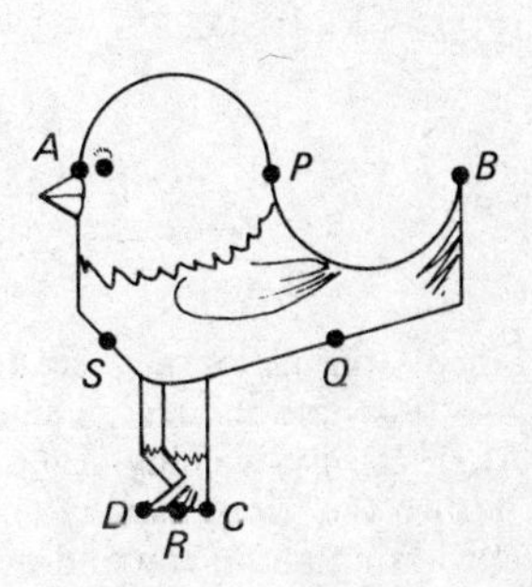

Fig. 10.6

the original in terms of $\overrightarrow{DB}$ and $\overrightarrow{CA}$ together with R(Q, 180°). It is a simple matter for the reader to complete the picture.

The symmetry exercises suggested by Fig. 10.3 are best summarised by the conversions of Fig. 10.6 for any quadrilateral $ABCD$ into a quadriside.

Exercise 10.1

Questions 1 and 2 refer to Fig. 10.2.

1. Describe in terms of $\overrightarrow{DB}$, $\overrightarrow{CA}$ and R(Q, 180°) how to place $ABCD$ on the following positions: (i) 10, (ii) 14, (iii) 13, (iv) 17.
2. State which position $ABCD$ is mapped onto by the following transformations:
   (i) $\overrightarrow{AC} + \overrightarrow{BD}$
   (ii) R(S, 180°)
   (iii) R(S, 180°) followed by $\overrightarrow{DB}$
   (iv) R(P, 180°) followed by $\overrightarrow{CA} + \overrightarrow{BD}$
3. A tile has five equal sides which make up the perimeter of a square, mounted by an equilateral triangle. Draw a tessellation using such a tile.
4. The domino-type diagram in Fig. 10.7 (i) consists of four small equal squares and has been lettered as quadrilateral $ABCD$ with mid-points P, Q, R, S. Tessellate a region with this tile to show the pattern for the whole region following the principles outlined in Fig. 10.2.

   The diagram of Fig. 10.7 (ii) consists of seven equal squares. Suggest four points $ABCD$ and corresponding mid-points with which to tessellate a region by translation and rotation only as discussed with Fig. 10.2. Describe the tessellations in terms of a basic rotation and translation.

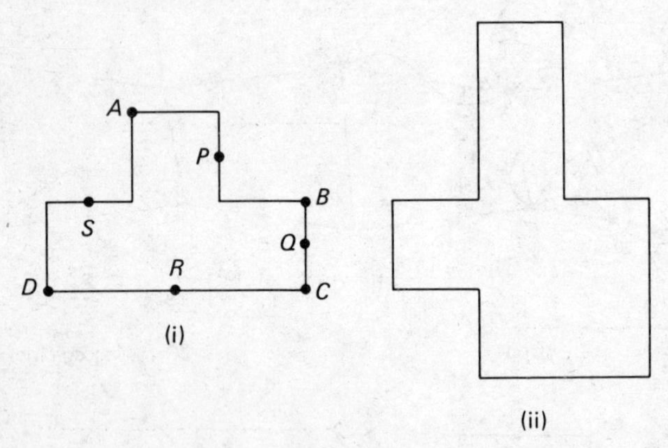

Fig. 10.7

Pentagons

Since a parallelogram will tessellate a region it is clear that any triangle will also tessellate the region since it may be considered as half the parallelogram. Alternatively, we may argue that because the sum of the angles of a triangle is 180° so the triangles may be tessellated along any straight line by the method of rotating about the mid-points P and Q, as seen in Fig. 10.8. Notice that on the straight line where three angles meet at K we have $C + A + B = 180°$.

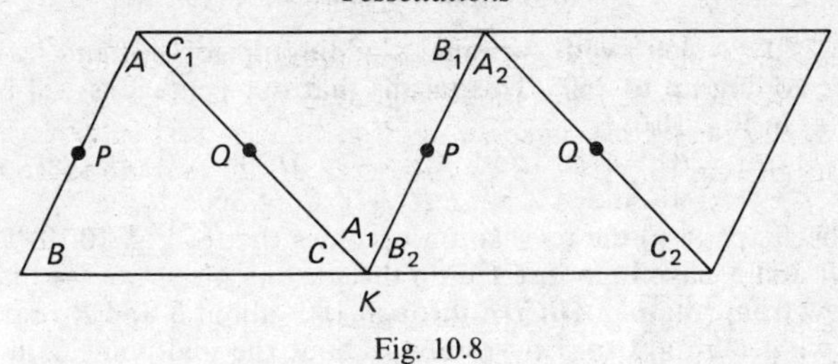

Fig. 10.8

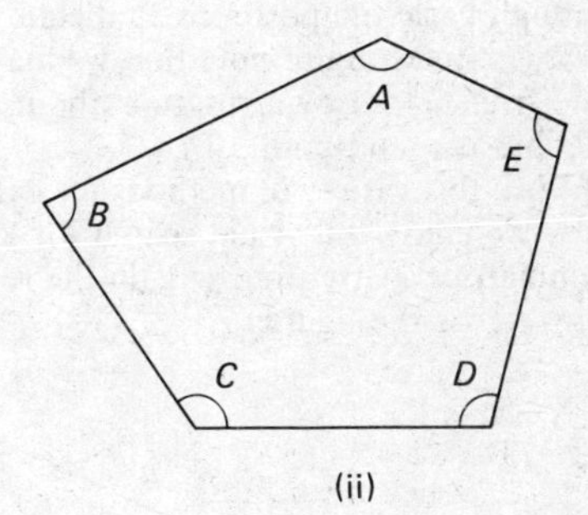

Fig. 10.9

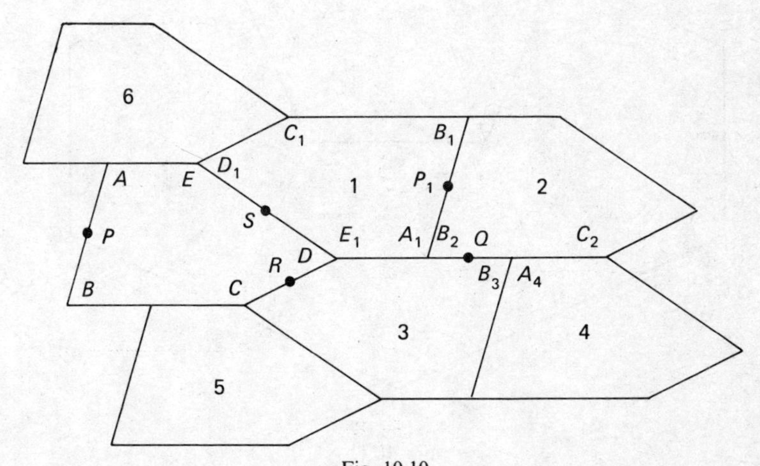

Fig. 10.10

Similar reasoning that a quadrilateral must tessellate a region by this rotation method is based on the sum of the angles of the quadrilateral being 360°.

But a five-sided figure has an angle sum of 540° and this makes tessellation impossible if we believe that all the five angles will meet at a point similar to $BC_1D_2A_3$ on Fig. 10.2. If we consider our effort (?) with Question 3 in the previous exercise this suggests that we should think of restricting our pentagon to $540° = 180° + 360°$, with two angles adding up to 180° and leaving the other three adding up to 360°. This means that our pentagons will be either like (i) or (ii) in Fig. 10.9.

In (i) $\angle A + \angle B = 180°$ and $\angle E + \angle D + \angle C = 360°$ while in (ii) $\angle B + \angle E = 180°$ and $\angle A + \angle D + \angle C = 360°$.

Using the shape in (i) the tessellation becomes that of Fig. 10.10. The mid-points P, R and S have been inserted on the starting tile so we see that tiles 1 and 3 follow from rotating $ABCDE$ through 180° about S and R respectively. Similarly, we get from 3 to 5 by rotating about the mid-point S in its new position. Position 2 may be obtained by translation and so we now have enough basic tile positions to dictate the rest of the tessellation.

It is interesting to note that we may get from 1 to 2 by a rotation about P_1 and from 2 to 3 by a rotation about Q, which is certainly not a mid-point of BC but is a mid-point of E_1C_2.

With this variety of methods for gaining the tessellation one can understand why we enjoy the visual search for key points like Q and the satisfaction of symmetrical patterning. (Again the reader must use a cut-out shape to map a way across the figure.)

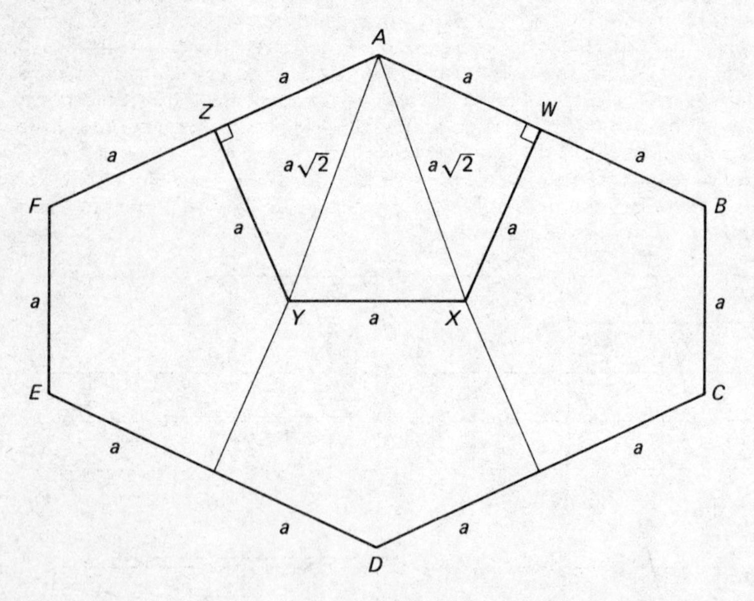

Fig. 10.11

Diagram (ii) of Fig. 10.9 will require some practical simplification and so we choose the simplest example $AWXYZ$ of Fig. 10.11 in which all the sides are of equal length a and angles W and Z are 90° each. Four of these pentagons form the hexagon $ABCDEF$ of Fig. 10.11, the tessellation of which is quite trivial, being formed by translations $n\overrightarrow{FB}$ (with n an integer) and $n\overrightarrow{EA}$. The four inner pentagons of $ABCDEF$ are related in pairs by reflection—for example, a reflection of $XYZAW$ in YX shows one pair. Alternatively, the presence of a 90°

angle suggests a possibility of rotating the pentagons in order to tessellate the region (the reader is asked for this in the next exercise).

Exercise 10.2

1. Tessellate a region using the hexagonal outline of Fig. 10.11.
2. Notice in Fig. 10.11 that the four pentagons of the hexagon are formed by joining A to the mid-point of ED and DC and similarly D to the mid-point of FA and AB. In your answer to Question 1 outline all the pentagons in the tessellation.
3. In Fig. 10.10 suggest basic vectors and a rotation which enable you to describe a mapping of $ABCDE$ onto the positions (i) 2, (ii) 4, (iii) 6, (iv) 3.
4. Draw a tessellation picture similar to Fig. 10.10 in which Fig. 10.9 (i) has a re-entrant angle at D, i.e. D is closer to AB than E.
5. Explain why a regular pentagon cannot be used to tessellate a region.

Hexagons

At first sight it might appear that increasing the number of sides increases the difficulty of finding a tile which can tessellate a region. In fact, a hexagonal tile is quite easy to use if we take note of the symmetry of the hexagonal Fig. 10.11.

We may consider that the basic cartesian grid has been distorted in the manner of Fig. 10.12 from (i) to (ii) to (iii), in which the tile of Fig. 10.11 is seen to appear but not necessarily with the same relationship between the lengths of its sides because the choice of the start point A is arbitrary. If the brickwork pattern is staggered out of (i) into (iv) we can produce the tessellation shown where we have another non-regular hexagon. (A regular hexagon has all its sides equal in length and all interior angles 120°.)

Notice that a rotation of a tile through 180° about the mid-point of any one of its sides will map it onto another tile, which again may be plotted in terms of the translation of one basic tile.

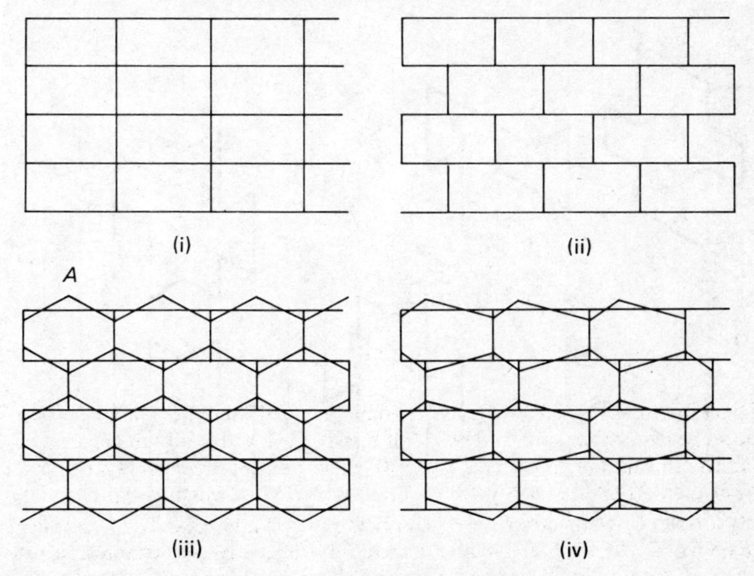

Fig. 10.12

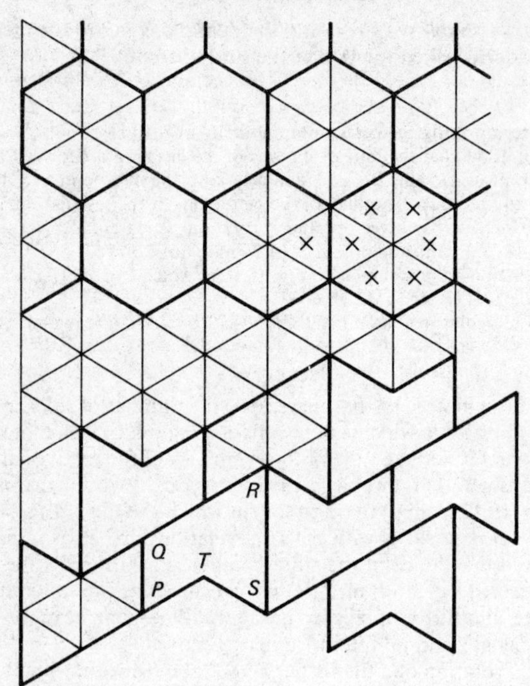

Fig. 10.13

Fig. 10.14

Naturally we would expect a regular hexagon to tessellate a region because each angle being 120° permits three vertices to meet and make up 360° at a point. However, this would be merely a special case of Fig. 10.12 (iii) in which all sides of the hexagon are of equal length. Interest in a regular hexagon tessellation lies in the variety of modification which can be made after the tessellation has been completed.

The regular hexagon tiling of the upper part of Fig. 10.13 is trivial and the use of the constituent equilateral triangles even more so. In fact, we may regard the regular hexagons as made up of six equilateral triangles deliberately arranged in hexagon outline.

Suppose we try to tile the region using five equilateral triangles. For this purpose we see that the pentiamond shape of $PQRST$ in Fig. 10.13 will tile successfully. This is not unexpected since a return to Fig. 10.9 shows that our pentiamond shape is like (i) with D moved towards AB, to create a re-entrant angle at D, and Fig. 10.10 (and Question 4, Exercise 10.2) showed that it was a simple matter to tessellate with this shape of tile. Again, putting two of the pentiamonds together, as in $ABCDEF$ in Fig. 10.14, shows them to form a hexagon, with two re-entrant angles at B and E. This hexagon will also tessellate.

Any tiling based on the equilateral triangle tends to be very versatile in creative outlines. For example, consider the large equilateral AXY of Fig. 10.15 (i), which consists of four equilateral triangles. Joining YB and XD locates the centre of the figure at C so that the $\triangle AXY$ can be seen as three quadrilaterals congruent to $ABCD$. Now divide each side of $\triangle BDW$ into three and connect up in the three-legged zig-zag of (ii) and we have three congruent quadrisides, but here we have the added attraction that we can move $ABCD$ into position 1 by a rotation through 120° about the point C. The bird picture is merely decoration.

Thus any tessellation, aside from the visual pleasure of its design content, presents us with the challenge of trying whenever possible to interpret its geometric structure in terms of the operations of translation, rotation and reflection.

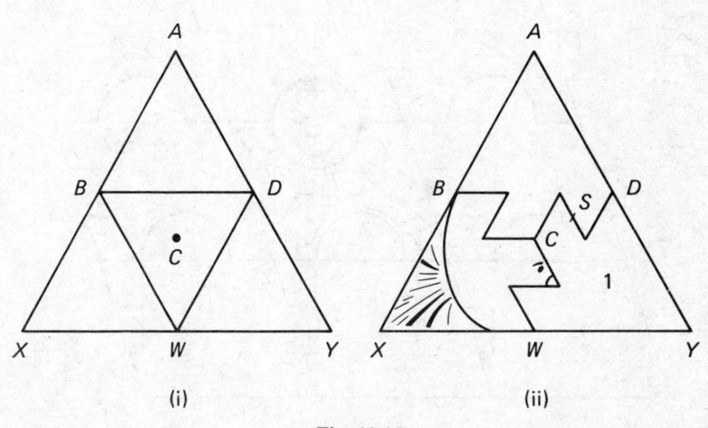

Fig. 10.15

Off Cuts

We may use any parallelogram to produce a new tiling simply by adding to one side whatever we remove from the opposite side, the two sides remaining congruent after the change. Taking the diagrams of Fig. 10.16 in order illustrates the method of producing the final tessellation of Fig. 10.17. The subsequent 'demon bowler' decoration is optional.

A little more variation may be obtained by attempting to make the decoration flow in two different directions. It is possible to be quite deliberate about this by starting with the two directional arrows, and distorting the sides of the quadrisides $ABCD$ by sides symmetrical about the mid-points to create the plaice in the Fig. 10.18.

The work has good practical uses in a mathematics course because it encourages us to search for those formal geometric relations which we might otherwise ignore.

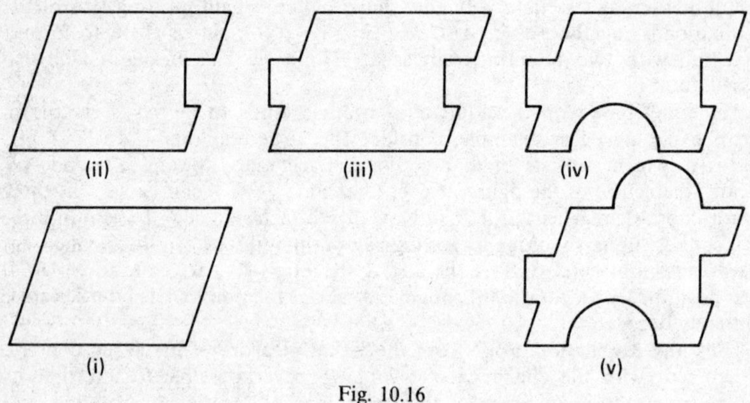

Fig. 10.16

Fig. 10.17

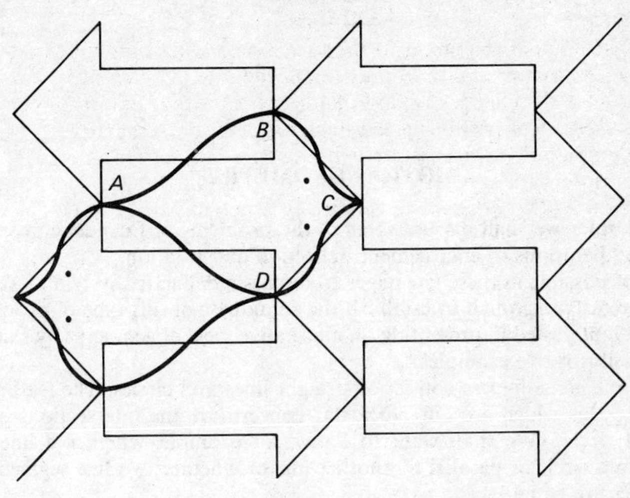

Fig. 10.18

Exercise 10.3

1. Show that the tile consisting of the seven equilateral triangles marked x in Fig. 10.13 will tessellate a region.
2. Consider a regular hexagon (e.g. as shown in Fig. 10.13) from which one of the six elementary triangles has been removed. Show that the remaining seven-sided figure of five equilateral triangles will tessellate.
3. In the top corner of Fig. 10.14 is a collection of four pentiamonds (in a different arrangement from the four pentiamonds in the bottom right of this figure). Tessellate a region using this collection of four pentiamonds as one single tile.
4. With the arrangement of Fig. 10.15 as a guide consider a square *ABCD* with centre *E*. Divide the square into four congruent quadrisides so that a rotation through 90° will map one quadriside onto another.
5. Show that the two plaice of Fig. 10.18 are really in the form of a hexagon, or hexiside.

11

MOTION GEOMETRY

In this chapter we shall discuss some of the problems of Euclidean geometry using the operations of enlargement, reflection and rotation.

It is not possible in these few pages to discuss a deductive system of axioms and theorems with which to establish the soundness of this type of proof. We are merely interested in presenting an alternative view of geometry by examining some illustrative examples.

All of our plane figures consist of straight lines and circles. The features of these diagrams which are our particular concern are the intersections of the lines and circles. We shall want to know, for example, whether a line is a tangent to a circle or parallel to another line or whether two line segments or two angles are equal.

When we transform a diagram by enlargement, reflection and rotation we need to know which features of the original are retained in the new diagram. For example, will two parallel lines remain parallel after the transformation or will two equal angles remain equal? Putting two such questions together we may ask more positively if a square can be transformed into a parallelogram with none of its angles being 90°?

Our main interest is in those features of the diagram which remain invariant under the transformation—that is, unchanged by the transformation.

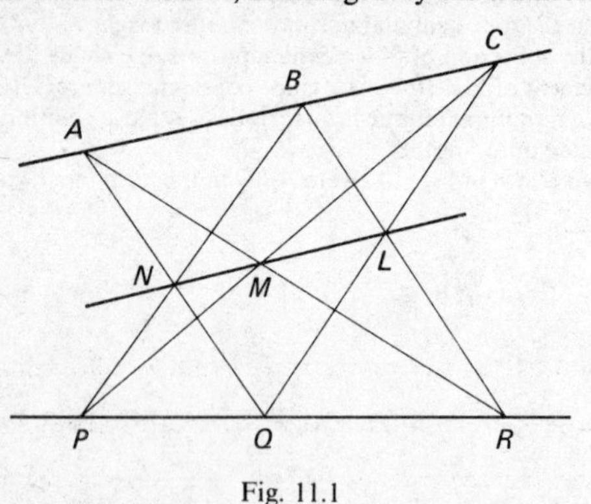

Fig. 11.1

Consider the diagram in Fig. 11.1, where we start with three distinct points *A*, *B*, *C* on one line and three distinct points *P*, *Q*, *R* on another line.

If the points are joined as shown in the diagram each pair of cross-joins $AQ.BP$, $BR.CQ$, $CP.AR$ will give the three points of intersection *N*, *L*, *M* respectively on a new straight line. Now any transformation of this diagram which moves the points *A*, *B*, *C* and *P*, *Q*, *R* into different positions on their straight lines will still retain the result of *N*, *M* and *L* being on a straight line—not the same line in general but still a straight line. All we require for this result to remain true is that the transformation allows straight lines to remain straight lines and their intersections to be preserved.

Radial Enlargement

This transformation forms the basis for perspective drawing. In a radial enlargement every length is the same multiple (k say) of the corresponding length in the original figure.

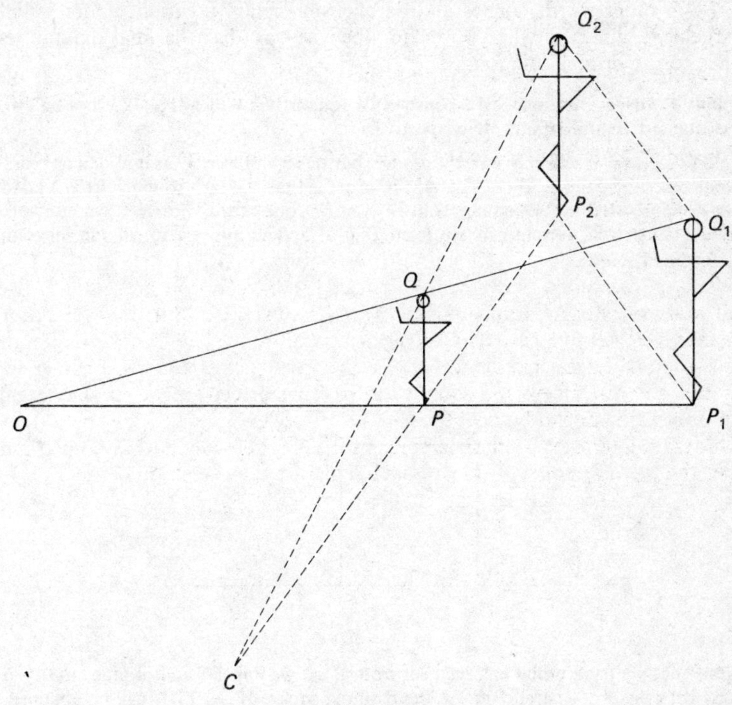

Fig. 11.2

Figure 11.2 shows O as the centre of the radial enlargement of PQ onto P_1Q_1 with an enlargement multiple of $k = \frac{5}{3}$, which makes $P_1Q_1 = \frac{5}{3}\,PQ$. Had there been another line segment of length 9 cm (say) in the plane then this radial enlargement from O would have mapped the line segment onto another line segment of length $\frac{5}{3} \times 9 = 15$ cm. Now suppose the man P_1Q_1 walks to P_2Q_2—that is, we have a translation from P_1Q_1 to P_2Q_2 (of course the figure $P_1Q_1Q_2P_2$ is a parellelogram). Now we join Q_2Q and P_2P, knowing that being two straight lines they must intersect at some point C, say, and this means that we may consider P_2Q_2 as a radial enlargement of PQ from the centre C.

This result may be more formally stated by saying that a radial enlargement followed by a translation is equal to a radial enlargement. Thus, the radial enlargement of PQ from centre O to P_1Q_1, followed by the translation from P_1Q_1 to P_2Q_2, is equal to a radial enlargement of PQ from centre C.

Naturally we should ask if the translation preceded the radial enlargement would we get the same final result? The answer is no, although the final result would still be a radial enlargement.

As a general result we may state that parallel lines will be transformed into parallel lines by a radial enlargement. Consequently, if two lines intersect at 60° in the original figure then they will still intersect at 60° in the enlarged figure—in other words, angles and parellelism remain invariant under radial enlargement, which is what we mean when we say that the final diagram is *similar* to the original diagram.

Example 1. Given a triangle ABC construct a square $WXYZ$ so that WX lies on BC, while Z and Y lie on AB and AC respectively.

Solution. A few trials will convince the reader that this is not as easy as it at first appears. But one of the nice things about this type of proof is that we learn so much from a good guess. Suppose we draw any square with WX on BC, one vertex Z on AB, Y not on AC. This is quite trivial as one can see from Fig. 11.3. The appearance of the diagram

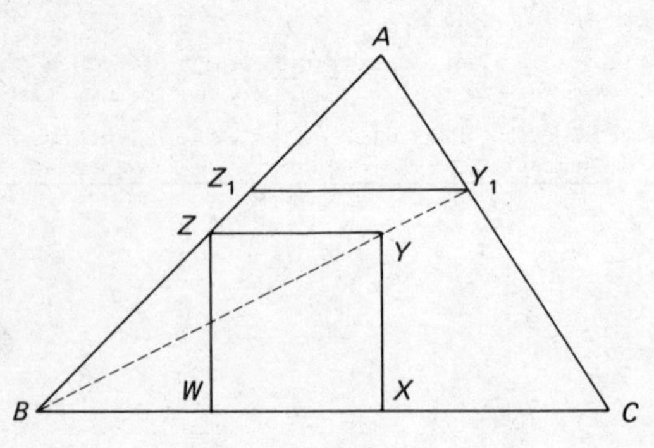

Fig. 11.3

suggests that we try a radial enlargement from B, so we join BY and produce to Y_1 on AC. By drawing Y_1Z_1 parallel to YZ the required square $W_1K_1Y_1Z_1$ will be obtained.

Answer

Example 2. Given any triangle ABC find a construction for the points P, Q on AB, AC respectively so that $BP = PQ = QC$.

Solution. Let the problem be represented by the diagram of Fig. 11.4. A scaled-down version of the solution is given by making $BX = XW = WZ$, where WZ is parallel to AC, so that if we can construct this part of the figure we shall obtain the final result by an enlargement of $BXWZ$, taking W to Q.

Now we know BX and WZ in direction and length but not the direction of XW, so instead of constructing the figure in the sequence B, X, W, Z we use the sequence B, X, Y, Z, where $XYZW$ is a rhombus and XY is parallel to AC.

Choose any point X on BA, then draw XY parallel to AC. Make $XY = BX$ using a pair of compasses with centre X, radius XB. Keeping the same radius but with centre Y, obtain Z on BC. With centre Z and the same radius as before, obtain W on the original arc centre X. The figure $WXYZ$ is now a rhombus with $BX = XW = WZ$ and $XY // WZ // AC$.

We now make a radial enlargement of $BXWZ$ onto $BPQC$ which gives $BP = PQ =$

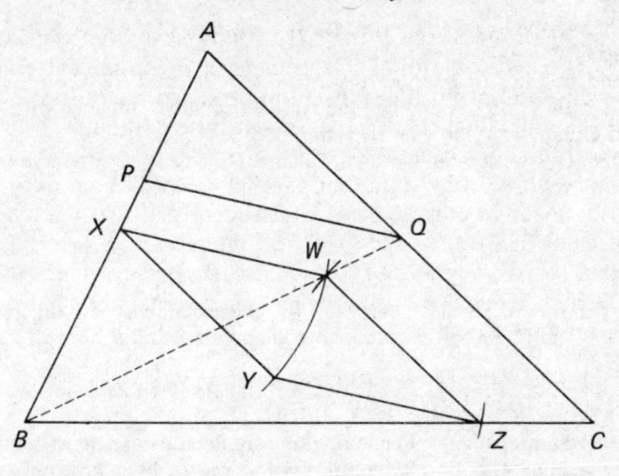

Fig. 11.4

QC. (An accurate construction is very difficult if the angle C makes the rhombus $WXYZ$ too flat, i.e. W close to Y.) *Answer*

Example 3. Construct a square in a sector of a circle having two vertices on the arc and one each on the boundary radii.

Solution. Let Fig. 11.5 represent the problem with the sector $OABC$ drawn with OB as the axis of symmetry (page 211). We know intuitively that the square will be symmetri-

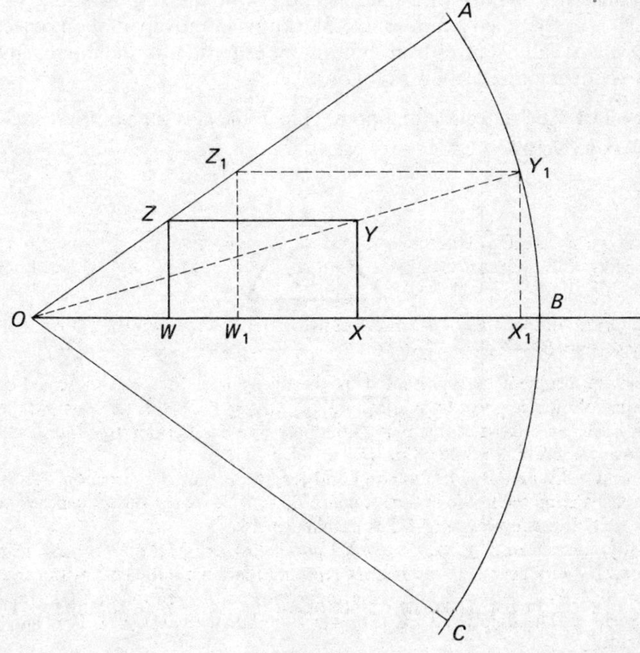

Fig. 11.5

cally placed across the line OB, so we start by drawing half a square $WXYZ$ as shown, in the manner of Example 1, with $ZY = 2ZW$.

Join OY and produce to intersect the arc of the sector at Y_1. From Y_1 draw Y_1Z_1 parallel to YZ with Z_1 on OA and half the required square is obtained with $W_1X_1Y_1Z_1$. The complete square will follow by reflection in OB.

Notice that we have used the fact that radial enlargement will preserve the ratios

$$\frac{XY}{YZ} = \frac{X_1Y_1}{Y_1Z_1} = \frac{1}{2}$$

Answer

Exercise 11.1

1. Using the diagram of Fig. 11.1, carry out the construction (i) with different distances between A, B and C, and (ii) interchanging the points P and R but still joining up the same letters.
2. Using the diagram of Fig. 11.3 construct a rectangle $WXYZ$ having Z on AB, Y on AC and W, X on BC such that $2WZ = ZY$.
3. Using the diagram of Fig. 11.3 construct any equilateral triangle with one vertex on each of the sides of ABC. Modify your construction to obtain an equilateral triangle with one side parallel to AC.
4. Given a segment of a circle, construct a square $WXYZ$ with WX on the chord of the segment and the other two vertices on the arc.
5. $ABCD$ is a square and two perpendicular lines not diagonals are drawn through its centre intersecting AB, BC, CD, DA in points W, X, Y, Z respectively. Construct a square passing through A, B, C, D and having the two perpendicular lines as diagonals.

Reflection

We are concerned mainly with the reflection of points and lines in a straight line, which is aptly called a mirror line. We know intuitively that an object and its image are equal distances from but on opposite sides of the mirror line but formally we define a reflection as follows.

Definition 11.1. The reflection of a point A in a line m is the point A_1 such that m bisects AA_1 at right angles.

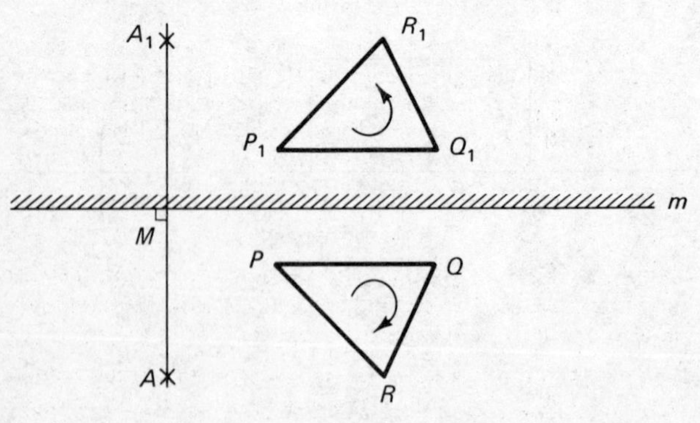

Fig. 11.6

In Fig. 11.6 the point A_1 is the reflection of the point A in the line m. Hence $m \perp AA_1$ and $MA = MA_1$.

One or two facts to be accepted about reflections are:

(i) Distances between points are unchanged (i.e. distances are invariant, e.g. $PR = P_1R_1$).

(ii) Angles between lines remain unchanged (i.e. angles are invariant, e.g. $\angle PRQ = \angle P_1R_1Q_1$). Thus parallel lines remain parallel after reflection and any object is congruent to its image.

Inspection of the reflection of $\triangle PQR$ in Fig. 11.6 shows that while the image $\triangle P_1Q_1R_1$ is congruent to $\triangle PQR$, the orientation of the triangle has reversed from clockwise to anticlockwise. (Note the usual convention of lettering the images after the reflection, i.e. we do not reflect the letter *A* to *∀*, etc.)

Sometimes it is necessary to select a suitable mirror line in order to declare one point to be the image of another. This is the same as finding points *A* and A_1 first in Fig. 11.6 and drawing the perpendicular bisector *m* afterwards.

An interplay of these ideas may be seen in the following example.

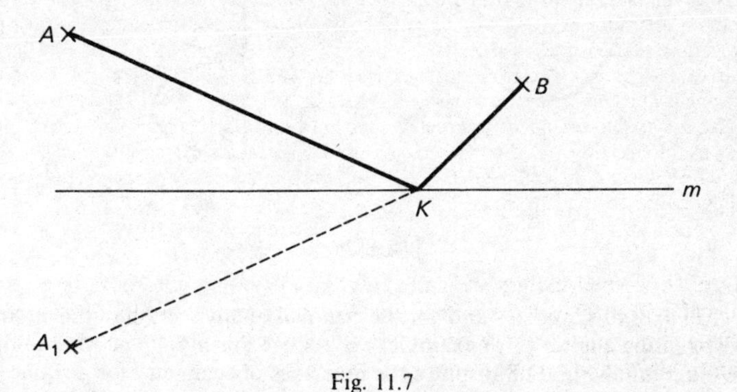

Fig. 11.7

Example 4. In terms of the diagram of Fig. 11.7 the problem is to find the shortest distance from point *B* to the line *m* and then on to the point *A*.

Solution. A sample path from *B* to *A* via the line is *BKA*, where *K* has been taken as any point on *m*. There is nothing about the path *BKA* to indicate whether or not it is the shortest possible path but an insight into the required solution is gained by inserting the image A_1 of *A* on the diagram and joining KA_1. Since KA_1 is the reflection of *KA* then $KA_1 = KA$. Now we see the length *BKA* as BKA_1. Where shall we put *K* in order to minimise $BK + KA$ (Question 7, Exercise 11.2 explains)? *Answer*

Axes of Symmetry

Inspection of $\triangle AKA_1$ in Fig. 11.7 shows that if we reflect the whole triangle in the line *m* the outline of the triangle would stay in the same position. Such lines *m* have a special significance for a figure.

Definition 11.2. If a figure contains its own reflection in a line *m* then the figure is said to be symmetrical about *m*. The line *m* is called an axis of symmetry of the figure.

We shall note that some figures have more than one axis of symmetry. With paper-drawn figures the existence of an axis of symmetry may be found by folding the figure about the line. A circle may be folded about any diameter to show one side of the fold coincident with the other, thus a circle has an infinite

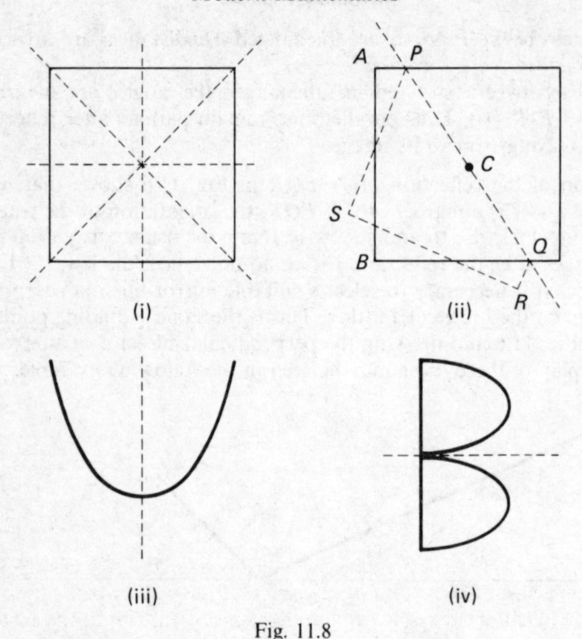

Fig. 11.8

number of axes of symmetry and yet the removal of any section of the circumference (creating a letter C for example) will reduce this infinite number to one.

The diagram of Fig. 11.8 displays the four axes of symmetry for a square in (i). In diagram (ii) notice that a general line through the centre of the square divides the square into congruent areas but this line is not an axis of symmetry of the square. If one half *PABQ* is cut out and turned round into the position given by *PQRS* then we have constructed a figure which is now symmetrical about *PQ*. Usually letters of the alphabet are taken as good familiar examples of symmetrical figures—like the letter U in (iii) and B in (iv). Much will depend on the style of printing—for example, a letter B is not always equilooped like the B in (iv).

Composite Reflections

Since a reflection of a figure reverses its orientation a second reflection in any other mirror will re-establish the original cyclic order of lettering.

With two such reflections, called say M_h and M_k for mirror lines called *h* and *k*, we should ask ourselves something about the composite mappings suggested by M_hM_k and M_kM_h. (M_kM_h means reflection in *h* first followed by reflection in *k* second.)

(i) Is it possible that these two mappings or reflections could result in the same final image?

(ii) Is it possible to arrange the mirror to map the object point onto any image we please?

For (i) we shall first arrange for the mirrors to be parallel, as suggested by

Fig. 11.9 in which the letter F_0 is taken as the object and F_1, F_2, F_3, F_4 are the images of F_0 in the different mirrors.

The upper part of the figure shows the reflection of F_0 in h first to give ꟻ$_1$ followed by reflection in k second to give F_2. Thus M_kM_h maps F_0 onto F_2. Notice that F_2 is facing the same way as F_0 after two reflections.

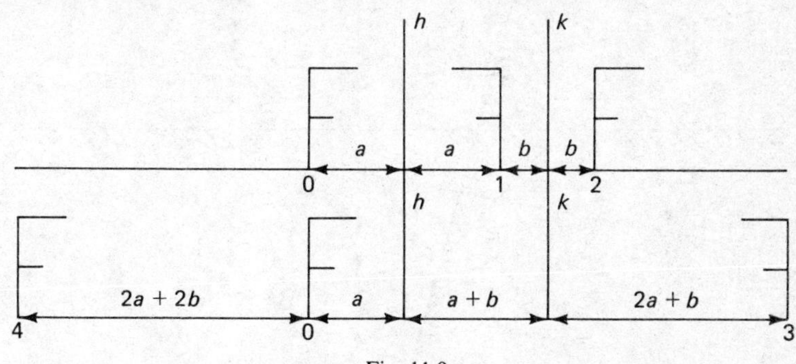

Fig. 11.9

The lower part of Fig. 11.9 shows the reflection of F_0 in k first to give ꟻ$_3$ followed by reflection in h second to give F_4. Thus M_hM_k maps F_0 onto F_4. As expected the reflections are not commutative since F_4 is not in the same place as F_2. Hence $M_hM_k \neq M_kM_h$. The distances entered on the diagram indicate that the distance between F_0 and F_2 is $2a + 2b$, which is the same as the distance between F_0 and F_4. So, the distance between the object and the final image is always $2(a + b)$, which is twice the distance between the mirrors.

We also notice that the two reflections are equivalent to a translation, since we could slide F_0 onto either position F_2 or F_4 directly.

Setting the two mirrors h and k at an angle a makes life more difficult but equally more interesting. The arrangement suggested in Fig. 11.10 again shows the object as F_0.

Taking the clockwise order first we see F_1 as the image of F_0 in h first, followed by F_2 as the image of F_1 in k second. Thus F_2 is the final image of F_0 in the mapping M_kM_h.

In the anticlockwise direction we see F_4 as the final image of F_0 in the mapping M_hM_k.

In the suggested arrangement of Fig. 11.10 F_2 is not in the same position as F_4, but it looks as though it could be if we changed the angle between the mirrors. Now the result in Fig. 11.9 suggests a connection between the placement of the mirrors and the relative positions of the object and the final image. We ask 'perhaps the angle between F_0 and F_2 or F_0 and F_4 is twice the angle a or $180° - a$ between the mirrors'.

Using the diagram of Fig. 11.10, we see that $2b$ is the size of the angle between F_0 and F_3, and $2c$ is the angle between F_3 and F_4. Therefore, the size of the angle between F_0 and F_4 is $2b + 2c$. But $a = b + c$ (being the exterior angle = sum of the two interior and opposite angles), which shows that the double reflection turned the object through an angle which is twice the angle a between the mirrors. Thus, one rotation through $2a$ about the point of intersection of the mirrors is equivalent to the two reflections M_hM_k. Similarly, the

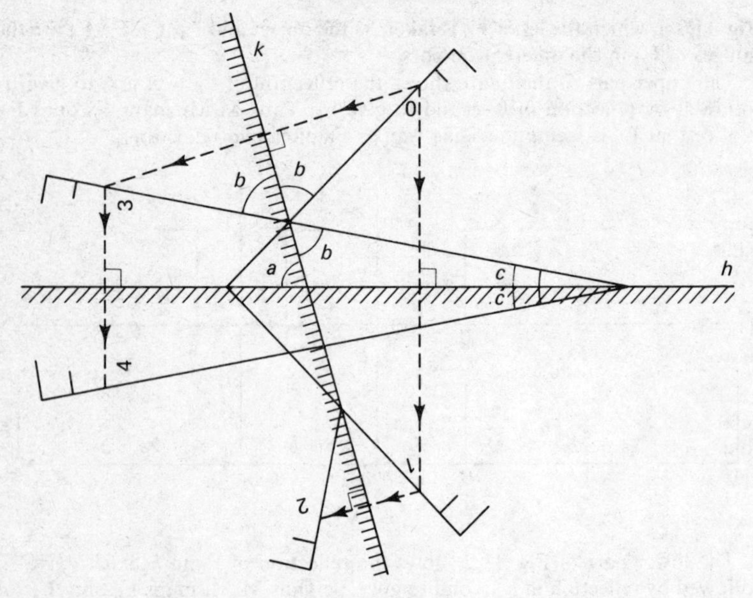

Fig. 11.10

angle between F_0 and F_2 is also $2a$, but in the clockwise direction. This result clearly suggests that if $a = 90°$ then F_2 will coincide with F_4.

This enquiry into the possible equivalent combinations of reflections which might lead to the same end result suggests that we try to see where to put the mirrors in order to map any two congruent triangles onto each other.

For an answer to this query consider the diagram of Fig. 11.11, in which we map the right-angled $\triangle$ ABC onto $\triangle$ $A_3B_3C_3$. Since the lettering of one

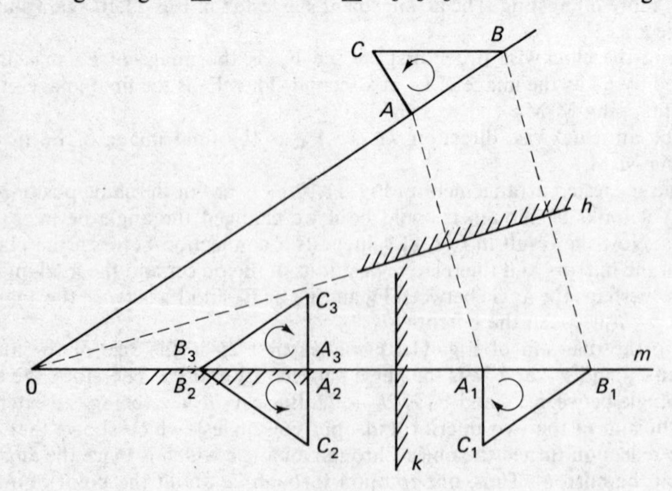

Fig. 11.11

triangle is in the reverse order to the other triangle we shall clearly need an odd number of reflections (in fact one or three).

The steps in the method are as follows:

(i) Produce sides BA, A_3B_3 to intersect at O. Bisect $\angle AOB_3$ and call the bisector line the first mirror line h. Reflect $\triangle ABC$ in h onto $\triangle A_1B_1C_1$.

(ii) This first reflection has made C_1A_1 parallel to C_3A_3 so that the required second mirror k is parallel to A_1C_1 and through the midpoint of A_1A_3. Reflect $\triangle A_1B_1C_1$ in k onto $\triangle A_2B_2C_2$.

(iii) Now take A_2B_2 as the third mirror m and reflect $\triangle A_2B_2C_2$ onto $\triangle A_3B_3C_3$. (Note that A_2 is the same point as A_3 and B_2 the same point as B_3.) The triple reflection given by $\mathbf{M}_m\mathbf{M}_k\mathbf{M}_h$ has mapped $\triangle ABC$ onto $\triangle A_3B_3C_3$ as required.

Had $\triangle ABC$ been positioned with B below A in the straight line OBA then a bisection of the obtuse angle between BA and A_3B_3 would have provided the first mirror for a two-reflection transformation mapping $\triangle ABC$ onto $\triangle A_3B_3C_3$.

Exercise 11.2

1. Which letters of the alphabet (according to the style of printing) have only (i) one, (ii) two, (iii) three, (iv) four axes of symmetry?
2. State the number of axes of symmetry of each of the following figures: (i) isosceles triangle, (ii) equilateral triangle, (iii) rectangle, (iv) regular hexagon.
3. Given a paper triangle ABC of general shape (scalene) explain how to find, without pencil or ruler: (i) the midpoint of AB, (ii) the perpendicular bisector of AB, (iii) the bisector of the angle ABC.
4. How can we obtain an angle of 45° by folding a piece of paper?
5. The coordinates of the vertices of $\triangle ABC$ are $A(3, 2)$, $B(7, 3)$, $C(5, 6)$. Find the final position of the $\triangle ABC$ after reflections in the line $y = 0$ followed by reflection in the line $x = 0$. Reverse the order of the reflections and show that the final image is the same in both cases.
6. Using $\triangle ABC$ of Question 5, find the position of the image after the sets of reflections in the following lines:

   (i) (a) $y = x$ (b) $y = 0$
   (ii) (a) $y = 0$ (b) $y = x$
   (iii) (a) $y = x$ (b) $x = 0$
   (iv) (a) $x = 0$ (b) $y = x$
   (v) (a) $y = x$ (b) $y = -x$
   (vi) (a) $y = -x$ (b) $y = x$

   Comment on any equalities in the above results.
7. Using Fig. 11.7, find the shortest distance BKA.

Rotations

To rotate a figure in its own plane we need (i) a point about which to perform the rotation, called the centre of rotation, and (ii) a stated direction of rotation of either clockwise or anticlockwise.

As shown in Fig. 11.12 (i) the centre of rotation O may not be a point in the figure to be rotated. All we need is some fixed base line such as AB in the figure so as to be able to measure the angle of rotation, between the position AB and A_1B_1 as shown in (i).

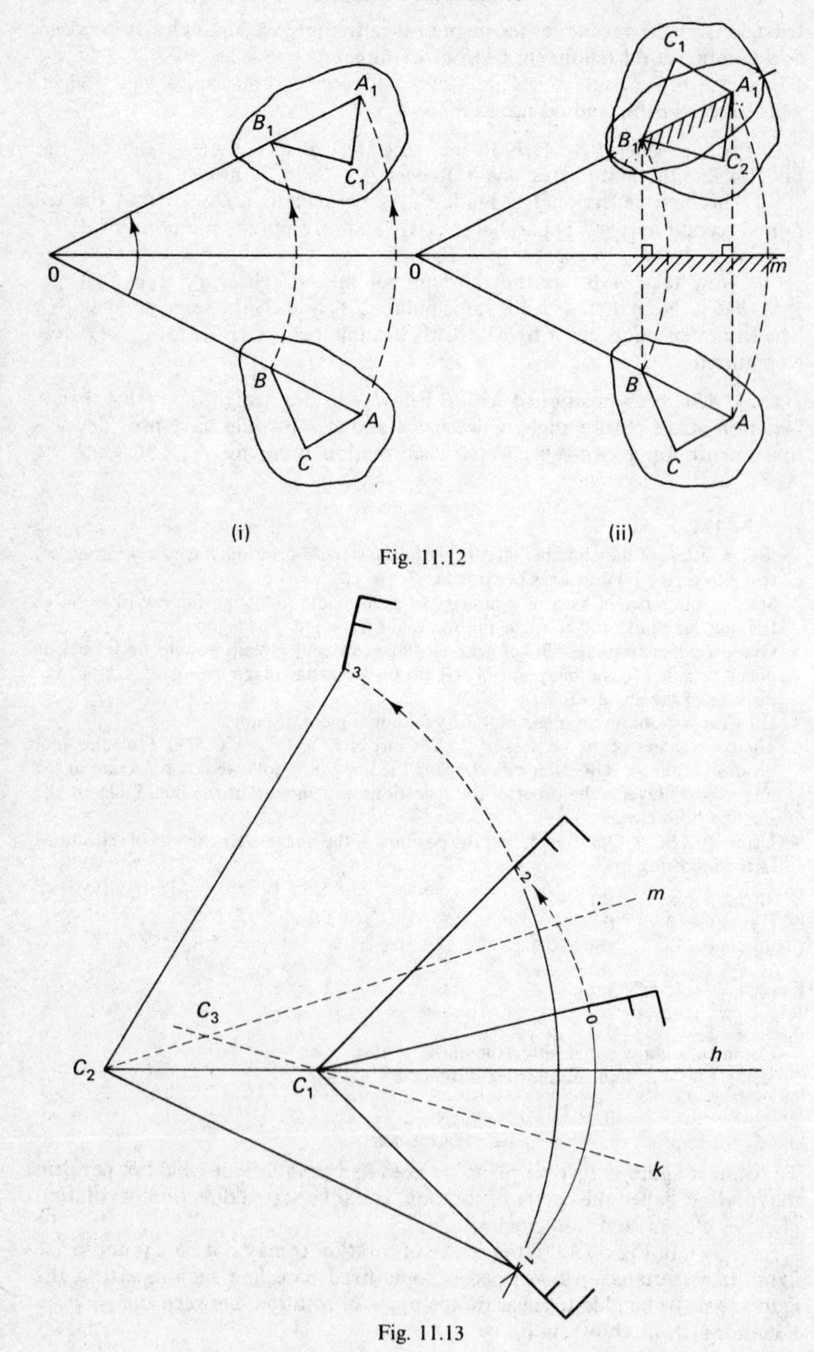

Fig. 11.12

Fig. 11.13

We saw (Fig. 11.10) that two successive reflections are equivalent to a rotation and Fig. 11.12 (ii) shows a repeat of the methods of the previous section. This was to bisect angle BOB_1 and use the bisector as a mirror line m to reflect ABC onto $A_1B_1C_1$ and then reflect the figure $A_1B_1C_1$ in the line A_1B_1 onto $A_1B_1C_2$ to make the two reflections in (ii) equivalent to the one rotation in (i). In fact, the representation of a rotation by two successive reflections is much more general than the method used above suggests, for we may choose any line we please through the centre of rotation O to act as one of the mirror lines. Figure 11.13 illustrates the intention.

The original object F_0 is rotated about C_1 onto F_2 and then rotated about C_2 onto F_3. These rotations are shown by the dotted circular arcs.

Now let us analyse the separate parts of this transformation. In Fig. 11.13 we take h as the line C_1C_2 and allow F_1 to be the reflection of F_2 in h. Taking k as the bisector of angle $F_0C_1F_1$ we may also allow F_1 to be the reflection of F_0 in k.

(i) The rotation from F_0 to F_2 is equal to the two successive reflections $F_0 \to F_1$ in k followed by $F_1 \to F_2$ in h. (h is the line C_1C_2 and k is the bisector of $\angle F_0C_1F_1$, k being obtained after finding F_1.) Thus M_hM_k maps F_0 onto F_2.

(ii) The rotation from F_2 to F_3 is equal to the two successive reflections $F_2 \to F_1$ in h followed by $F_1 \to F_3$ in m where m is the bisector of angle $F_3C_2F_1$. Thus, M_mM_h maps F_2 onto F_3.

Now bearing in mind that $M_hM_h = I$, the identity transformation leaving everything as it is, we now see that the two rotations about C_1 and C_2 that we started with are equal to M_hM_k followed by M_mM_h—that is, $M_mM_hM_hM_k$.

But $M_mM_hM_hM_k = M_m(M_hM_h)M_k = M_mIM_k = M_mM_k$, and since this result is reflection M_k followed by M_m this is equal to a rotation about the intersection of m and k, the point C_3 in Fig. 11.13. Hence the rotation of F_0 about C_1 onto F_2, followed by the rotation of F_2 onto F_3 about C_2, is equivalent to the one rotation of F_0 about C_3 onto F_3.

This result is a typical ingredient of transformation geometry where we are concerned about the equivalent movements of figures from one position to another in the same plane. As an illustration of the use of rotation in solving problems in transformation geometry consider the following examples.

Example 5. A variable size right-angled isosceles triangle ABC with $AB = AC$ and angle $BAC = 90°$ rotates about its vertex A and keeps another vertex B on a fixed line m. Find the path traced out by the vertex C.

Solution. Always draw a sample diagram as shown in Fig. 11.14 with $\triangle AB_1C_1$. Receiving no insight from this, we have drawn two more samples $\triangle AB_2C_2$ and $\triangle AB_3C_3$, to reveal the promising discovery that $C_1C_2C_3$ is a straight line perpendicular to the fixed line m, but can we prove this is always true?

Turning our attention to the beginning of the problem, we see the line segment AB_1 touching m at B_1 in Fig. 11.15 (i) with the point A a fixed distance of d from m. Rotate this figure 90° clockwise about A and we obtain Fig. 11.15 (ii), in which m has been mapped onto m_1, B_1 has been mapped onto C_1 with $AB_1 = AC_1$ and m_1 is therefore at a distance d from A.

We see therefore that in Fig. 11.14 all the points C are the images of B in the rotation $R(A, -90°)$ and so C will always lie on a line m_1 perpendicular to m and the same distance d from A.

Answer

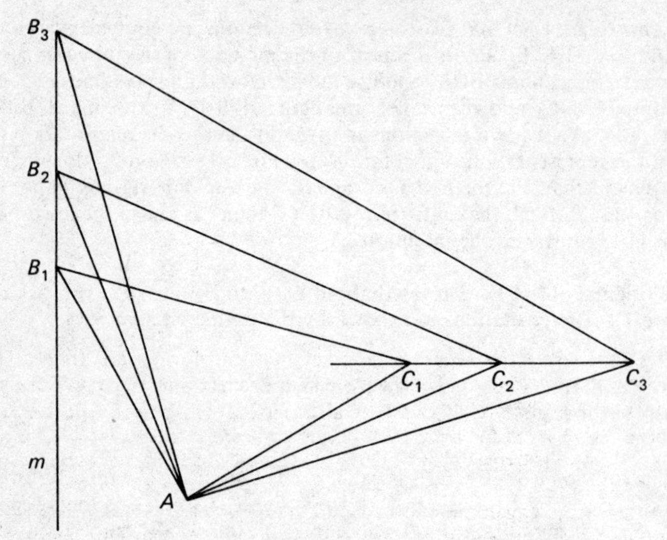

Fig. 11.14

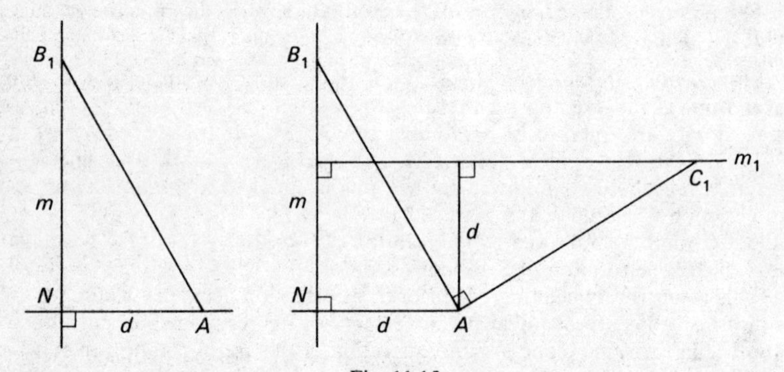

Fig. 11.15

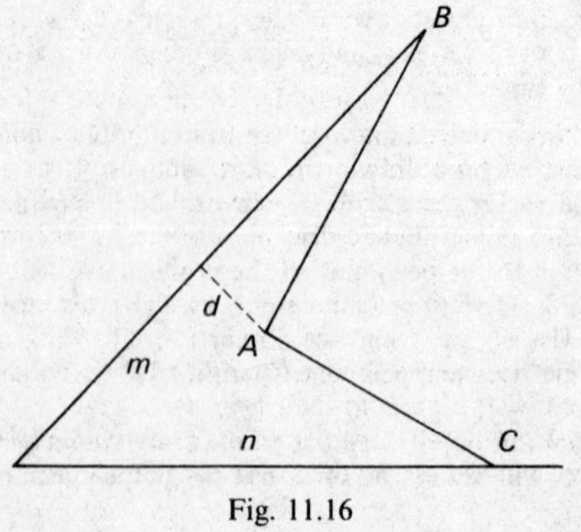

Fig. 11.16

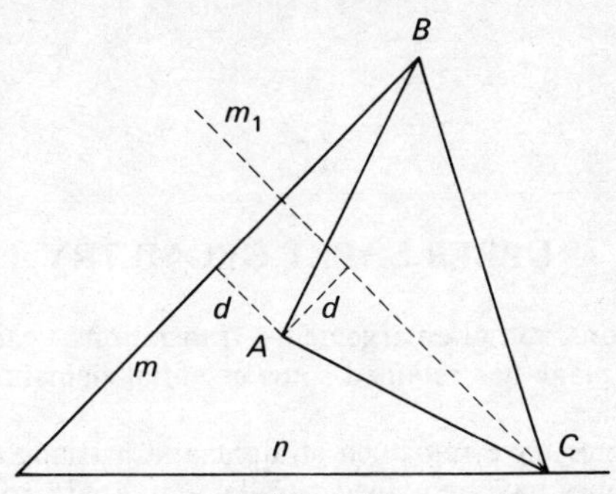

Fig. 11.17

Question. Suppose $\triangle$ *ABC* is a variable equilateral triangle, what is the path of *C* under the same conditions (Question 6, Exercise 11.3)?

Example 6. Two lines *m* and *n* include an acute angle and a fixed point *A*. Construct a right-angled isosceles triangle *ABC* with $AB = AC$ and $\angle BAC = 90°$ with *B* on *m* and *C* on *n*.

Solution. A sample diagram to see the nature of the problem is shown in Fig. 11.16. Attempts to suggest another such triangle indicate that there is only one solution for the position shown in the diagram.

Experience in Example 5 suggests that we try $R(A, -90°)$, in which case *C* is the image of *B* and we have the same situation as in Example 5—namely, that *C* must be on a line m_1 perpendicular to *m* and the same distance *d* from *A*, as shown in Fig. 11.17.

The intersection of m_1 with *n* gives the position of *C* and we can complete the $\triangle$ *ABC* as shown. *Answer*

Exercise 11.3

1. Find the image of the line $y = 5$ after each of the following rotations:
   (i) $R(A, 90°)$, $A(0, 0)$ (iii) $R(C, -90°)$, $C(1, 2)$
   (ii) $R(B, 90°)$, $B(2, 2)$ (iv) $R(O, -45°)$, $O(0, 0)$
2. A square *OABC* lettered clockwise has *OB* as diagonal where *O* and *B* are the points (0, 0), (5, 5) respectively. Find the coordinates of the vertices *A* and *C*.
   The square is now given a positive rotation of 90° about *B*. Find the new coordinates of the other three vertices.
3. A letter F_0 stands at a point *O* and is rotated through 90° about a point *C* to F_2. Obtain the two reflections which map F_0 onto F_2 the second reflection being in the line *CO*.
4. The letter F_0 stands at the point *O* on the end of the diameter C_2C_1O of a circle centre C_1. The letter is rotated through 45° about C_1 to position F_2, which is then rotated through 45° about C_2 to position F_3. Using Fig. 11.13 for guidance obtain the mirror lines *m* and *k* and describe each of the two reflections which give the rotations. Find also the angle of rotation about the point C_3 which maps F_0 onto F_3.
5. (i) Using Fig. 11.14 sketch the path of *C* when *B* is moved down the line *m*.
   (ii) Suppose the sample diagram had drawn both *B* and *C* below *A*, what path would *C* have traced out?
6. Again using Fig. 11.14, what would have been the path of *C* had $\triangle$ *ABC* been equilateral?
7. Using Fig. 11.17, determine if a second triangle may be constructed to satisfy the given conditions but with *B* on *m* to the left of *A*.
8. *A*, *B* and *C* are three given non-collinear points in a plane. Show how to construct a square with centre *A* and having two of its adjacent sides, produced if necessary, passing through *B* and *C*.

12

RUBBER SHEET GEOMETRY

Reflections, rotations, radial enlargements, translations—each of these transformations of the plane has retained some essential elements of the diagrams they have transformed.

Thus straight lines have remained straight and a triangle has remained a triangle. Parallel lines have remained parallel and, apart from radial enlargement, the distances between points remain unchanged. Each transformation applied itself to its own particular type of problem, but all of these problems are essentially problems of straight line intersections.

Suppose we introduce a transformation in which lines in general were not straight, so that only their intersections would be retained, and all our quadrilaterals would become quadrisides, what you may ask would there be left to talk about and discover and what geometrical 'properties' would be left to explore?

Well certainly this transformation must be flexible if it produces any shapes like quadrisides. It is so flexible that the only way to transform a diagram with this much freedom is to put it on a perfectly elastic rubber sheet which will stretch as much as we please without tearing. Such transformations will be called **topological transformations.** A typical set of diagrams arising from these transformations is given in Fig. 12.1.

Some curves may be drawn with one stroke or movement of the pencil, from the beginning of the curve to the end without taking the pencil off the paper; such a curve is called a continuous curve. A curve like the circle, which is continuous and without loops and where the starting point is the same as the end point, is called a simple closed curve because when drawn on our paper or rubber sheet it separates the surface into two regions. These two regions are intuitively called the inner and outer regions. Intuitively because we may consider the effect of drawing the equator on the surface of the earth; which of the two regions of the earth shall we call the inside and which the outside?

When we examine the diagrams of Fig. 12.1 we notice the following facts about a rubber sheet transformation:

(i) The continuity of a curve is preserved—that is, an unbroken curve remains unbroken.

(ii) The inner region of a simple closed curve remains an inner region. For example, the point D remains inside the region ABC, which in turn remains inside the region FSG.

(iii) Similarly to (ii), the outer region of a simple closed curve remains an outer region. For example, the point E remains outside ABC, and H remains outside FSG.

(iv) Points of intersection are preserved. For example, T will still be the intersection of WX and YZ after the transformation on the rubber sheet.

(v) The order of the points on a curve is preserved (unlike reflection). For example, ABC anticlockwise remains anticlockwise.

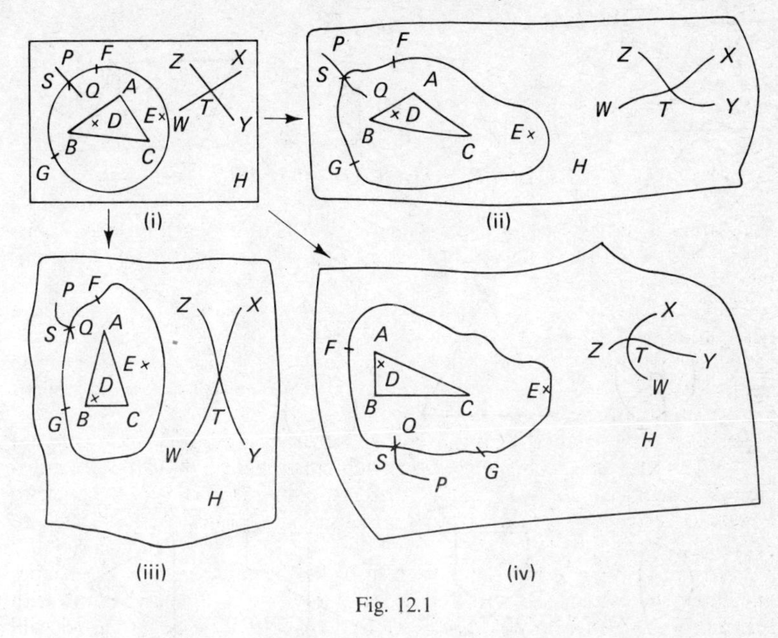

Fig. 12.1

All these five facts are invariants and a transformation which has those invariant properties is a topological transformation. If we break continuity by disconnecting or cutting a curve we create a new figure—that is a new topological relationship. Similarly, if we perforate the surface or fill in a hole we create a new figure.

Definition 12.1. A topological transformation of a diagram preserves continuity and is made without cutting, folding or perforating. A topological transformation is a rubber sheet transformation.

Topological Equivalence

In Fig. 12.1 the diagrams of (ii), (iii), (iv) are derived from (i) and we use the following definition to refer to this relation.

Definition 12.2. Diagrams which are derived from each other by a topological transformation are said to be topologically equivalent.

The question of topological equivalence is not one about angles, corners or curvature but merely of connection—that is, if A is connected to B then their images must be connected after the transformation. The examples of Fig. 12.2 illustrate what is meant by saying that two figures are topologically equivalent.

Before reading on, examine these diagrams and divide them into sets of topologically equivalent figures. Think of the figures drawn on a rubber sheet and decide if that figure could be stretched onto the other figure.

The sets of topologically equivalent figures are as follows:

(a) {(ii), (iii), (iv), (v), (vi)}. Each figure is a simple closed curve and they are therefore topologically equivalent.

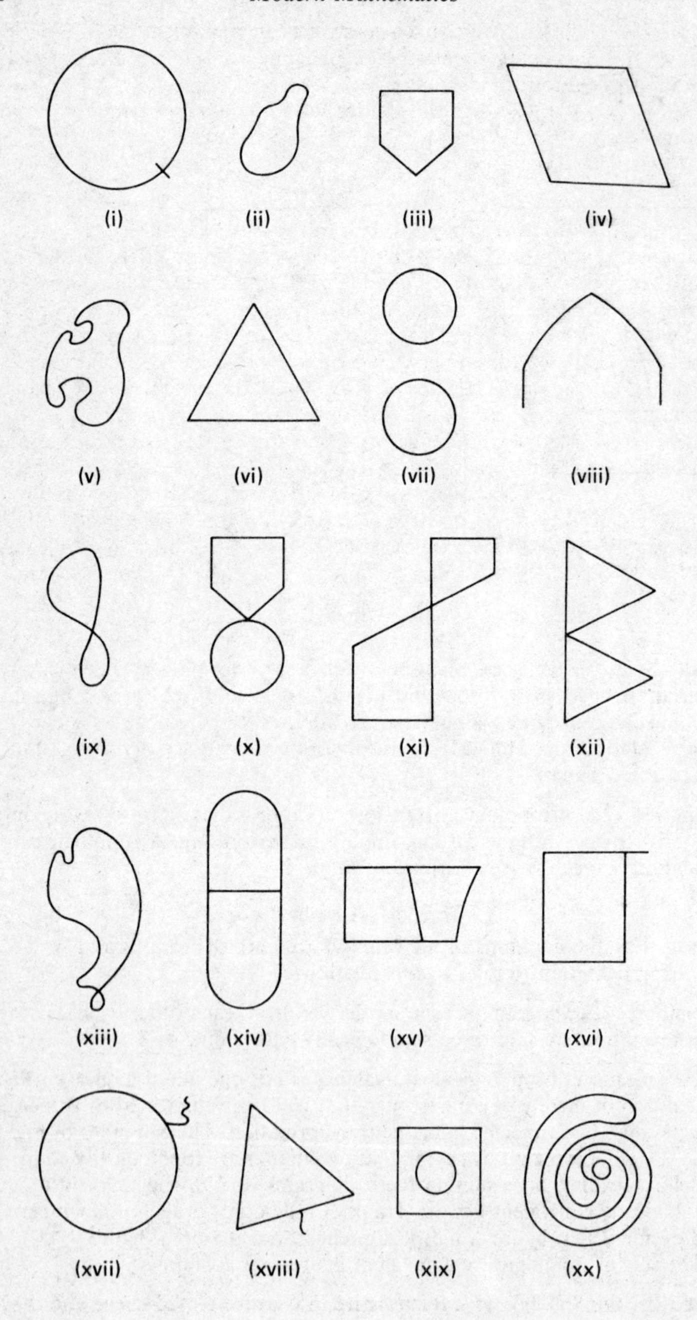
(i)
(ii)
(iii)
(iv)
(v)
(vi)
(vii)
(viii)
(ix)
(x)
(xi)
(xii)
(xiii)
(xiv)
(xv)
(xvi)
(xvii)
(xviii)
(xix)
(xx)

Fig. 12.2

(b) {(ix), (x), (xi), (xii), xiii)}. Each figure has two simple closed loops with one and only one point of intersection in common. Notice that they each divide the plane into three regions.

(c) {(xiv), (xv)}. Unlike (b) these figures have a common boundary of more than one point.

(d) {(xvi), (xvii), (xviii)}.

(e) {(xix), (xx)}.

Diagrams (i)—topologically equivalent to the usual capital letter *Q*—(vii)—two disjoint regions—and (viii)—an arc and not a closed curve, being topologically equivalent to the usual capital letter *C* or *L*—are the odd ones out in the given set in Fig. 12.2. Understand that it might appear that the diagram (viii) could be joined by pulling our rubber sheet together to close the gap, but this not so because we agree that what is broken or disconnected to begin with must remain so. Similarly, for (vii) we may *not* cut (ix) into two in order to get (vii) because cutting or tearing is not allowed.

Whenever we make a true sketch we are creating a topologically equivalent diagram. An aerial photograph of an area might look like Fig. 12.3, which is very nice for showing the relative widths of the roads but is probably less convenient from the point of view of instructing someone to get from the points *A* to *C* (say) than the topologically equivalent line diagram of Fig. 12.4.

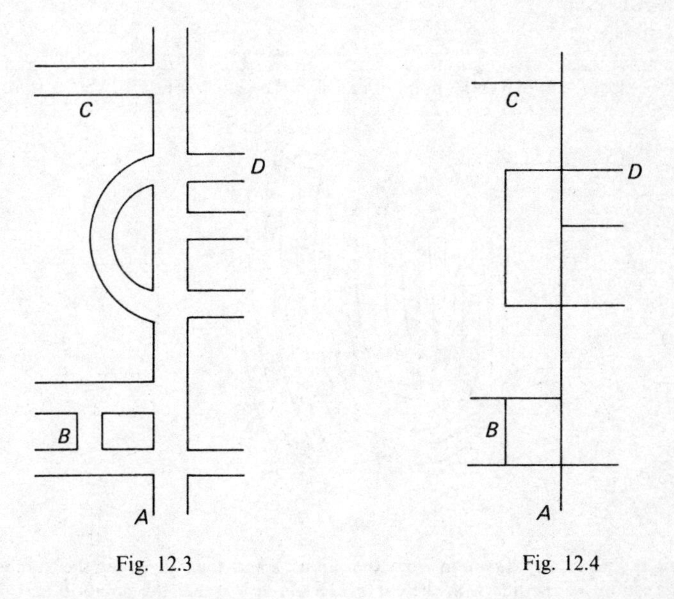

Fig. 12.3

Fig. 12.4

For example, the instructions to someone at the point *B* wishing to get to the point *D* would be no more than 1st right, 1st left, 3rd right. The angles between the roads or turnings are irrelevant to the person walking from *B* to *D*, and they do not need to be told any distances, just as long as the diagram of Fig. 12.4 is topologically equivalent to the true road map of the region.

Exercise 12.1

1. Using Fig. 12.4, give the necessary directions for someone trying to get from (i) A to C, (ii) C to A, (iii) D to A, (iv) C to D.
2. Using Fig. 12.2, which of these figures is topologically equivalent to the following?

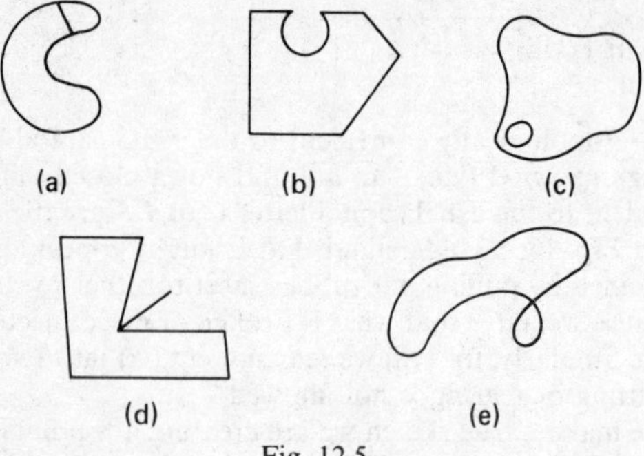

Fig. 12.5

3. Which of the diagrams in Fig. 12.2 are simple closed curves?
4. What would make the following pairs of figures topologically equivalent in Fig. 12.2:
   (a) (i), (xvi)
   (b) (iv), (xviii)
   (c) (iii), (xiv)

5. A simple closed curve separates a plane into two regions. Fig. 12.6 is a diagram of a simple closed curve and the point A is inside the curve. A test for this is to draw a

Fig. 12.6

straight line in any direction from the point A and then to count the number of intersections of the line with the curve. Determine whether the points B and C are inside the curve.

Networks

Any set of points on a surface where each point is connected to at least one other point, is called a network. We can clarify what this means by examining the illustrations in Fig. 12.7.

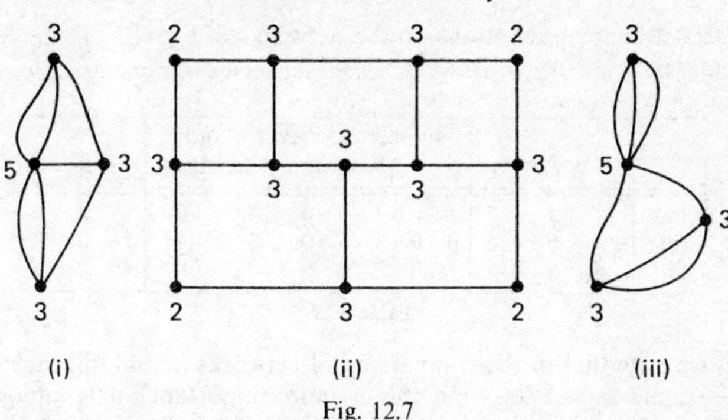

Fig. 12.7

The points in the network are called its **vertices** and it is necessary to emphasise their presence by marking them as • (vertices are also called nodes, i.e. places of rest). Any join of two vertices will be called an **arc** even when the join is a straight line, and every arc will start and end at a vertex of the network. The network in Fig. 12.7 (ii) has 12 vertices connected by 16 arcs and the network of (iii) has 4 vertices connected by 7 arcs. When we compare the different vertices of these two networks it becomes clear that it is possible to classify the vertices according to the number of arcs which leave the vertex. This number will be called the **order of the vertex** and will sometimes be written at the side of the vertex as shown in Fig. 12.7. For example, network (iii) has one vertex of order 5 and three vertices of order 3; so has network (i) but they certainly do not 'look' like the same network. All the vertices of network (ii) are either of order 2 or of order 3.

A vertex will be referred to as either odd or even according to its order. Thus all the vertices of network (i) are odd.

After vertices and arcs the third feature of a network is the regions formed by the arcs on the surface. The network (ii) has six regions, the sixth region being described, for want of a better word, as the 'exterior' of the network. Hence, any simple closed contour such as a circle separates a plane surface (i.e. like this sheet of paper) into two regions although it is not always clear which of those regions we should call the exterior—the equator on the earth's surface being an example of this uncertainty.

Two points will be said to be located in the same region of a network if they can be joined by an arc which does not cut any arc of the network. Clearly it is only possible to pass from one region to another in a network by crossing at least one of the arcs of the network.

If we are to compare networks in order to see if they have any special qualities in common we shall have to take note of the number of

(i) vertices (V), and their order,
(ii) arcs (A),
(iii) regions (R).

The first relationship we shall discover is

$$V + R - A = 2$$

for any network and this will remain true under any rubber sheet transformation. This result is called an invariant under the rubber sheet transformation.

We tabulate our observations on the networks of Fig. 12.7 in the following type of table:

	V	R	A	Number of vertices of order 5	4	3	2	1	Odd	Even	$V+R-A$
(i)	4	5	7	1	0	3	0	0	4	0	2
(ii)	12	6	16	0	0	8	4	0	8	4	2
(iii)	4	5	7	1	0	3	0	0	4	0	2

Table 12.1

Again we note that in the comparison of networks (i) and (iii) each entry in the table is the same. Yet—and this is more important—it is impossible to transform one network into the other, because the joins are not the same: that is, in (i) no vertex of order 3 has more than two joins to the vertex of order 5, yet in (iii) one vertex of order 3 has three joins to the vertex of order 5.

When confronted with any network the slightest touch of curiosity usually leads to doodling which, aside from creating a more artistic outline, will probably reach the stage of trying to draw the network in one continuous line without repeating any of the arcs. At first we may not mind where we start or finish but afterwards it may occur to us that perhaps some networks can be drawn by beginning and ending at the same vertex, yet still without repeating an arc.

Testing any of the networks in Fig. 12.7 will show that it is not possible to draw them in one continuous line without repeating an arc, let alone starting and finishing at the same vertex.

Definition 12.3. A network is said to be **traversable** if the complete network can be drawn in one continuous line without repeating any arc. If the network can be traversed by starting and ending at the same vertex then the network is said to be **unicursal**.

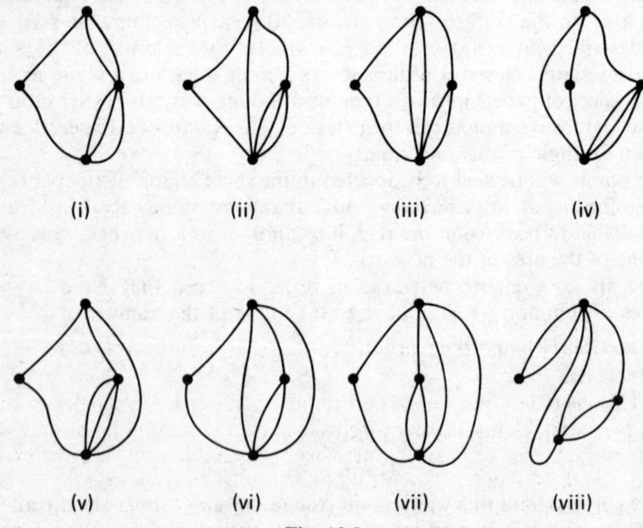

Fig. 12.8

Exercise 12.2

1. Examine the networks of Fig. 12.8 and enter the information as in Table 12.1. (a) Which of these networks is topologically equivalent? (b) Which of these networks are traversable?
2. Suppose that the networks of Fig. 12.8 had pieces of string for the arcs, knotted to give the vertices. If the strings are lifted off the surface and put down anywhere else which of the networks has the same knotted string structure?
3. In Table 12.1 find a relation between A and the sum of the orders of the vertices.
4. Disallowing a vertex being connected to itself by a single arc loop, show that it is not possible to have a three-vertex network with one and only one odd vertex.
5. Draw two networks each having $V = 4$, $R = 2$, $A = 4$ but not being topologically equivalent.

Euler's Results

The invariant result $V + R - A = 2$ is one of Euler's results which we shall meet again later. His other results here concern the construction of networks, and they are as follows:

(a) There must always be an even number of odd vertices. This includes having no odd vertices at all.

(b) For a network to be traversable it must have either two odd vertices or no odd vertices.

(c) For a network to be unicursal it must have all even vertices.

For (a) we may argue that in any network there is a different vertex at each end of an arc (i.e. to any arc there are two vertices). This means that the sum of the orders of the vertices is equal to $2A$, where A is the number of arcs. Clearly $2A$ is an even number. We can appreciate this result with Fig. 12.9 (i) and (ii).

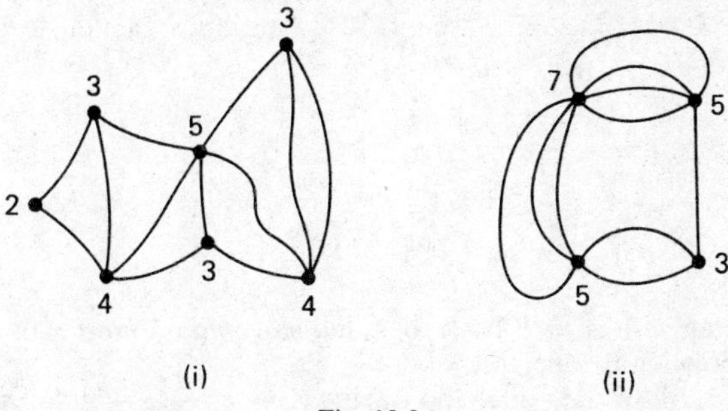

Fig. 12.9

In (i) the sum of the orders of the vertices is $2 + 3 + 5 + 3 + 4 + 3 + 4 = 24 = 2 \times 12$ since there are 12 arcs.

In (ii) the sum of the orders of the vertices $5 + 7 + 5 + 3 = 20 = 2 \times 10$ since there are 10 arcs.

We have seen elsewhere that odd numbers must come in pairs in order to get an even sum. Hence there must always be an even number of odd vertices.

(b) First consider traversing a network and entering an odd vertex in Fig. 12.10 (i) for the first contact with that vertex. This leaves an even number of unused arcs connected to this vertex, each pair being drawn thereafter by going out and back to this vertex until we must finally end at this vertex.

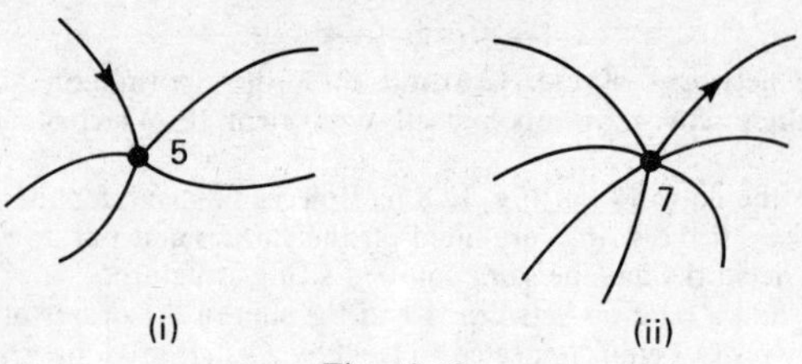

Fig. 12.10

Therefore, in any traversable network if we enter an odd vertex for the first contact with that vertex then we must eventually end there.

Similarly, in Fig. 12.10 (ii) we see that we are starting at an odd vertex. This leaves an even number of arcs connected to this vertex, each pair being drawn thereafter by coming in and going back out. Therefore, in any traversable network if we start at an odd vertex we cannot end there.

It follows that a traversable network may only contain two or no odd vertices.

(c) For an unicursal network one must accept sensible behaviour. An obviously unicursal network like Fig. 12.11 (i) is expected to be drawn unicursally

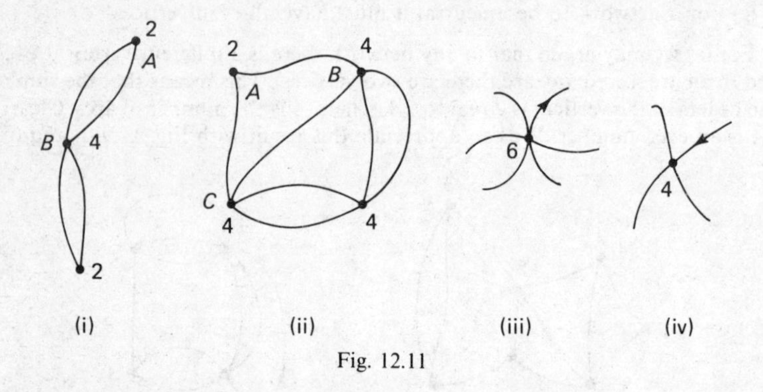

Fig. 12.11

so that starting with *A* to *B* back to *A*, full stop, would not be an acceptable attempt at drawing this network.

Similarly, in (ii) starting at *A* and tracing $A \rightarrow B \rightarrow C \rightarrow A$, full stop, is not really trying.

Using (iii) it can now be seen that if we start at any even vertex we must always end at that vertex. Similarly, in (iv) if we make first contact with this vertex by entering it, then we must (and always can) leave that vertex.

Therefore, provided we use all the arcs available, any all even vertex network will be unicursal.

These results now enable us to discover by inspection if a network is traversable or unicursal.

Neither of the two networks of Fig. 12.9 is traversable because they have more than two odd vertices. Both of the networks in Fig. 12.11 are unicursal because all the vertices are even.

Networks for Maps

Any map divides a surface into regions which might be called countries each separated from the others by borders. Without getting too close to the examples of an atlas, a typical map is shown in Fig. 12.12 (i) with six countries *A*, *B*, *C*, *D*, *E*, *F* and their borders.

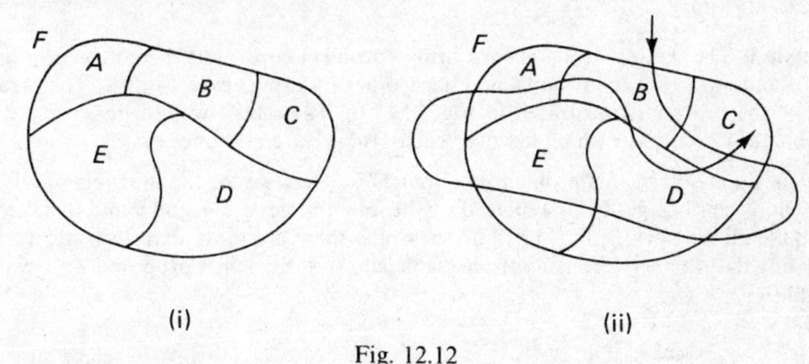

Fig. 12.12

The question of visiting each country by crossing each border once and once only is closely related to the traversability of a network. All we need is some method of reading the same problem into the two situations.

A sample visiting route is shown in (ii) and reveals that we have missed two borders. Readers should try to see if they can improve on this attempt (it is possible to miss only one border): it will help towards seeing the clue to the dual relationship between networks and maps. The clue lies in seeing:

(a) the regions of the map as vertices of a network,
(b) the borders separating regions as arcs connecting the vertices.

With the lettering of Fig. 12.12 (i) we construct the dual network in Fig. 12.13 (i). We see that *F* has a common border with each of the other five countries so this means that in the dual network in Fig. 12.13 (i) the vertex *f* will be connected to each of the other vertices *a*, *b*, *c*, *d*, *e* by one arc. The other connections are: *a* is connected to *f*, *b* and *e*; *b* is connected to *a*, *f*, *c*, *d* and *e*; and so on until we produce the complete dual network from which we see immediately that with more than two odd vertices this network cannot be traversed and so the original map cannot be travelled by crossing each border once and once only.

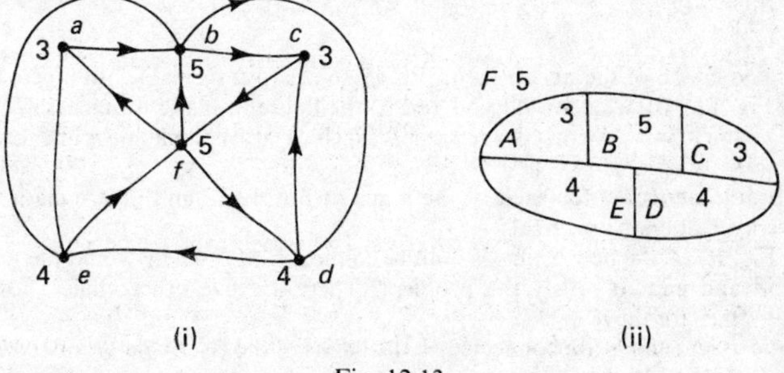

Fig. 12.13

Since this information depends entirely on the order of the vertices there is little point in drawing the dual network. We could merely write the number of borders in the regions and deduce the result from Fig. 12.13 (ii).

The arrows of the network diagram in Fig. 12.13 (i) record the map journey in Fig. 12.12 (ii) in the order *fbcfdefabdc*. Note the omission of travelling the arcs *ae* and *eb*.

Example 1. The famous Koenigsburg bridge problem concerned the connection of two river islands with the river banks and each other by using seven bridges. The arrangement of the bridges is illustrated in Fig. 12.14 (i). The puzzle was to find out if it was possible to walk a path which led over each bridge once and once only.

Solution. Euler solved this problem in about 1736. Here we see the four regions *A*, *B*, *C* and *D* and their bridges being replaced by the four vertices *a*, *b*, *c* and *d* and their arcs, to give the dual network in Fig. 12.14 (ii), in which there are more than two odd vertices, indicating that the network is not traversable and so the original proposed walk was not possible.

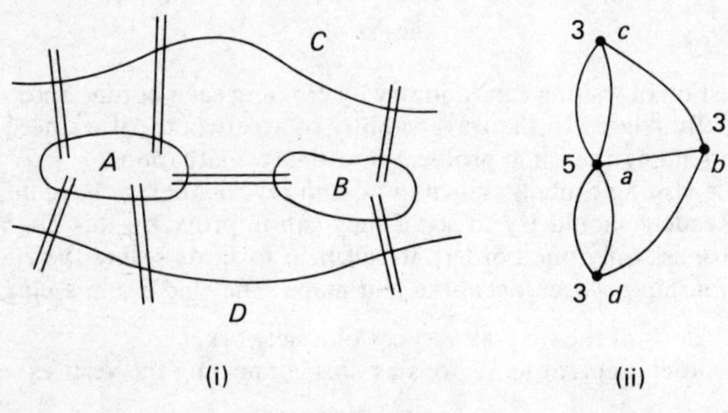

Fig. 12.14

Example 2. Threadwork problems are similar in nature to the above problem. An electrician has to see if he can thread one single piece of wire once and once only through each of the walls as marked in Fig. 12.15 (i).

Solution. Although a few hit or miss sample runs may be made as in (iii) and (iv) we shall not reach a firm conclusion without using Euler's results.

Converting the six regions *A*, *B*, *C*, *D*, *E* and *F* to vertices and noting their order in (ii) we observe that there are more than two odd vertices in the dual network. We now know that the proposed threading is not possible.

Exercise 12.3

1. Determine which of the networks in Fig. 12.16 are (i) traversable, (ii) unicursal.
2. Using Fig. 12.9 (i), what would you add to the figure to make it unicursal?
3. (i) Is a unicursal network traversable? (ii) Is a traversable network necessarily unicursal?
4. Obtain dual networks for each of the maps in Fig. 12.17 and determine if they are either traversable or unicursal.
5. Using Fig. 12.14, if a new bridge is built to connect *C* to *B* where would the walk have to begin and end to satisfy the problem? There are five other choices for such a bridge. What are they?
6. Two mid-river islands are connected to the banks of the river by a set of 12 bridges as shown in Fig. 12.18. (i) Determine whether a traversable or a unicursal walk is pos-

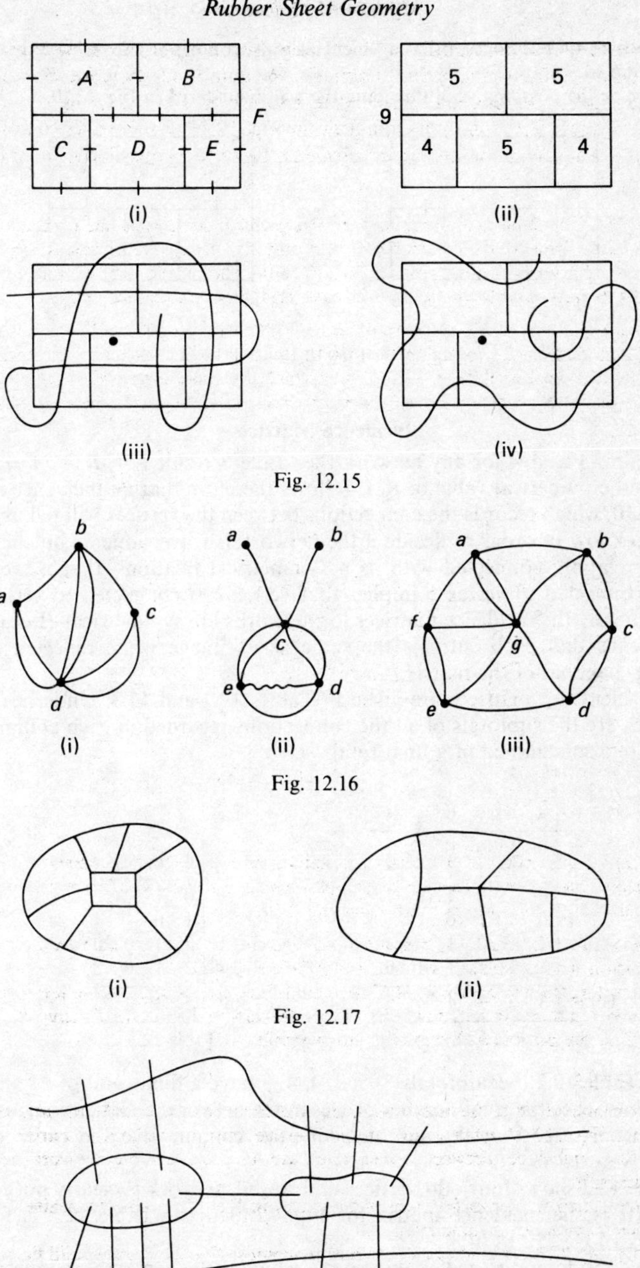

Fig. 12.15

Fig. 12.16

Fig. 12.17

Fig. 12.18

sible over these bridges. (ii) If a unicursal walk is not possible, remove bridges to make it so.

7. Examine the possibilities of threading the wall boundaries in Fig. 12.19.

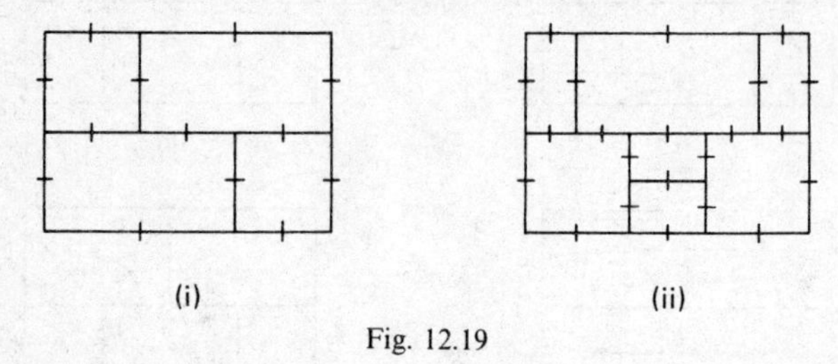

Fig. 12.19

Incidence Matrices

If we know V and A for any network then Euler's result $V + R - A = 2$ will give us the numerical value of R. It follows therefore that an incidence matrix (page 110) which records the connections between the vertices will tell us all we need to know in order to decide if the network is traversable or unicursal.

The relation 'connected with' is a symmetrical relation—that is, vertex a being connected to vertex b implies that vertex b is connected to vertex a—consequently the incidence matrices for networks are symmetrical. (Recall that this means that each entry is the same as its image when reflected in the leading diagonal of the matrix.)

Two incidence matrices are given in Tables 12.2 and 12.3. Underneath the matrices are the subtotals of all the connections recorded in each column and these are also summed in a final total.

	a	b	c	d	e	f	
a	0	1	0	0	1	1	
b	1	0	1	1	1	1	
c	0	1	0	1	0	1	
d	0	1	1	0	1	1	
e	1	1	0	1	0	1	
f	1	1	1	1	1	0	
Total	3	5	3	4	4	5	= 24

Table 12.2

	a	b	c	d	
a	0	1	2	2	
b	1	0	1	1	
c	2	1	0	0	
d	2	1	0	0	
Total	5	3	3	3	= 14

Table 12.3

For Table 12.2 the subtotals 3, 5, 3, 4, 4, 5 have a final total of 24 and this represents twice the total number of arcs in the network, a fact already used on page 227. Notice that the subtotals of each column give the order of the respective vertices.

Table 12.2 shows four odd vertices and so this network is clearly not traversable. (It is the incidence matrix for Fig. 12.13 (i), as Table 12.3 is for the network of Fig. 12.14 (ii).)

If we consider a directed connection in which vertex a may be 'connected *to*' b without b being connected *to* a, it is possible to have a non-symmetrical incidence matrix. We shall not only see such matrices in the next example but also see how their multiplication may produce meaningful results.

Example 3. Table 12.4 is a directed incidence matrix recording the connections of vertices a, b to vertices p, q, r. Table 12.5 is a directed incidence matrix recording the connections of vertices p, q, r to vertices x, y.

A suggested meaning for the matrix information is given in terms of island ferries.

from \ to	p	q	r
a	2	3	1
b	2	1	0

Table 12.4

from \ to	x	y
p	2	1
q	2	1
r	1	2

Table 12.5

Consider that a, b, p, q, r, x and y are islands which are served by boat ferries plying from one island to another. The directional sense of the connection arises from noting that the boats do not necessarily go back and forth, but tour around.

In this interpretation, Table 12.4 shows that there are three ferries from a to q and Table 12.5 shows that there are two ferries from q to x and one from q to y. There must of course be some connections to a and b but this matrix has been omitted.

If we detach the matrices from their headings we can produce the product.

$$\begin{pmatrix} 2 & 3 & 1 \\ 2 & 1 & 0 \end{pmatrix} \begin{pmatrix} 2 & 1 \\ 2 & 1 \\ 1 & 2 \end{pmatrix} = \begin{pmatrix} 11 & 7 \\ 6 & 3 \end{pmatrix} = \text{from} \begin{matrix} a \\ b \end{matrix} \overset{\text{to}}{\overset{x \quad y}{\begin{pmatrix} 11 & 7 \\ 6 & 3 \end{pmatrix}}}$$

which shows among other pieces of information that there are 11 ferries from a to x via p, q and r. If we interchange the matrices and suggest that the ferry connections are 'two-way' then we may introduce a further matrix multiplication (see Exercise 12.4)

$$V + R - A = 2$$

It is possible to see this result become obvious by examining the build-up of a network bearing in mind that no vertex may exist without connection to at least one other vertex. Therefore all networks must start with the basic elements of two vertices and one arc. This is the first entry in the following table, which shows the possible progression towards a final network.

		V	R	A	E	$V + R - A = E$
1		2	1	1	2	basic element
1*a*		3	1	2	2	add one vertex
1*a*		2	2	2	2	add one arc
1*b*		3	1	2	2	add one arc and one vertex
2*a*		3	2	3	2	add one further arc
2*b*		3	2	3	2	add one further arc

Each type of addition now builds up the network and continues to preserve the Euler result.

Exercise 12.4

1. Write out the incidence matrices for the networks of Fig. 12.16.
2. Rearrange the matrix results of the above worked example to show the number of alternative routes from x, y to a, b via p, q and r, assuming that all ferry connections are two-way.
3. There are two airports a, b serving two cities p and q. From p, q it is possible to reach two further places x, y. The matrices for the number of two-way connections are

	p	q
a	3	4
b	2	1

	x	y
p	2	2
q	1	1

(i) Find the matrix expressions for the connections between a, b and x, y via p, q.
(ii) If *all* the connections are reduced by one, what is the matrix for the connections from x, y to a, b via p, q?

4. Take any network whatsoever and say what happens to the values of V, R, A if: (i) an arc is removed, (ii) two extra vertices are added to one arc, (iii) an extra vertex is placed on two adjacent arcs and then joined.

$V + F - E = 2$

The topological networks and figures discussed so far have all been drawn in a plane and we have discovered that Euler's result of $V + R - A = 2$ is invariant for any topological transformation in the plane. Now consider Euler's result applied to some of the standard three-dimensional figures.

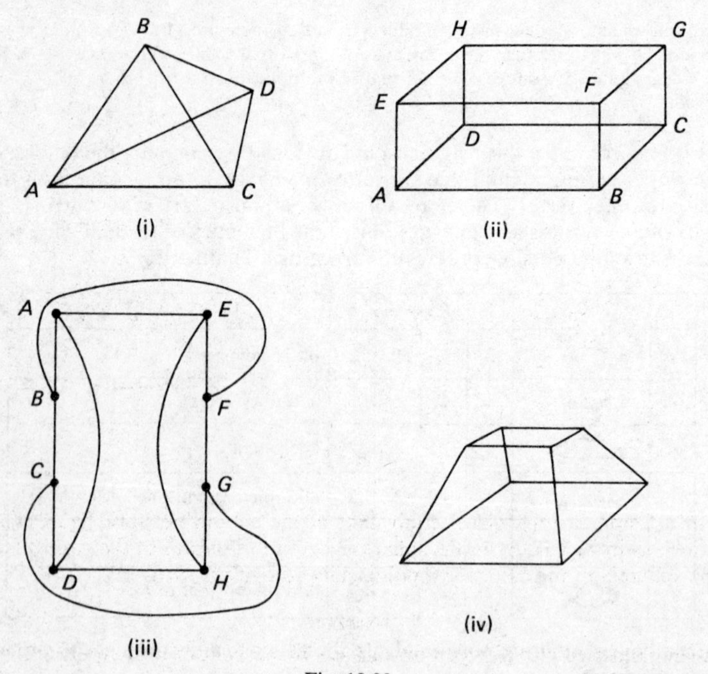

Fig. 12.20

The joins of four points in space form a tetrahedron diagram (i) in Fig. 12.20. This figure has an inside and an outside, so that it divides space into two sections, but remember that Euler's result is for networks with the regions formed by the joins or connections in the networks.

In three dimensions the 'regions' of the figures become 'faces' of the figures and so we change R to F.

The arcs will now become 'edges' of the three-dimensional figure and so we change A to E. Hence, $V + R - A$ has now become $V + F - E$. For the tetrahedron $ABCD$ in Fig. 12.20 (i) we have

$$V = 4, E = 6, F = 4, V + F - E = 2$$

Fig. 12.20 (ii) represents a cuboid, and here we have

$$V = 8, E = 12, F = 6, V + F - E = 2$$

Both of these figures are examples of what are called polyhedra, and like two-dimensional figures they may undergo rubber sheet transformation if we consider them to be drawn on an inflated rubber balloon. Thus, diagrams (ii), (iii) and (iv) in Fig. 12.20 are topologically equivalent, although the restriction of drawing three-dimensional networks on two-dimensional paper is difficult to overcome visually.

Topological equivalents of the cuboid $ABCDEFGH$ may also be drawn on simply connected surfaces, such as the sphere. (A simply connected surface is a surface which is separated into two separate regions by a simple closed curve.)

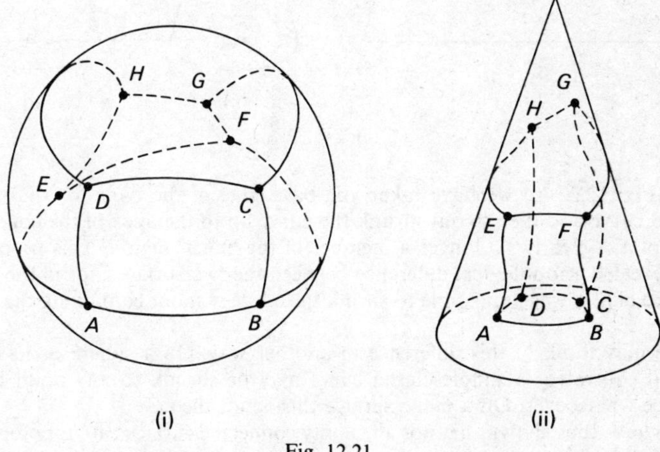

Fig. 12.21

For example, a topological equivalent of the cuboid network figure may be drawn as in Fig. 12.21 on the sphere (i) or on the surface of the cone (ii). Both the sphere and the cone are topologically equivalent surfaces.

Surfaces

A simple closed curve drawn on surfaces like a balloon will separate the surface into two regions; one inner region and one outer region. Strictly speaking, we could consider the surface to be divided into three regions or sets of points—one inside the curve, one on the curve, and one outside the curve—but we are content with our two-region description here.

A surface which is always separated in this manner by a simple closed curve is called a **simply-connected** surface. The two surfaces of Fig. 12.21 were simply connected, and so is the plane surface of this page. At first sight it appears that these surfaces are all topologically equivalent, until we actually try to obtain a cuboid or sphere from the page which you are reading at the moment. Some of the edges would have to be stitched together and this is not allowed in a topological transformation since no edges were stitched before the transformation.

The surfaces in Fig. 12.22 offer an interesting illustration of the difficulty of deciding which region is inside or outside a simple closed curve. A simple closed curve is drawn in the plane surface of Fig. 12.22 (i) and a point A is taken in the inner region. If we now make a topological transformation which shrinks the curve it will always contain A in its inner region.

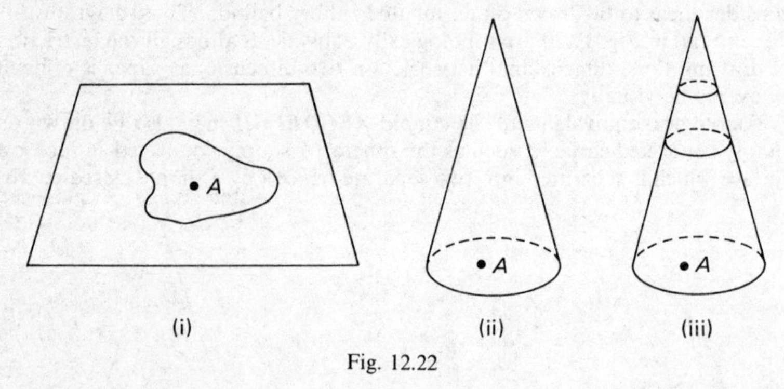

Fig. 12.22

In Fig. 12.22 (ii) we have taken the base edge of the cone for the simple closed curve and then in (iii) shrunk the curve up to the apex of the cone with the point A clearly no longer a member of the 'inner' region. This procedure has revealed a topological difference between the two surfaces, for in the plane surface of (i) it was impossible to shrink the curve without containing the point A.

We may think of this difference in another way. On a sphere or its topological equivalent a simple closed curve may be shrunk to any point of the surface whatsoever. On a plane surface this is not the case.

We have thus shown that not all simply connected surfaces are topologically equivalent.

Definition 12.4. A surface is said to be simply connected whenever any simple closed curve on the surface can be shrunk to an interior point without leaving the surface.

In the above definition we are only suggesting that we may shrink the simple closed curve to be as small as we please; we are not suggesting that it becomes the point—for this would not be an allowable topological transformation.

On the basis of this definition a sphere is a simply connected surface, and so is any other surface which is topologically equivalent to the sphere, such as the cone and cuboid.

A torus is an example of a surface which is not simply connected. This is the doughnut-shaped Fig. 12.23 (i). The dotted simple closed curve could not be shrunk to a point in the surface, and neither could the dashed curve.

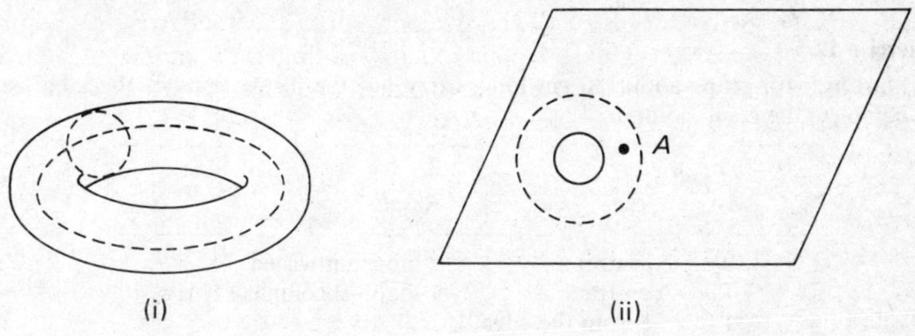

Fig. 12.23

The surface of a piece of paper punctured by a hole shown in Fig. 12.23 (ii) is not a simply connected surface, because the dotted simple closed curve cannot be shrunk to the point A without leaving the surface and crossing the hole.

Experiments in Topology: The Moebius Strip

For all the surfaces we have used so far we have restricted our attention to one side only. In the case of a sphere we only dealt with the outer side, and when speaking of a plane sheet of paper we only referred to one side as being the surface under consideration. At first sight it would appear that all surfaces are two-sided, but, as Moebius (1790–1868) showed, there are surfaces with only one side. The simplest of such surfaces is the Moebius Strip. (For all of the following experiments the reader would find it an advantage to buy a roll of gumstrip paper 5 cm wide.)

Cut a 30 cm strip and stick the ends together after giving the strip half a twist. The diagrams of Fig. 12.24 should make the required arrangement clear.

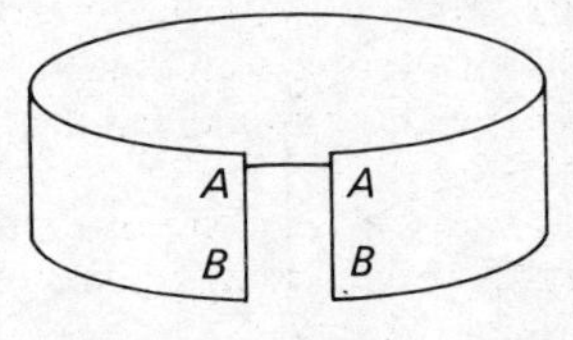

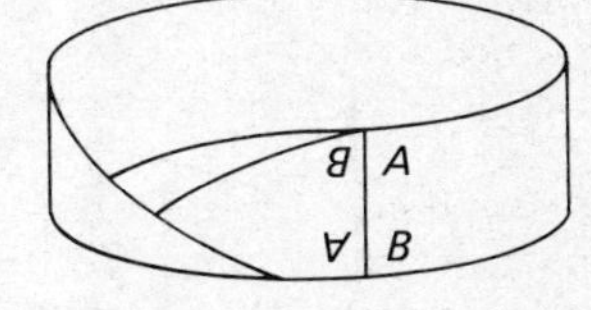

Fig. 12.24

Letter the ends as shown on both sides of the strip (this will mean 4 A's and 4 B's); then join the ends so that the A's and B's come together as shown in the final figure. This is a Moebius Strip.

Make a starting mark on the edge and then follow the edge round the strip as far as possible. You will arrive back at the starting mark. Only one edge!

Now draw a line down the middle of the strip as far as possible and again you will come back to the starting mark—only one side! The Moebius strip is therefore a onc-edged surface which has only one side.

Now take a pair of scissors and make a perforation in the strip, then cut the strip down the middle as far as possible and so return to the starting cut. You should obtain a single loop having two edges and two whole twists. A further cut down the middle of this strip will give—try it first—two separate but interlocked loops.

There are many combinations of cuts and twists which yield surprising results.

Exercise 12.5

1. Starting with strips about 30 cm long, carry out the instructions in the table and complete the result column.

No. of half twists	*Type of cut*	*Result*
0	central	2 loops untwisted
1	central	1 loop—2 complete twists
1	$\frac{1}{3}$ from the edge	
2	central	
2	$\frac{1}{3}$ from the edge	
3	central	
3	$\frac{1}{3}$ from the edge	
4	central	
4	$\frac{1}{3}$ from the edge	

2. Give each of the strips in Fig. 12.25 a half twist in the same direction, then connect up. Cut round the hole as shown by the dotted line.
3. Repeat Question 2, but this time give *A* a twist in the opposite direction to *B*. Cut round the hole as before.

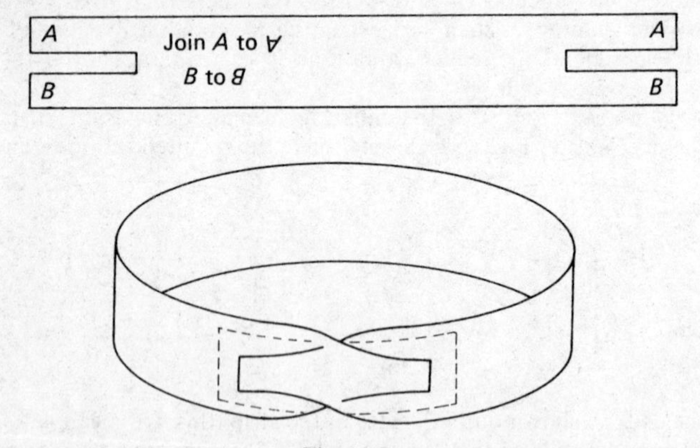

Fig. 12.25

The Punctured Torus

The best example for a punctured torus is the inner tube and valve of a car tyre. Imagine that we have stripes painted round the inner hole on both sides of the torus surface. By threading the tube through the valve (not very likely, but a good topological transformation) we shall produce the change in the direction of the stripes shown in Fig. 12.26.

The following exercise is an easy demonstration of this transformation.

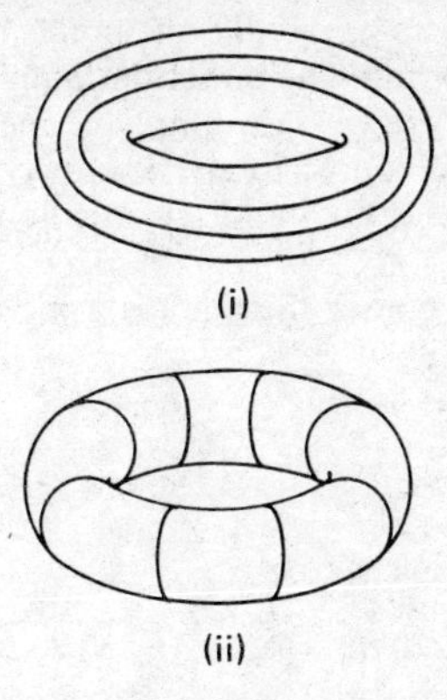

Fig. 12.26

Exercise 12.6

1. Take a pair of striped pyjamas with a draw-string top and stitch the bottoms of the legs together. You now have a punctured torus with the draw-string controlling the valve opening, as in Fig. 12.27 (i). Reach through the valve down the leg and grab where the two ends have been stitched. Pull up through the valve. Close the valve by drawing the string tight. You will now have Fig. 12.27 (ii), which is the topological equivalent of Fig. 12.26 (ii).

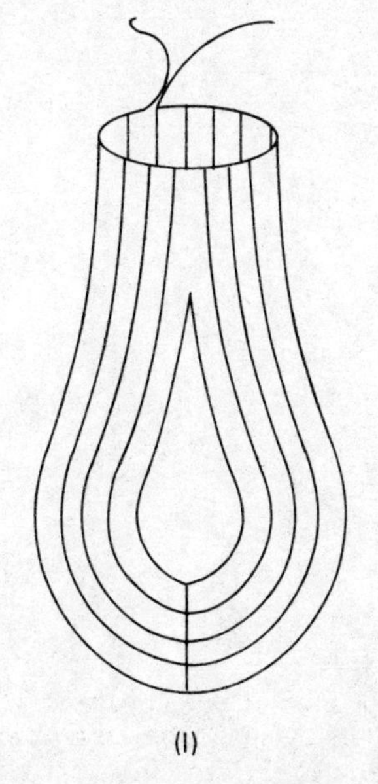

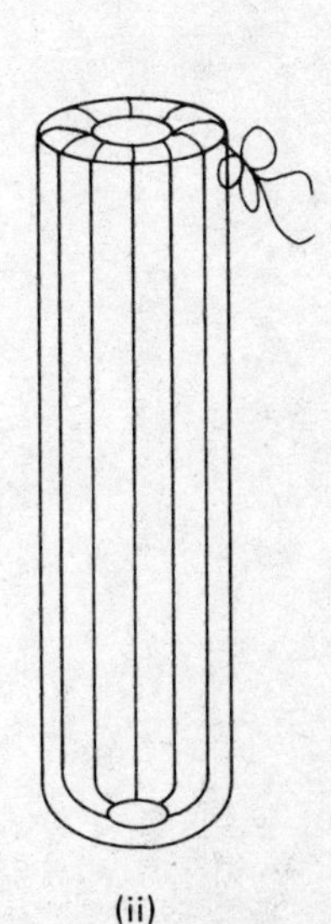

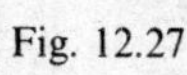

Fig. 12.27

The Four-colour Problem

In colouring a map, two countries or regions which have a common frontier or boundary must be given two different colours. The same colour may be used for countries whose frontiers have only one point in common, as in Fig. 12.28 (i) below. We enquire now into the minimum number of colours necessary to colour any map which is drawn on a simply connected surface.

The first map has been coloured to give the idea of what is meant by colouring the map. Copy the other two and try to colour them using four colours only.

It is always possible to colour these maps using only four colours.

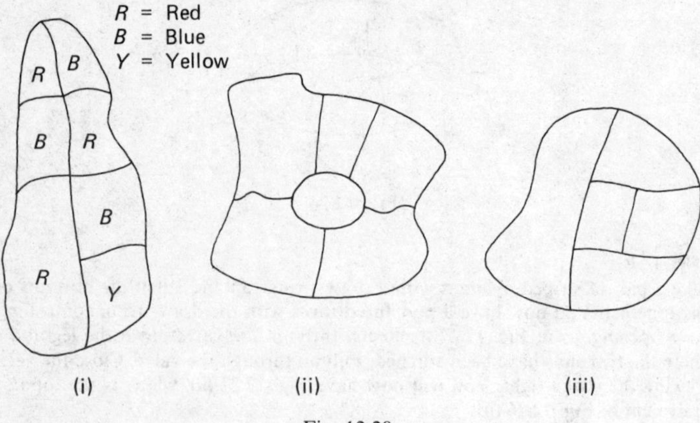

Fig. 12.28

SOLUTIONS

Exercise 1.1

1, 3, 5, 7 well defined. **2, 4, 6** not well defined. **8.** $P = \{b, e, g, i, n\}$
9. $Q = \{g, i, n, e, r\}$ **10.** $R = \{l, e, t, s\}$ **11.** $T = \{e, g, i, n\}$
12. $U = \{b\}$ **13.** $V = \{r\}$ **14.** $W = \{1, 3, 4, 9\}$, i.e. 134.999999 . . .
15. {All the months of the year whose names begin with J}.
16. {All the even numbers between 9 and 17}.
17. {All the odd numbers between 2 and 12}, see Question 1.
18. {The squares of the first five natural numbers}, i.e. $\{1^2, 2^2, 3^2, 4^2, 5^2\}$.
19. {The first five multiples of five}, i.e. $\{5 \times 1, 5 \times 2, 5 \times 3, 5 \times 4, 5 \times 5\}$.
20. {All the rearrangements of the three letter word, *pet*}.
21. $\in$ **22.** $\notin$ **23.** $\notin$ **24.** $\notin$ **25.** $\in$ **26.** $\notin$ **27.** $\in$ **28.** $\in$ **29.** $\notin$ **30.** $\notin$

Exercise 1.2

1. Match $K = \{a, b, c\}$ to set A. $a \to p, b \to q, c \to r$.
Match $T = \{a, b, c, d\}$ to set B. $a \to k, b \to l, c \to m, d \to n$.
No, because $n(K) \neq n(C)$, $n(T) \neq n(C)$.
2. There are six different ways:
$p \to 1, q \to 2, r \to 3; q \to 1, p \to 2, r \to 3; q \to 1, r \to 2, p \to 3;$
$r \to 1, q \to 2, p \to 3; r \to 1, p \to 2, q \to 3; p \to 1, r \to 2, q \to 3.$
3. $1 \to r, 2 \to q, 3 \to p$. **4.** (i) 2, (ii) 24.
5. Each element of the set {1, 2, 3, 4, 5} is half the corresponding element in the set {2, 4, 6, 8, 10}.
6. The most obvious match is given by the following plan. $1 \to 5$, $2 \to 10$, $3 \to 15$, $4 \to 20$, $5 \to 25$ so that each element of the set {1, 2, 3, 4, 5} is one fifth of the corresponding element in the set {5, 10, 15, 20, 25}.
7. $Y = \{7, 8, 9, 10\}$ and the matching is given by $1 \to 7$, $2 \to 8$, $3 \to 9$, $4 \to 10$.
8. $n(E) = 16$. **9.** $n(F) = 33$. **10.** $n(G) = 1$, i.e. $G = \{4\}$.
11. $n(H) = 0$, i.e. $H = \phi$. **12.** $n(J) = 2$, i.e. $J = \{0, 1\}$. **13.** 13, 16, 19.
14. 16, 14, 12. **15.** 80, 85, 90. **16.** 40, 48, 56.
17. 64, 100, 144, i.e. $64 = 8^2$, $100 = 10^2$, $144 = 12^2$.
18. 21, 34, 55, i.e. except for the first two elements each element is the sum of the two previous elements.
19. 13, 17, 19, i.e. prime numbers in order of size. **20.** 5/6, 6/7, 7/8.

Exercise 1.3

1. $\phi, \{a\}$ **2.** $\phi, \{a\}, \{b\}, \{a, b\}$ **3.** ϕ
4. $A = \{1, 3, 5, 7, 9, 11, 13, 15, 17, 19\}$ **5.** $B = \{3, 6, 9, 12, 15, 18\}$
6. $C = B$ **7.** $D = \{5, 10, 15\}$ **8.** $E = \{1, 2, 3, 5, 7, 11, 13, 17, 19\}$
9. $X \not\subset Y$ **10.** $Z \subset Y$ **11.** $X \not\subset Z$ **12.** $\phi \subset Z$ **13.** $\{1, 3\} \subset X$
14. $\{5\} \subset Z$ **15.** $\{1, 5\} \not\subset X$ **16.** T **17.** F **18.** T **19.** F
20. T **21.** F **22.** T **23.** T **24.** F **25.** T **26.** F **27.** F
28. F **29.** T **30.** F

Exercise 1.4

True statements are **1, 2, 3, 5, 6, 7, 8, 10.** Statements **4** and **9** are false.
11. All hardworking people are intelligent. **12.** All students are intelligent people.
13. Not all children are hardworking people.

14. Not all males are students. **15.** k is a hardworking male.
16. p is a student who is not intelligent and is not hardworking, or p is an unintelligent student who is not hardworking.
17. All males are hardworking. **18.** Not all males are intelligent people.
19. Not all intelligent people are students.
20. Not all people who are not students are unintelligent.
21. {children}. **22.** {violet, blue, red}. **23.** {0}.
24. {numbers which are greater than or equal to 7}.
or {numbers $\geqslant$ 7}. **25.** $U' = S' = \phi$

Exercise 1.5

1. {1, 2, 3, 5, 7} **2.** {2} **3.** {4, 8} **4.** {4, 8} **5.** {1, 2, 3, 5, 7} $= A \cap B$
6. B **7.** D' **8.** {11} **9.** {2} **10.** {4, 8} **11.** ϕ **12.** Yes **13.** Yes
14. $B =$ {all prime numbers from 1 to 11 inclusive}.
$C =$ {all even numbers from 2 to 10 inclusive}.
$B \cap C = \{2\} =$ {all even prime numbers}.
15. (i) {green parrots}.
(ii) {talking parrots}.
(iii) {green parrots which talk}.
(iv) {green birds which do not talk}.
(v) {non talking green birds which are not parrots}.
16. (i) Fig. 1.5 or Fig. 1.6.
(ii) Fig. 1.6 and possibly $A = B$, i.e. $A \subseteq B$.
(iii) Fig. 1.4.
(iv) Same as for (i) but with A and B interchanged.
(v) $A = B$

Exercise 1.6

1. {1, 2, 4, 6, 8} **2.** {1, 3, 5, 7, 9} **3.** {2, 3, 4, 5, 6, 7, 8, 9}, $F' = \{1\}$
4. {2, 4, 5, 6, 8}, $G' = \{1, 3, 7, 9\}$ **5.** {4, 5, 8} **6.** U **7.** A' **8.** C
9. $A \cup B = \{2, 3, 4, 5, 6, 7, 8, 9\}$; $A \cap B = \phi$ so that $A \cup B = A \triangle B$.
10. $B \cup C = \{2, 3, 4, 6, 8, 9\}$, $B \Delta C = \{2, 3, 4, 8, 9\}$.
11. {3, 5, 7, 9}, {2, 4, 6, 8}, $H \cup K = A \cup B$.
12. $B \Delta A = H \cup K$. **13.** {green apples}. **14.** {lemons or cherries}.
15. {apples or lemons}, $A \cup L = A \Delta L$ because $A \cap L = \phi$.
16. {lemons or ripe fruit} note that 'lemons' includes both ripe and unripe lemons.
17. {unripe lemons or ripe fruit excluding ripe lemons}.
18. {cherries or unripe fruit} note that this includes ripe cherries.
19. $K =$ {ripe apples}, $K \cup G =$ {ripe apples or green fruit}.
20. $H =$ {unripe apples}, $H \cup G =$ {unripe apples or green fruit}.

Exercise 1.7

1. {1, 3, 5, 7, 9} **2.** {0, 5, 6, 7, 8, 9} **3.** {0, 1, 2, 6, 7}
4. {1, 2, 3, 4, 5, 9} **5.** {4} **6.** ϕ **7.** U **8.** {0, 1, 2, 3, 4, 6, 7, 8}
9. $\{2, 4\} \cup C = \{2, 3, 4, 5, 8, 9\}$ **10.** $A \cap \{1, 2, 3, 4, 5, 8, 9\} = \{2, 4, 8\}$
11. {5, 9} **12.** {0, 2, 3, 4, 6, 8} **13.** {2, 4, 8} **14.** {5, 7, 9}
15. {0, 1, 3, 5, 6, 7, 8, 9} **16.** {1} **17.** U **18.** R_5, R_7 **19.** R_6, R_7
20. R_5 (Think of: in A and in B and outside C.)
21. R_3 (Think of: outside A and outside B and inside C.) **22.** R_8.

Exercise 1.8

1. (i), (ii), (iv), (v) no solution. (iii) and (vi) {0, 2, 3, 4, 5, 6}.
2. (i) {0, 1, 2 and any selection from 5, 6, 7}, (ii) and (iv) no solution. (iii) {3, 4}, (v) {3, 4 and any selection from 0, 1, 2}, (vi) {3, 4}.
3. (i), (ii), (iv), (v) no solution. (iii) and (vi) {7, 3, 4}.

4. (i) and (v) no solution. (ii) {4, and any selection from 1, 2, 3}, (iii) and (vi) {4}, (iv) {1, 2, 3 and any selection from 0, 5, 6, 7}.
5. (i) and (v) no solution. (ii), (iii) and (vi) {1, 2}.
(iv) {any selection of elements not including 1 or 2}.
6. $X \subseteq A'$, i.e. any subset of A'. **7.** $X = \phi$ **8.** X may be any subset of U.
9. $X = A$ **10.** $X = \phi$

Exercise 1.9

1. 4 **2.** 1 **3.** 15
4. 63 and 38. Insufficient information to gain an exact number because we are not told that *all* members of the group are either blue-eyed or fair-haired.
5. 10 or 9. The non-eater of cheese is either a male or female. Knowing that the number of guests is 25 is irrelevant.
6. (a) 60%. (b) 9% + 7% + 11% = 27%. Use a n(set) Venn diagram and start by filling in $n(A \cap B \cap C) = 1$.
7. (i) 7; (ii) 2. Use the formula

$$n(H \cup D \cup B) = 100 = n(H) + n(D) + n(B) - n(H \cap D) - n(D \cap B) - n(B \cap H) + n(H \cap D \cap B)$$

8. Put $n(C \cap M \cap S) = x$ on the n(set) Venn diagram and produce the equation $51 + x = 56$. (i) 5, (ii) 3.

Exercise 2.1

1. (i) $1 + 3 + 5 + 7 + 9$
$1 + 3 + 5 + 7 + 9 + \ldots + 21$
$2^2, 3^2, 4^2, \ldots$ the tenth line gives 11^2.
(ii) $1 + 4 \times 6 = 5^2$
$1 + 10 \times 12 = 11^2$
(iii) $5^2 - 4^2 = 3^2$, $13^2 - 12^2 = 5^2$, $25^2 - 24^2 = 7^2$, $41^2 - 40^2 = 9^2$, $61^2 - 60^2 = 11^2, \ldots, 221^2 - 220^2 = 21^2$
The pattern may be seen as increasing the numbers on the left-hand side by 8, 12, 16, 20, 24, 28, 32, 36, 40 to start the lines with 13, 25, 41, 61, 85, 113, 145, 181, 221, etc.
(iv) Similar to (iii) but producing squares of even numbers on the right-hand side.
$5^2 - 3^2 = 4^2$, $10^2 - 8^2 = 6^2$, $17^2 - 15^2 = 8^2$, $26^2 - 24^2 = 10^2$, $37^2 - 35^2 = 12^2, \ldots 122^2 - 120^2 = 22^2$
(v) We see this as $1 \to 2^1, 2 \to 2^2, 3 \to 2^3, 4 \to 2^4, 5 \to 2^5, \ldots 10 \to 2^{10}$
2. The values of the lines are 1, 3, 6, 10. The formula gives the value of the line. Thus $n = 4$ gives $\frac{4}{2} \times 5 = 10$, which is the value of the fourth line. The value of the 200th line is $\frac{200}{2}(201) = 20100$.
3. The values of the lines are $1, 3^2, 6^2, 10^2$, i.e. the squares of the results in question 2. Therefore the value of the 20th line is $\left(\frac{20}{2} \times 21\right)^2 = 210^2 = 44100$.
4. $n = 1, 2, 3, 4$ gives the values of 1^2, $1^2 + 2^2$, $1^2 + 2^2 + 3^2$, $1^2 + 2^2 + 3^2 + 4^2$, respectively. Putting, for example, $n = 12$ gives $650 = 1^2 + 2^2 + 3^2 + \ldots + 12^2$.
5. The expression gives a prime number for all values of n from 0 to 40 inclusive but not for $n = 41$, i.e. the expression does not always give a prime number for n = whole number.
6. Prime numbers can be put in the form $6n \pm 1$.

Exercise 2.2

1. Informally let each number be represented by a rectangle of width 3 and any length. Put two rectangles together to make a longer rectangle but still of width 3.

Formally, put $3n$ for one number and $3k$ for the other number. The sum is $3n + 3k = 3(n + k)$, which is a multiple of 3. (Note that the difference $3(n - k)$ is also a multiple of 3.)

2. The second player has only two choices: (i) fill a corner square (which leads to a draw) or (ii) fill a non-corner square, in which case he loses.

3. (i) All elch's are trids. No trid is a quid. ∴ No elch is a quid.
(ii) All pards are quids. No elch is a quid. (We have just proved this.) ∴ No elch is a pard.

4. n^2 must end in 0, 1, 4, 5, 6, 9. $2m^2$ must end in 0, 2, 8.
∴ If $2m^2 = n^2$ then $2m^2$ and n^2 must end in 0.
∴ m^2 must end in 0, 5 and therefore m must end in 0, 5.
Since n^2 ends in 0 then n must end in 0.
∴ The only possibility is that m ends in 0, 5 and n ends in 0, and this will mean that 5 will be a common factor of m and n, which is not allowed. Hence it is not possible to have m and n so that $2m^2 = n^2$.

Exercise 2.3

1. T **2.** F **3.** not a statement **4.** T **5.** F **6.** not a statement
7. not a statement **8.** not a statement **9.** F **10.** T **11.** $\{-4, 4\}$
12. $\{-1\}$ **13.** ϕ **14.** $\{-2, 4\}$ **15.** $\{0, 1\}$

Exercise 2.4

1, 2. x(T), y(T), $x \wedge y$(T), $x \vee y$(T) **3.** x(T), y(F). $x \wedge y$(F), $x \vee y$(T)
4. x(F), y(F), z(T), $x \wedge y \wedge z$(F), $x \vee y \vee z$(T)
5. This is an interesting example because as outlined in the definition of a statement none of the sentences x, y, z is a statement, yet the compound sentence is a statement. $x \wedge y \wedge z$(F), $x \vee y \vee z$(T) because one and only one of the sentences x, y, z must be true.
6. As for 5. **7.** x(T), y(T), $x \wedge y$(T), $x \vee y$(T)
8. x(T), y(T), z(T), $x \wedge y \wedge z$(T), $x \vee y \vee z$(T) **9.** $\{2, 3\}$
10. Any number between 5 and 10.
11. $(x \wedge y) \vee (z \wedge y) = (x \vee z) \wedge y$; this is similar to the result in sets $\{(A \cap B) \cup (C \cap B) = (A \cup C) \cap B\}$, and shows that the distributive law holds for $\wedge$ over $\vee$ (page 19).
12. (i) Operation $\vee$ is associative (page 17).
(ii) This proves that the distributive law holds for $\vee$ over $\wedge$ (page 19). The truth values for $x \vee (y \wedge z)$ will read F, F, F, T, T, T, T, T.

Exercise 2.5

1. (i) F, F, F, T
(ii) tautology
(iii) F, T, T, F
(iv) F, T, F, F
(v) T, T, F, F

2. (i) T, T, T, F
(ii) F, F, F, T
(iii) tautology
(iv) contradiction

3. F, T, T, F: $\overline{x \equiv y}$

Exercise 2.6

1. T **2.** T **3.** T **4.** T **5.** T **6.** F **7.** F **8.** T
9. F. A rectangle which is not a square satisfies this condition.
10. T. But not sufficient.

Exercise 2.7

1. Th(T), Con(F), (e.g. a bucket of water), In(F), Conpos(T).
2. Th(T), Con(F), (e.g. 15), In(F), Conpos(T).
3. Th(T), Con(F), (e.g. $10 \times 2 = 20$), In(F), Conpos(T), (e.g. $2.4 \times 2 = 4.8$)

4. Th(F), Con(T), In (T), Conpos(F).
5. Th(T), Con(T), In(T), Conpos(T).
6. Exactly as for Example 21 (page 46).

Exercise 2.8

1. T, T, F, T 2. T, T, F, T 3. tautology 4. tautology
5. F, F, T, F, F, F, T, F 6. F, T, T, F 7. T, F, T, F, T, T, T, T
9. x: price goes up, y: people will not buy, $\bar{x}$: price comes down. $\{(x \rightarrow y) \wedge (y \rightarrow \bar{x})\} \rightarrow (\bar{x} \rightarrow \bar{y})$, has truth values T, F, T, T.
10. It is cooking which makes the room dangerous: therefore cooking in the bedroom converts the bedroom into a dangerous kitchen.

Exercise 3.1

1. $1 + 1 = 1, 0 + 0 = 0; 1.1 = 1, 0.0 = 0$ 2. $\bar{x}, \bar{x}$
3, 4. $x.x + x.y = x + x.y = x$ 5. $y.x + y.y = y.x + y = y$
6. $z.x + z.y + z.\bar{z} = z.x + z.y = z.(x + y)$ simplest form.
7. y because $x + \bar{x} = 1$. 8. y because $x.\bar{x} = 0$.
9. y easier to see by tracing the circuit for the original form.
10. Draw circuit for original form and note that current flows if $x = y = 1, x = y = 0$.
11. 1 since $x + \bar{x} = 1, y + \bar{y} = 1$.

Exercise 3.2

1. These are examples of the distributive law for . over +.
   (i) $y.x + z.x$, which may be written as $x.y + x.z = x.(y + z)$.
   (ii) $y.y + z.y = y + z.y$.
   (iii) $y.z + z.z = y.z + z$.

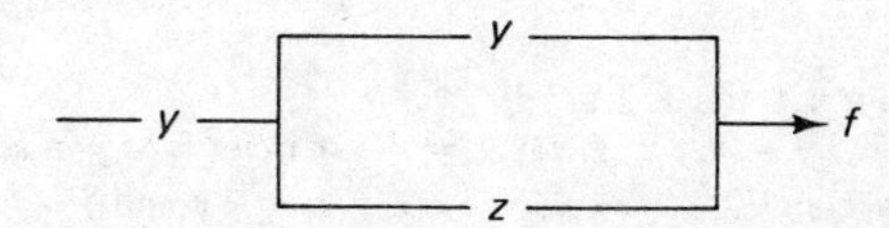

2. From the circuit we see that the flow of current is entirely dependent on y. Thus $f = y$, i.e. $y.(y + z) = y$.
   Return to Question 1 (iii) and note $(y + z).z = z$. This is called the absorption law.
3. (i) $x.x + x.z + y.x + y.z = x + y.z$ (i.e. x absorbs $x.z$, $x.y$).
   (ii) $x.z + x.y + z.z + z.y = x.y + z$ (i.e. z absorbs $x.z$, $z.y$).
4. (i) $x.\bar{y} + x.z + x.y + x.\bar{z} + x = x$ (i.e. x absorbs all of the other four terms).
   (ii) $\bar{x}.y + \bar{x}.\bar{z} + x.\bar{y} + x.\bar{z} = \bar{x}.y + x.\bar{y} + \bar{z}.(x + \bar{x}) = \bar{x}.y + x.\bar{y} + \bar{z}$
5. (i) $f_1 = x.\bar{x} + x.y + \bar{x}.\bar{y} = x.y + \bar{x}.\bar{y}$
   (ii) $f_2 = x.y + \bar{x}.\bar{y}$

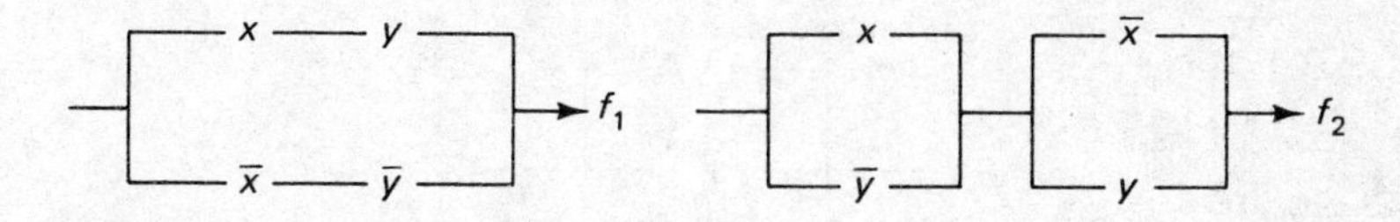

Exercise 3.3

1. (i) $f_1 = \bar{x}.\bar{y}.\bar{z} + \bar{x}.\bar{y}.z + x.y.\bar{z}$
   $= \bar{x}.\bar{y}.(\bar{z} + z) + x.y.\bar{z} = \bar{x}.\bar{y} + x.y.\bar{z}$
   (ii) $f_2 = x.y.z + x.y.\bar{z} + x.\bar{y}.\bar{z} = x.y + x.\bar{y}.\bar{z} = x.(y + \bar{y}.\bar{z})$
   (iii) $f_3 = 0$ is the same as $f_2 = 1$.
   (iv) $f_4 = \bar{x}.\bar{y}.\bar{z} + \bar{x}.y.\bar{z} + x.y.\bar{z} = \bar{x}.\bar{z} + x.y.\bar{z}$
   $= \bar{z}.(\bar{x} + x.y)$

2. (i) $x \not\equiv y = \bar{x}.y + x.\bar{y}$

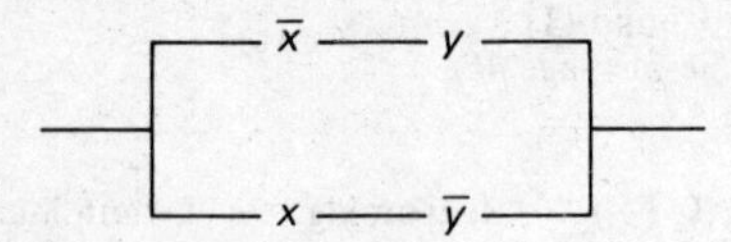

(ii) $x \rightarrow y = y + \bar{x}$

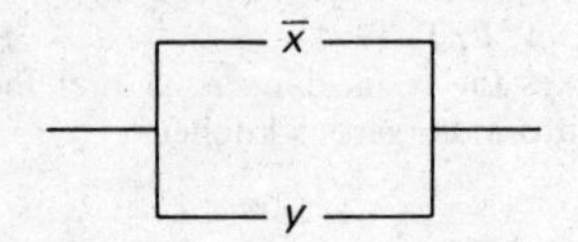

3. If $y = 0, f_1 = 0$, but $f = 1$ (Fig. 3.17) if $x = 1$; hence $f_1 \neq f$.

4. (i) $\bar{x}.y + x.(\bar{y} + \bar{z})$

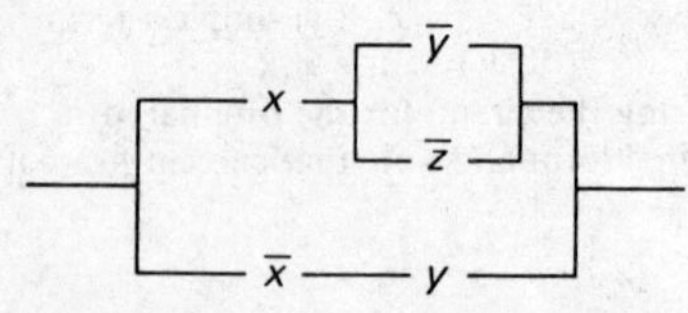

(ii) $(\bar{x}.y.z + x.\bar{z}).(\bar{x}.y.\bar{z} + x.\bar{y}) = x.\bar{y}.\bar{z}$, this result is easily gained by inspecting the current flow in the circuits

x —— $\bar{y}$ —— $\bar{z}$

Exercise 3.4

1. (i) $x.\bar{y}$; (ii) $\bar{x}.y$; (iii) $x.y$; (iv) $\bar{x}.\bar{y}.\bar{z}$.

2. (i) $f = 1$ on lines 7, 8. $f = x.y.z + x.y.\bar{z} = x.y$. Circuit is x, y in series.

(ii) $\bar{f} = 1$ on lines 1 and 2 $\therefore \bar{f} = \bar{x}.\bar{y}.\bar{z} + \bar{x}.\bar{y}.z = \bar{x}.\bar{y}$ and $f = \bar{\bar{f}} = \overline{\bar{x}.\bar{y}} = x + y$. Circuit is x, y in parallel.

3. $\bar{f} = 1$ on lines 1, 3, 5 only (similar to worked Example 11, page 60)

$$\therefore \quad \bar{f} = \bar{x}.\bar{y}.\bar{z} + \bar{x}.y.\bar{z} + x.\bar{y}.\bar{z}$$
$$= \bar{x}.\bar{z}.(\bar{y} + y) + x.\bar{y}.\bar{z}$$
$$= \bar{x}.\bar{z} + x.\bar{y}.\bar{z}$$
$$\therefore \quad f = \bar{\bar{f}} = (x + z).(\bar{x} + y + z) = x.\bar{x} + x.y + x.z. + z.\bar{x} + z.y + z$$
$$= x.y + z \text{ (by absorption law)}$$

Circuit: Interchange x and z in Fig. 3.10.

4.

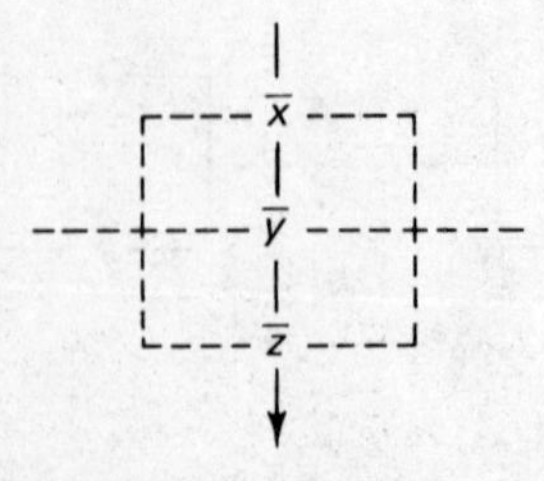

This is DeMorgan's result $\overline{(x + y + z)} = \bar{x}.\bar{y}.\bar{z}$.

5. This question has written x for $\bar{x}$. By inspection $f = \bar{x} + \bar{y}.\bar{z}$.

6. (i) $f = \bar{y}.\bar{z}.(\bar{y} + z) = \bar{y}.\bar{z}$; circuit $\bar{y}$ and $\bar{z}$ in series.

(ii) $f = (\bar{x} + \bar{z}).z = \bar{x}.z$; circuit $\bar{x}$ and z in series.

7. 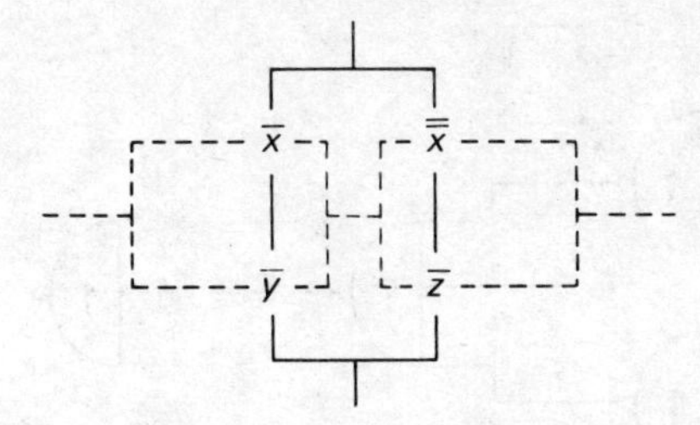

The circuit for $\bar{f} = (x + y).(\bar{x} + z)$ is in dotted outline. The circuit for f is in solid line. By inspection $f = \bar{x}.\bar{y} + x.\bar{z}$.

Exercise 3.5

1. (i) $x + y$; (ii) $\bar{x} + y$; (iii) $\bar{x} + \bar{y}$; (iv) $\bar{x} + y + \bar{z}$; (v) $x.y + 1 = 1$; (vi) $x.y + x = x$; (vii) $x.y + y = y$; (viii) $\bar{x}.\bar{y} + \bar{z}$.
2. (i) $\bar{x}.\bar{y}$; (ii) $x.\bar{y}$; (iii) $x.y$; (iv) $x.\bar{y}.z$; (v) 0; (vi) $\bar{x}$; (vii) $\bar{y}$; (viii) $z.(x + y)$.
3. (i) $x.y$; (ii) $\bar{x}.y$; (iii) $\bar{x}.\bar{y}$; (iv) $\bar{x}.y.\bar{z}$; (v) $x.y.1 = x.y$; (vi) $x.y.x = x.y$; (vii) $x.y.y = x.y$; (viii) $\bar{x}.\bar{y}.\bar{z}$.
4. (i) $\bar{x} + \bar{y}$; (ii) $x + \bar{y}$; (iii) $x + y$; (iv) $x + \bar{y} + z$; (v) $\bar{x} + \bar{y}$; (vi) $\bar{x} + \bar{y}$; (vii) $\bar{x} + \bar{y}$; (viii) $x + y + z$.
5. Two AND gates, one each for $x.z$ and ( ).( ).
   Two OR gates, one each for (+) and (+).
   Suggested circuit

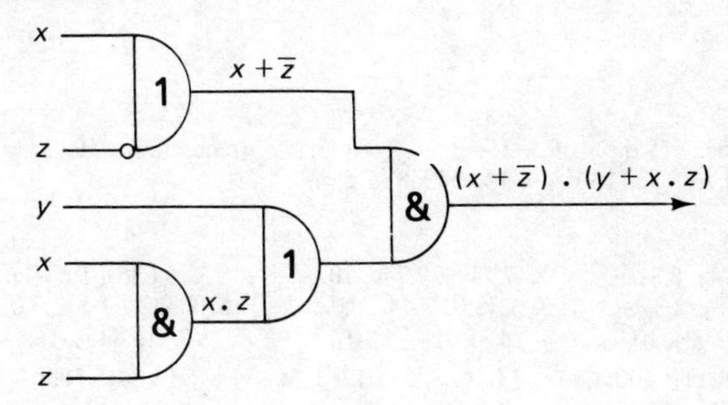

6. $f = x + \bar{y}.\bar{z} = x + \overline{y + z}$

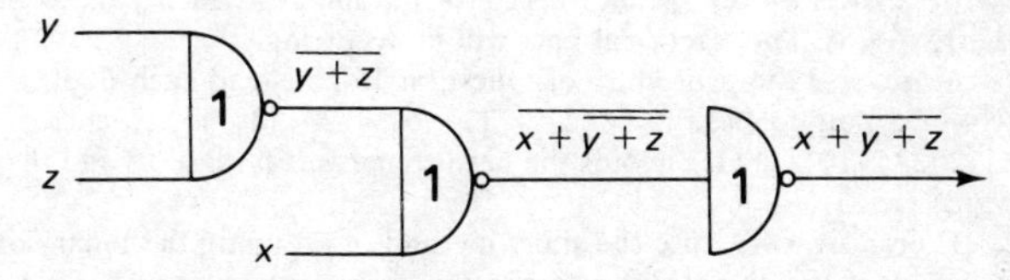

The last NOR gate is used as a negater.

7. $f = x + \bar{y}.\bar{z} = \overline{\bar{x}.(\overline{\bar{y}.\bar{z}})}$ by DeMorgan's law.

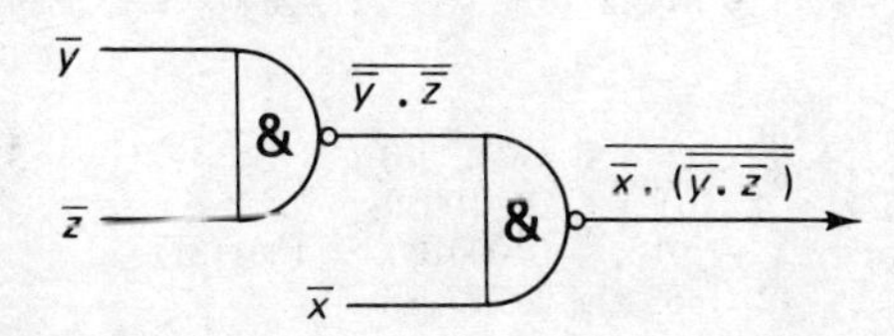

8. (i) $x.\bar{y} = \overline{\bar{x} + y}$, $\bar{x}.y = \overline{x + \bar{y}}$

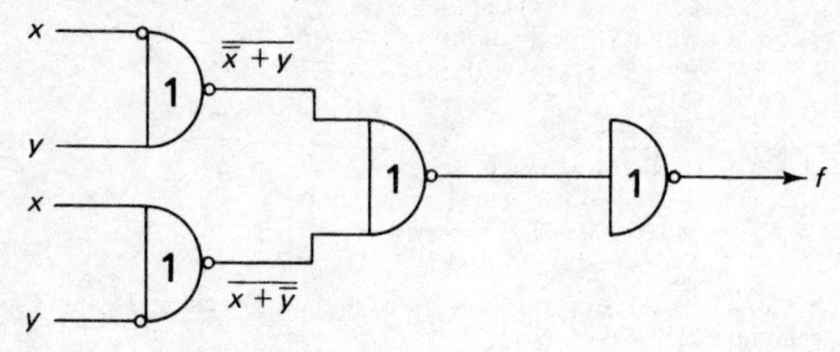

(ii) $A + B = \overline{\bar{A}.\bar{B}}$; $f = \overline{(\overline{x.\bar{y}}).(\overline{\bar{x}.y})}$

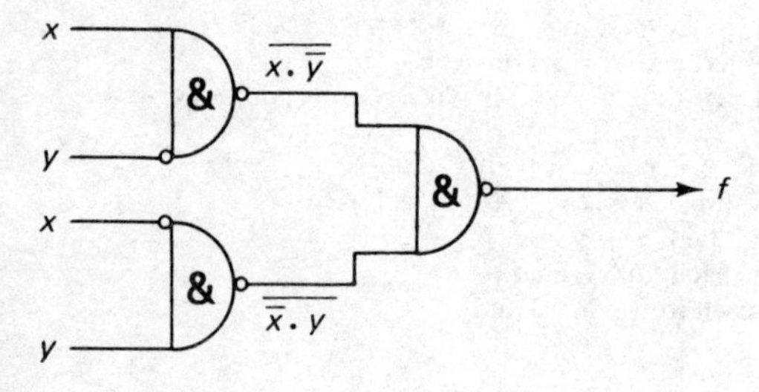

Exercise 4.1

1. 26 **2.** 20 **3.** 14 **4.** 81 **5.** 95 **6.** 31 **7.** 6.8
8. $9\frac{13}{49} = 9.26531$ (5 d.p.) **9.** $33 + \frac{1}{6} + \frac{4}{36} = 33.2777$ (recurring) **10.** $5\frac{3}{8} = 5.375$
11. $19\frac{19}{25} = 19.76$ **12.** $130\frac{28}{81} = 130.34568$ (5 d.p.)

Exercise 4.2

1. 8.835_9, 8.84_9, 8.8_9. **2.** 7.744_8, 7.74_8, 10.0_8. **3.** 5.636_7, 5.64_7, 6.0_7.
4. 2.233_6, 2.23_6, 2.3_6. **5.** 1.433_5, 1.44_5, 2.0_5. **6.** 2.332_4, 3.00_4, 3.0_4.
7. 10.000_3, 10.00_3, 10.0_3. **8.** 1.011_2, 1.10_2, 1.1_2. **9.** 16.835_9, 16.84_9, 16.8_9.
10. 11.111_2, 11.11_2, 100.0_2. **11.** 415_7 **12.** $0.\overline{3630}_7$ **13.** $415.\overline{3630}_7$
14. $120.\overline{1012}_3$
15. $16.\overline{35}_9$, write answer 14 as 01, 20.10, 12, 10, 12, and read each pair to base 3, i.e. $10_3 = 3_9$, $12_3 = 5_9$. The fractional part will be recurring.
16. Here we can reverse the procedure of Question 15 and read each digit in 468.735_9 separately in a pair to base 3, i.e. $4_9 = 11_3$, $6_9 = 20_3$, $8_9 = 22_3$, etc., to gain the answer 112022.211012_3. This avoids the need to convert to base 10 and then to base 3 afterwards.
17. 122.312_4; 32.66_8, $1A.D8_{16}$, use the short method of grouping the binary digits.
18. Make a group of the three binary digits for each digit to base 8, i.e. $6_8 = 110_2$, $7_8 = 111_2$, etc. 110111.100011_2, 313.203_4.
19. $234.30\bar{3}_5$ **20.** $94\frac{7}{16}{}_{10} = 94.4375_{10} = 234.2343_6$
21. $143\frac{22}{49}{}_{10} \approx 143.4490_{10} \approx 1033.21\overline{103}_5$. **22.** $26.\bar{8}_{10} \approx 26.8889_{10} \approx 32.7071_8$.

Exercise 4.3

1. 7, 6, 5, 4 **2.** 6, 5, 4, 3, 2 **3.** 1121_6, 1010_7 **4.** 1221_5 **5.** 400_5
6. 660_7 **7.** 1667_8 **8.** 1012_3 **9.** 10010_2
10. The sum becomes $22_4 + 1010_4 + 01000100_4 = 1001132_4$
11. 1013.12_6, 41.20_6 **12.** 1463.53_8, 263.77_8

Exercise 4.4

1. 23400 in each base. **2.** 3434 in each base.
3. $11 \times 11 = (11 \times 10) + (11 \times 1) = 110 + 11 = 121$
4. All bases greater than 4. **5.** 0.0152_8 **6.** 0.00143_8
7. base $7: 6 + 6 = 15$, $6 \times 6 = 51$
base $6: 5 + 5 = 14$, $5 \times 5 = 41$
base $5: 4 + 4 = 13$, $4 \times 4 = 31$.
8. $2 = 10_2$, $2^2 = 100_2$, $2^3 = 1000_2$, etc.
$S = 11\,111\,111_2 = 100\,000\,000_2 - 1 = (10_2)^8 - 1.\ \ 2^{10} - 1.$
9. $4 \times 5 = 32_6$, $5 \times 6 = 42_7$, $6 \times 7 = 52_8$, $7 \times 8 = 62_9$, $8 \times 9 = 72_{10}$, $9 \times A_{11} = 82_{11}$.
See each result as $(10 - 2)(10 - 1) = 100 - 20 - 10 + 2$ in the relevant base.
10. $21_7 = 3 \times 5$, $36_8 = 2 \times 3 \times 5$, $42_6 = 2 \times 21_6$, $302_5 = 12_5 \times 21_5$.
11. $82_5 = 42_{10}$, $410_5 = 105_{10}$, $240_5 = 70_{10}$, LCM $= 210_{10} = 1320_5$.

Exercise 4.5

1. 110_3 **2.** 110_6 remainder 1 **3.** 1011.1_2 **4.** 1515_7
5. 11_7 remainder 1 **6.** 111_8 remainder 3 **7.** 40_9 **8.** 41.03_7
9. 241_5 **10.** 432_5 remainder 1 **11.** 111_7 remainder 11_7
12. 234_7 remainder 22_7 **13.** 0.024_7 **14.** 2
15. even, even, odd, even. (Use the place values and the fact that odd + odd = even.)

Exercise 4.6

1. $\dfrac{132_5}{444_5} = \dfrac{42_{10}}{124_{10}} = \dfrac{21_{10}}{62_{10}}$ **2.** $\dfrac{132_4}{333_4} = \dfrac{30_{10}}{63_{10}} = \dfrac{10_{10}}{21_{10}}$

3. $\dfrac{5}{3}$ **4.** $\dfrac{25_7}{100_7 - 1} = \dfrac{25_7}{66_7} = \dfrac{19_{10}}{48_{10}}$ **5.** $\dfrac{1}{9}$ denary.

6. Convert to base 5, to get $0.\bar{1}$ and $10_5S = 1 + S$; $S = \dfrac{1}{4}$.

7. Convert to $0.\overline{10}_7$; $S = \dfrac{7}{48_{10}}$. **8.** Convert to $0.\overline{100}_4$; $S = \dfrac{16_{10}}{63_{10}}$.

Exercise 4.7

1. (i) $11_{10} = 102_3 = 110_3 - 1$
(ii) $22_{10} = 211_3 = 1011_3 - 100_3$
(iii) $26_{10} = 222_3 = 200_3 + 20_3 + 2 = 1000_3 - 100_3 + 100_3 - 10_3 + 10_3 - 1$
$= 1000_3 - 1$
(iv) $35_{10} = 1022_3 = 1000_3 + 100_3 - 10_3 + 10_3 - 1 = 1100_3 - 1$
(v) $32_{10} = 1012_3 = 1010_3 + 10_3 - 1 = 1020_3 - 1 = 1100_3 - 11_3$
2. (i) $45_{10} = 1200_3 = 2000_3 - 100_3 = 10000_3 - 1100_3$, $(81_{10} - 36_{10})$
(ii) $59_{10} = 2012_3 = 10100_3 - 1011_3$
(iii) $72_{10} = 2200_3 = 10000_3 - 100_3$, $(81_{10} - 9_{10})$
3. No. Change to 6, 5, 3. **4.** No. Change to 15, 8, 7. **5.** Yes.
6. No. Change to 13, 9. 6, 2. **7.** No. Change to 5, 4, 3, 2 or 5, 4, 2, 2, 1, etc.
8. 16_{10} light; 19_{10} heavy; 18_{10} light.

Exercise 5.1

1. 3; 3 **2.** 3; 3 **3.** 4; 4 **4.** 2; 5 **5.** 5; 3 **6.** 3; 3 **7.** 1; 0
8. $(3 + 4) \times (2 \times 5) = 1 \times 4 = 4$; $(3 \times 4) + (2 + 5) = 0 + 1 = 1$
9. 4 **10.** 3 **11.** 2 and 4 (two solutions) **12.** 0, 2, 3 (three solutions)

13. Since odd + odd = even, addition on S is not closed, but because odd × odd = odd, multiplication on S is closed. On E both operations are closed.

+	0	1	2
0	0	1	2
1	1	2	0
2	2	0	1

×	0	1	2
0	0	0	0
1	0	1	2
2	0	2	1

14. 1 **15.** 1 **16.** 0 **17.** (i) 1 or 2; (ii) no solution. **18.** 2

Exercise 5.2

1. $1 - 5 = 1 + 1 = 2$ **2.** 4 **3.** 0, 0 **4.** 2 **5.** $3 + 1 + 2 + 4 = 4$
6. 2 **7.** $2 - 5 = 3$ **8.** $(2) - (0 + 1) = 1$ **9.** $(3 + 2) - 1 = 4$
10. $0 - (2) - (2 + 2) = 0 - 2 - 4 = 0 + 4 + 2 = 0$
11. 0, 6, 5, 4 respectively (i) 4, (ii) 6, (iii) 5, (iv) 2.

Exercise 5.3

1. 1 **2.** 2^{-1} does not exist. **3.** $3 \div 5 = 3 \times 5^{-1} = 3 \times 5 = 3$
4. Does not exist. **5.** Does not exist. **6.** Does not exist.
7. $4 \times 2 \times 5 = 4$ **8.** $2 - 5 = 2 + 1 = 3$ **9.** 1
10. 2, 5 **11.** 2, 5 **12.** 1, 4 **13.** 1, 3, 5 **14.** 4

Exercise 5.4

1. All. **2.** 1, 3, 5, 7
3. Yes, because whenever the multiplicative inverse exists, it is unique.
4. $6 + 1 = 7$ **5.** $6 \times 1 = 6$
6. $3 \times 3 = 1 = 3 \times 6 + 3 \times 5 = 2 + 7 = 1$—an example of the distributive law.
7. $5 \times 3 = 7$ **8.** $3 \div 7 = 3 \times 7 = 5$
9. (i) does not exist, (ii) does not exist, (iii) 2, (iv) does not exist, (v) $7 \times 5 = 3$, (vi) $5 \times 7 = 3$
10. (i) 3, 7; (ii) 2, 6; (iii) 1, 3, 5, 7; (iv) 0, 2, 4, 6; (v) 2, 6; (vi) 3, 7.

Exercise 5.5

1. —mod 6; $3 + 5 = 0$ mod 8 **2.** —mod 6; $1 - 4 = 1 + 4 = 5$ mod 8
3. —mod 6; 0 mod 8 **4.** —mod 6; 6 mod 8
5. Does not exist in mod 6 or mod 8; solutions $2n = 4$; 2, 5 mod 6; 2, 6 mod 8
6. 0, 1, 3, 4 mod 6; 0, 1 mod 8 **7.** 1, 2, 4, 5 mod 6; 1, 2 mod 8
8. —mod 6; 4, 5 mod 8 **9.** —mod 6; 2, 6 mod 8 **10.** 2, 4 mod 6; 2 mod 8
11.

+	0	1	2	3	4
0	0	1	2	3	4
1	1	2	3	4	0
2	2	3	4	0	1
3	3	4	0	1	2
4	4	0	1	2	3

×	0	1	2	3	4
0	0	0	0	0	0
1	0	1	2	3	4
2	0	2	4	1	3
3	0	3	1	4	2
4	0	4	3	2	1

(i) 0, 1; (ii) $1^{-1} = 1, 2^{-1} = 3, 3^{-1} = 2, 4^{-1} = 4, 0^{-1}$ does not exist; (iii) no;
(iv) (a) $2 + 4 = 1$, (b) $1 - 3 = 1 + 2 = 3$, (c) $2 - 3 = 2 + 2 = 4$,
(d) $2 \times 3 \times 4 = 4$; (v) 1, 3, 2

12. (i) $365 \equiv 1 \pmod 7$; Wednesday, because we take Tuesday as day 0; (ii) $366 \equiv 2 \pmod 7$; Thursday.
13. Concentrate on the 1st of March which is a Saturday to be taken as day 0. In the

next leap year 1st March is $365 \times 3 + 366 = 1461$ days away. $1461 \equiv 5 \pmod 7$, a Thursday. The next 29th February is a Wednesday.

Exercise 5.6

1. $a * b = a, b * a = b, \therefore\ a * b \neq b * a$ for all $a, b \in S$, i.e. $*$ not commutative on S. $(a * b) * c = a * c = a$, $a * (b * c) = a * b = a$, i.e. $*$ is associative on S.
2. Neither commutative nor associative.
3. Commutative and associative, $a * b * c = a + b + c - 2$.
4. Commutative, not associative,

$$(a * b) * c = [2(a + b)] * c = 4a + 4b + 2c$$
$$a * (b * c) = a * [2(b + c)] = 2a + 4b + 4c$$

5. Commutative and associative, $a * b * c = \sqrt{a^2 + b^2 + c^2}$; as written this means the positive square root.
6. $a = 0, b = 1$; the table may also be used to register the idea of even $\times$ even = even, even $\times$ odd = even, odd $\times$ odd = odd.
7. $a = -1, b = 1$
8.

+	E	O
E	E	O
O	O	E

9.

$$\begin{aligned} a \circ (b * c) &= a \circ (b + c + 2) = ab + ac + 2a + 2a + 2b + 2c + 4 + 2 \\ &= ab + ac + 4a + 2b + 2c + 6 \\ (a \circ b) * (a \circ c) &= (ab + 2a + 2b + 2) * (ac + 2a + 2c + 2) \\ &= ab + 2a + 2b + 2 + ac + 2a + 2c + 2 + 2 \end{aligned}$$

i.e. $$a \circ (b * c) = (a \circ b) * (a \circ c)$$

Exercise 5.7

1. (i) V; (ii) H; (iii) V; (iv) V; (v) I; (vi) E.
2. (i) I; (ii) $I^{-1} = I$, $V^{-1} = V$, $H^{-1} = H$, $E^{-1} = E$, i.e. each element is its own inverse.
3. (i) $*$ is commutative and associative, identity element is I, $\mathrm{I}^{-1} = \mathrm{I}$, $\mathrm{A}^{-1} = \mathrm{C}$, $\mathrm{B}^{-1} = \mathrm{B}$, $\mathrm{C}^{-1} = \mathrm{A}$.

*	I	A	B	C
I	I	A	B	C
A	A	B	C	I
B	B	C	I	A
C	C	I	A	B

(ii) Same structure as + in arithmetic (mod 4): I = 0, A = 1, B = 2, C = 3.

4.

*	1	2	3	4
1	1	2	3	4
2	2	4	1	3
3	3	1	4	2
4	4	3	2	1

(i) Commutative and associative.
(ii) $1^{-1} = 1$, $2^{-1} = 3$, $3^{-1} = 2$, $4^{-1} = 4$.

5. $*$ is not commutative, not associative, e.g. $a * (c * b) = a * a = c$; $(a * c) * b = a * b = b$.

$\circ$ is commutative and associative, $\circ$ corresponds to + arithmetic (mod 3) on set $\{0, 1, 2\}$; $*$ has no identity element (but note that $a * c = a$, $b * c = b$, $c * c = c$, i.e. c is an identity element on the right but not the left); hence no inverses.

o has identity element a, $a^{-1} = a$, $b^{-1} = c$, $c^{-1} = b$.

(i) $a \times a = c$; (ii) $b \circ a = b$, i.e. $\times$ is not distributive over o; (iii) c; (iv) a; i.e. o is not distributive over $\times$.

6. Not defined for P $\times$ P, S $\times$ S, K $\times$ K. If this is described as, no go, 0, then we have a new element given in the set.

$\times$	P	S	K
P		P	K
S	P		S
K	K	S	

Exercise 5.8

1. $\{0, 3\}$, $\{0, 2, 4\}$. 2. $\{I, H\}$, $\{I, V\}$, $\{I, E\}$.
3. Each element must be its own inverse. Tables are always of the same diagonal form.
4. $\{0, 4\}$, $\{0, 2, 4, 6\}$.
5. Concentrate on Table 5.14. Associativity follows from ordinary addition; inverses are $0^{-1} = 0$, $1^{-1} = 3$, $2^{-1} = 2$, $3^{-1} = 1$. Since Table 5.14 represents a group, so does Table 5.15 from the one-to-one correspondence which we found.
6. The element 4 does not have an inverse, i.e. we are unable to find a number of the set to satisfy $4 \times ? = 1$ the identity element. $\{1, 3\}$, $\{1, 5\}$, $\{1, 7\}$; $\{1, 3, 5, 7\}$.
7. $\{a, b, c\}$ and $\times$ do not form a group because $\times$ is not associative and there is no identity. $\{a, b, c\}$ and o do form a group, associative, a is the identity, $a^{-1} = a$, $b^{-1} = c$, $c^{-1} = b$.
8. Yes. $a \rightarrow I$, $b \rightarrow A$, $c \rightarrow C$. 9. $\{0, 1\}$, $\{0, 2\}$, $\{0, 3\}$, $\{0, 4\}$.
10. (i) $4 \times 3 = 2$; (ii) $3 \times 2 = 4$, which shows that $\times$ is not associative on S. By Lagrange's theorem and Question 9, subgroups of order 2 in a system of order 5. 2 is not a factor of 5.

Exercise 6.1

1. (i) (1, 2), (2, 4), (3, 6), (4, 8), (6, 12), (8, 16)
   (ii) (1, 4), (2, 8), (3, 12), (4, 16), (16, 64)
   (iii) (64, 8), (16, 2), (8, 1)
   (iv) (4, 1), (6, 3), (9, 6), (12, 9)
   (v) (1, 1), (1, 3), (3, 1), (1, 9), (9, 1), (3, 3), (3, 9), (9, 3), (9, 9)
   (vi) ϕ
2. (i) (2, 1), (3, 1), (3, 2)
   (ii) (1, 2), (1, 3), (2, 3). This relation is the inverse of (i).
   (iii) (1, 1), (2, 2), (3, 3)
3. $a = 6$, $b = 15$, $y = 5x$
4. Nine ordered pairs (0, 0), (0, 1), (0, 2), (1, 1), (1, 2), (1, 0), (2, 2), (2, 0), (2, 1).
5. The three sets are $A = \{6, 12, 36\}$ factors 2, 3.
   $B = \{21, 63\}$ factors 3, 7.
   $C = \{22, 44, 88\}$ factors 2, 11.

   R consists of 9 ordered pairs from each of A and C and 4 from B, making 22 ordered pairs in all.

Exercise 6.2

1. (i)

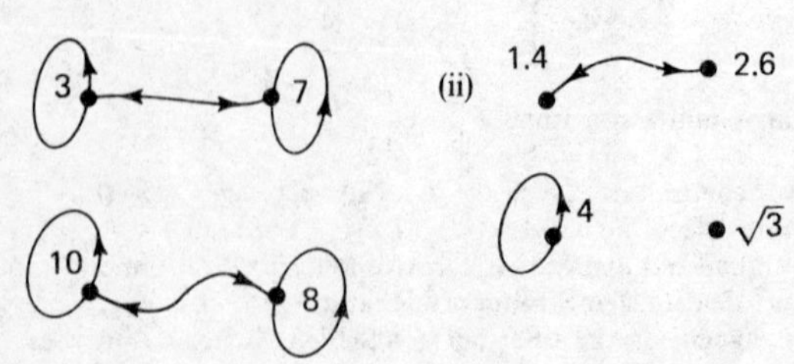

2. $(a, a), (a, c), (b, a)$ **3.** $(a, b), (b, a), (a, c), (d, c), (b, d), (c, c)$

4. (i) Each element would be related to itself like c.
(ii) c cannot be greater than itself.
(iii) If $c + a$ is divisible by 3 then so is $a + c$, i.e. two arrowheads required.

5. aRc does not imply that cRa, i.e. there are not two arrowheads on each join. No, for the same reason.

6. (i)

	a	b	c	d
a	0	1	0	0
b	1	0	0	0
c	1	0	1	1
d	0	1	0	0

(ii) The inverse relation, i.e. all the arrowheads reversed in direction.

Exercise 6.3

1. R, S, T, E **2.** S, T **3.** A, T, O **4.** A, T, O
5. A, T (i.e. not allowing a number to be a factor of itself).
6. S. Consider 2, 4, 5 to prove R is not transitive.
7. R, S, T, E; reflexive because $a - a = 0$ is divisible by 3; transitive because $a - b = 3k$, $b - c = 3m$ means $a - c = 3(m + k)$, i.e. divisible by 3.
8. S **9.** S, T
10. S, not transitive for consider 1, 2, 4, i.e. (1, 4), (4, 2) belong to R but (1, 2) does not. Note that R would be transitive on $\{2, 3, 4, 5, 6\}$.
11. A
12. (i) (4, 4)
(ii) (3, 1)
(iii) none required
(iv) none required
(v) (4, 4), (3, 1)
13. R_1 is R, S, T, E; R_2 is S only.

Exercise 6.4

1. 54; (i) (3, 6), (3, 7), (3, 8), (3, 9); (ii) $\{(6, 3), (7, 3), (8, 3), (9, 3)\}$; (iii) (1, 1), (2, 2), (3, 3), (4, 4), (5, 5), (6, 6); (iv) (b, a) is in R whenever (a, b) is in R.
2. Nine ordered pairs in all. (i) R = $\{(1, 1), (2, 1), (3, 1)\}$; (ii) (2, 1), (3, 1), (3, 2); (iii) $\{(1, 2), (1, 3), (2, 3)\}$.
3. (i) (1, 1), (2, 4), (3, 9); (ii) (1, 1), (2, 4), (3, 9), (4, 16), (5, 25).
4. (i) (1, 1), (1, 3), (2, 2), (2, 4), (3, 1), (3, 3), (4, 2), (4, 4); (ii) R, S obvious but for transitive $x - y = 2k$, $y - z = 2n$, where k or n are negative or positive integers $x - z = 2(k + n)$, which is even. (iii) 2 subsets of odd and even numbers.
5. Reflexive $\frac{a}{b} = \frac{a}{b}$ for every ordered pair (a, b)

Symmetric since $\frac{c}{d} = \frac{a}{b}$ implies $\frac{a}{b} = \frac{c}{d}$

Transitive $\frac{a}{b} = \frac{c}{d}$ and $\frac{c}{d} = \frac{e}{f}$ implies $\frac{a}{b} = \frac{e}{f}$

R.S.T. therefore E.

6. Reflexive: $(a, a) \in R$ for all $a \in A$. $\therefore$ $(a, a) \in R^{-1}$ for all $a \in A$.
Symmetric: $(a, b) \in R, (b, a) \in R$. $\therefore$ $(b, a) \in R^{-1}, (a, b) \in R^{-1}$.
Transitive: $(a, b) \in R, (b, c) \in R \Rightarrow (a, c) \in R$.
$\therefore$ $(b, a) \in R^{-1}, (c, b) \in R^{-1} \Rightarrow (c, a) \in R^{-1}$.

Exercise 6.5

1. $S(3, 1)$; $T(4, 2)$.
2. $\{(x, y) \mid x < 0, y < 0, x, y \in \mathbb{R}\}$; $(-1, -1), (-2, -21), (-3, -7)$.
3. (i) $y < x + 4$, (ii) $y < x + 8$.
4. $y = -2$ was intended $\{(x, y) \mid y > -2, x, y \in \mathbb{R}\}$, $(1, 1), (-11, 10), (5, -1)$.
5. (i) $\{(x, y) \mid y < -\frac{5x}{3} - 5, x, y \in \mathbb{R}\}$; (ii) $\{(x, y) \mid y < -\frac{5x}{3} - 10, x, y \in \mathbb{R}\}$.
6. $(1, 2), (2, 2), (1, 3), (1, 4)$.
7. As an inequality $0 < \frac{1}{2}x < y < 2x$ or $0 < x < 2y < 4x$ will do; alternatively, $R = \{(x, y) \mid 0 < x < 2y < 4x, x, y \in \mathbb{R}\}$, $(1, 1\frac{1}{2}), (3, 2), (10, 15)$.
8. The interior of the square with the line joining $(0, 0)$, $(2, 2)$ as diagonal. $(\frac{1}{2}, \frac{1}{2}), (\frac{1}{4}, \frac{1}{3})$, $(1, 1.4)$; $(0, 0), (0, 1), (0, 2), (2, 2), (2, 1), (2, 0), (1, 0), (1, 1), (1, 2)$.

Exercise 6.6

1. (i) $\{3, 4, 5, 6, 7\}$, function.
   (ii) $\{0, 2, 6\}$, function.
   (iii) $\{-2, -\sqrt{3}, -\sqrt{2}, -1, 0, 1, \sqrt{2}, \sqrt{3}, 2\}$, not a function, e.g. $(0, -\sqrt{2})$ and $(0, \sqrt{2})$ are in the relation.
   (iv) $\{-2^{\frac{1}{3}}, 0, 1, 2^{\frac{1}{3}}\}$, function.
   (v) Not a function, e.g. the relation contains $(0, 1), (0, 2), (1, 1.4)$.
   (vi) Not a function, range consists of $(2, y)$ for all $y < 2$.
2. (i), (iv) not a function; (ii), (iii) functions; (ii) $y = 1$, a line parallel to the x axis; (iii) $y = x + 1$.
3. $R = \{$all pairs of the form $(x, 3)\}$, e.g. $(0, 3), (-1, 3), (1, 3), (2, 3), (-10, 3), (-2, 3)$, etc., with no duplication of x.
4. R contains $(3, 1), (3, 5), (3, 9), (3, 10)$, etc., so cannot be a function. This line is parallel to the y axis. Straight lines parallel to the y axis are the only straight lines which do not represent functions.
5. (i), (vii) are the only functions.
6. (i), (ii) represent functions, i.e. one value of x gives rise to one value of y. In (iii) and (iv) one value of x can give rise to two or three values of y. It is useful to remember that if a line parallel to the y axis drawn everywhere in the domain cuts the curve graph in one and only one point, then the graph defines a function.

Exercise 6.7

1. (i) R is not a function; $R^{-1} = \{(1, 0), (2, 0), (3, 1), (5, 2), (7, 6)\}$ is a function.
   (ii) R is a function; $R^{-1} = \{(1, 1), (1, 2), (4, 3), (6, 5)\}$ not a function.
   (iii) $R = R^{-1}$ is a function.
   (iv) $R = R^{-1}$ not a function.
2. (i) $\{(x, y) \mid y = \frac{1}{2}(x - 1)\}$, i.e. interchange x and y in the original $y = 2x + 1$ to get $x = 2y + 1$.
   (ii) $\{(x, y) \mid y^2 = x + 2\}$ not a function because $y = \pm\sqrt{(x + 2)}$; if $x + 2 < 0$ then y is not defined in $\mathbb{R}$.
   (iii) $\left\{(x, y) \mid x = \frac{9}{y}\right\}$ a function, excluding $y = 0$ for which x is not defined.
   (iv) $\left\{(x, y) \mid x = 5 + \frac{6}{y}, y \neq 0\right\}$ a function.
3. (i) $x = 2x + 1$ when $x = -1$, $x = -1$ is an invariant element.
   (ii) $x = x^2 - 2$, i.e. $x^2 - x - 2 = (x - 2)(x + 1) = 0$, $x = 2$ or -1 are the two invariant elements.
   (iii) $x = \frac{9}{x}$, i.e. $x^2 = 9$, $x = \pm 3$ are the two invariant elements.

(iv) $x = 5 + \frac{6}{x}$, i.e. $x^2 = 5x + 6$, $x^2 - 5x - 6 = 0$, $x = 6$ or -1 are the two invariant elements.

4. Only (iv).

5.

Not a function

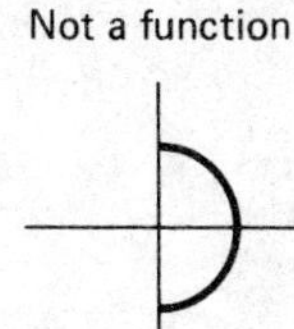

A function

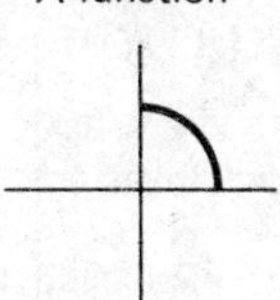

A function

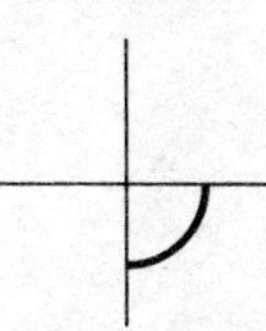

6. No.

7. $g \circ f = \{(1, 12), (2, 13), (3, 14)\}$; $f \circ g$ is not defined for any member of the domain of g, e.g. 12, 13, 14 are not in the domain of f.

Exercise 7.1

1. Sloped lines are $x + y = 1$, $\frac{x}{2} + \frac{y}{4} = 1$. Constraints are $0 < x$, $0 < y$, $1 - x < y$, $y < 4\left(1 - \frac{x}{2}\right)$. The point (1, 1) is in this region.

2. (i) $(1, 1), (\frac{1}{2}, \frac{1}{2}), (1, \frac{1}{4})$
(ii) $(-3, 1), (-3, 2), (-3, 3)$
(iii) $(1, 1), (1, 2), (2, 1)$

3. (i) $\frac{x}{4} + \frac{y}{6} = 1$; (ii) $\frac{x}{5} + \frac{y}{7} = 1$; (iii) $\frac{x}{6} + \frac{y}{8} = 1$. They all slope in the direction left to right down, i.e. negative slope. They are not parallel.

4. (i) $\frac{x}{8} + \frac{y}{16} = 1$; (ii) $\frac{x}{4} + \frac{y}{8} = 1$; (iii) $\frac{x}{1} + \frac{y}{2} = 1$. The three lines are parallel to each other. Note that: (i) $2x + y = 16$; (ii) $2x + y = 8$; (iii) $2x + y = 2$. The further the line from (0, 0), the greater $2x + y$ becomes.

5. $5x + y = 20$, C increases, i.e. as the line is moved away in the first quadrant so we obtain $5x + y = 25$, $5x + y = 30$, $5x + y = 35$, etc.

6. $0 < x$, $0 < y$, $y < 16 - 2x$, $8 - 2x < y$; (4, 2), (4, 4), (6, 2), (2, 6), (2, 8), (2, 10).

Exercise 7.2

1. (i) 20; (ii) 28. **2.** (i) 8; (ii) 7. **3.** 38

4. $P(38)$, $Q(64)$, $R(55)$ for $5x + 7y$; P(10), Q(16), R(17) for $3x + y$.

5. The value of C at each point is $P(32)$, $Q(31)$, $R(30)$, $S(29)$. The line reaches S first.

Exercise 7.3

1. (i) $P(-4)$, $Q(2)$, $R(4)$, $S(0)$, $T(-8)$; optimum values at R and T.
(ii) $P(-9)$, $Q(0)$, $R(6)$, $S(3)$, $T(-11)$; optimum values at R and T.
(iii) $P(8)$, $Q(14)$, $R(4)$, $S(-12)$, $T(-12)$; optimum values at Q and S, T. (Any point on line segment ST.)
(iv) $P(3)$, $Q(-6)$, $R(-6)$, $S(3)$, $T(13)$; optimum values at Q, R and T. (Any point on line segment QR.)
(v) $P(-16)$, $Q(-10)$, $R(4)$, $S(12)$, $T(-4)$; optimum values at P and S.

2. $Q(4, 2\frac{1}{2})$, $R(-3, 3)$, $S(-2\frac{1}{2}, -4\frac{1}{2})$, $P(-3, 5)$.
Values at extreme points (i) $Q(13)$, $R(0)$, $S(-14)$, $P(4)$; maximum at Q minimum at S.
(ii) $Q(3)$, $R(-12)$, $S(4)$, $P(-16)$; maximum at S minimum at P.

3. The extreme points of the region are $P(0, 2)$, $Q(0, 5)$, $R(5, 5)$, $S(2, 2)$. (i) max. $R(50)$, min. $P(6)$; (ii) max. $Q(15)$, min. $R(-20)$; (iii) max. $R(5)$, min. 0; (iv) max. 5, min. 2.

4. The region is unbounded.

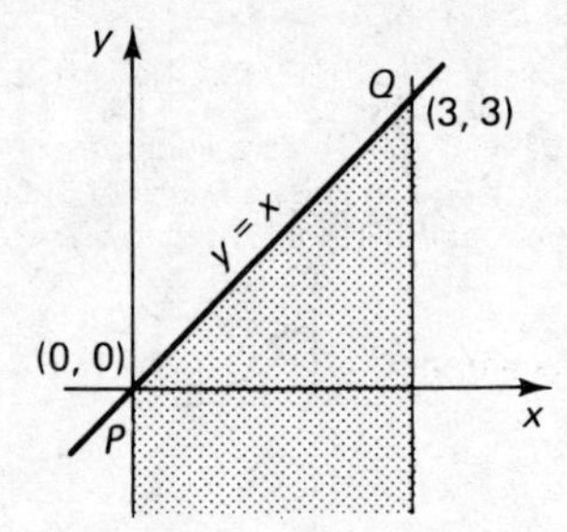

(i) Max. 3, min. 0.
(ii) Max. $Q(3)$, min does not exist.
(iii) Max. $P(0)$, min does not exist.
(iv) Max. $Q(3)$, min does not exist.

Exercise 7.4

1. (i) Profit line is $0.4x + 0.6y = C$. Max. P(£6) for an outlay of £40 on 10 boxes of brand Y.
(ii) Profit line is $0.55x + 0.22y = C$. Max. R(£4.40) for an outlay of £72 on eight boxes of brand X.
(iii) Profit line is $0.4x + 0.3y = C$. Max. at (6, 4) of £3.60 for an outlay of £70, i.e. 5.14%.

Reservation. Take five boxes of each with a profit of £3.50 for an outlay of £65, i.e. 5.4%.

2. (i) Profit line is $1000x + 800y = C$, which simplifies to $10x + 8y = C/100$. Max. at (4, 5) of £8000.
(ii) Profit line is $600(x + y) = C$, which simplifies to $x + y = C/600$. Max. at (3, 6) of £5400, same profit at (2, 7) but greater packaging costs.
(iii) Profit line is $400x + 500y = C$, which simplifies to $4x + 5y = C/100$. Max. at (2, 7) of £4,300.

3. (i) Max. 46 at (14, 4), min. 36 at (10, 6).
(ii) Max. 28 at (10, 6), min. 24 at (12, 4).
(iii) Max. 10 at (14, 4), min. 4 at (10, 6).

4. (i) $S(12)$; (ii) $R(10)$; (iii) $Q(49)$; (iv) $P(10)$.

5. For sowing: $150x + 90y \leqslant 2700$, i.e. $5x + 3y \leqslant 90$.
For labour: $4x + 5y \leqslant 100$.
For profit: maximise $400x + 480y = C$ or $5x + 6y = C/80$.
The extreme points are (0, 20), (18, 0), $(11\frac{7}{13}, 10\frac{10}{13})$; the point (11, 11) is inside the solution region, giving a maximum of 121 for $5x + 6y$ and a maximum $C = 121 \times 80 =$ £9680 for an outlay of £2640 in sowing.

The points (12, 10) and (10, 12) are also acceptable, giving profit of £9600 on £2700 outlay and £9760 on £2580 outlay. But note at (0, 20) a profit of £9600 arises from an outlay of £1800—clearly a better percentage proposition.

Notice that the limitation of resources means that not all 25 hectares may be sown.

6. Material: $x + 2y \leqslant 400$; $3x + 2y \leqslant 600$. Extreme points (0, 200), (200, 0), (100, 150).
A: Profit line $3x + 3y = C$, maximum £750 at (100, 150).
B: Profit line $2.5x + 3.5y = C$, maximum £775 at (100, 150).
C: Profit line $2.75x + 3.25y = C$, maximum £762.5 at (100, 150).

Reservation. It looks as though he could do better by selling 200 coats to wholesaler B to get a profit of £700 using only the coat making process, i.e. one type assembly operation as an easier way out at less profit but the problem is whether they would be able to sell so many coats, especially in the wrong season.

7. Bet £x on A and £y on B.

A wins: $y < \frac{5x}{2} - 10$; B wins: $y > \frac{1}{5}(10 + x)$; C wins: $10 > x + y$.

Obtain graphs similar to Fig. 7.10. Observe solutions obtainable inside triangular region. For example: $x = 6$, $y = 3.50$ gives a return of £1.50 if A wins, £1.50 if B wins and 50p if C wins. Of course, another horse might still win the race.

Exercise 7.5

1. (i) Max. $Q(11)$ (already found), min. (-6) as tangent to curve between O and P $(-5\frac{1}{3}$ by calculation).
(ii) Max. $Q(2)$, min. $P(-4)$.
(iii) Min. $O(0)$, max. $P(4)$.
(iv) Min. $P(-12)$, max. $Q(9)$.
(v) Min. (-1) as tangent to curve between O and P, max. $Q(3)$.

2. (i) Min. 1.2 at R, where $y = \frac{4}{x}$ cuts $y^2 = 9x$. R is the point (1.2, 3.3), max. $S(8)$ approx.
(ii) Min. $S(0.5)$, max. $P(6.5)$.
(iii) Min. 4 very near Q at (2, 2), max. 11.3 as tangent to curve in between P and S at (5.7, 5.7) approx.
(iv) Max. $2\frac{1}{4}$ at approx. $(2\frac{1}{4}, 4\frac{1}{2})$ between P and R, min. $S(-7.5)$.

3. (i) Min. $O(0)$, max. 1.
(ii) Min. $O(0)$, max. 1.
(iii) Min. $O(0)$, max. 2.
(iv) Max. $\frac{1}{4}$ at the point $(\frac{1}{4}, \frac{1}{2})$, min. $(-\frac{1}{4})$ at the point $(\frac{1}{2}, \frac{1}{4})$.

Exercise 8.1.

1. (i) $50 \div 8 = 6.25$ minutes; (ii) Lands at N where $QN = 50 \tan 37° = 37.7$ m (1 d.p.); alternatively produce PC until it meets the opposite bank at N, and measure to scale to get 37.5 m.
2. (i) $50 \div 6 = 8.3$ minutes (1 d.p.); (ii) Lands at N where $QN = 50$ m. Notice that the resultant direction in which the boat travels is 45° to the banks.
3. $50 \div 5.3 = 9.4$ minutes (1 d.p.).
4. (i) $\sin \angle CPA = 6/10 = 0.6$, angle $CPA = 36.9°$ (1 d.p.).
(ii) $CP^2 = PA^2 - AC^2 = 100 - 36 = 64$, $CP = 8$, i.e. $\overrightarrow{PC}$ represents 8 m/min from P to Q.
(iii) $60 \div 8 = 7.5$ minutes.
5. $7^2 + 5^2 = 74$, $\sqrt{74} = 8.6$ (1 d.p.). $\therefore$ $|\overrightarrow{AC}| = 8.6$; $\tan \angle BAC = 0.7143$; angle $BAC = 35.5°$ (1 d.p.).
6. $|\overrightarrow{AC} = 10.4$ (1 d.p.); $\angle BAC = 24.5°$ (1 d.p.).

Exercise 8.2

1. (i) 5 parallel to AB; (ii) 5 parallel to AD; (iii) $\overrightarrow{AC}$; $5\sqrt{2}$; 45° to $\overrightarrow{AB}$; (iv) $\overrightarrow{DB}$; $5\sqrt{2}$; 45° to $\overrightarrow{DA}$; (v) $\overrightarrow{AE}$, $\frac{5\sqrt{5}}{2}$; 26.6° to $\overrightarrow{AB}$; (vi) same as (v).
2. (i) 13; 22.6° to $\overrightarrow{AB}$; (ii) 13; 22.6° to $\overrightarrow{AB}$.
3. (i) Circumference of a circle centre O, radius 10.
(ii) Surface of a sphere centre O, radius 10.
4. $\mathbf{x} = 0$ **5.** $\mathbf{x} = 0$ **6.** $\mathbf{x} = \mathbf{a} - \mathbf{b}$ **7.** $\mathbf{x} = 2\mathbf{b} - \mathbf{a}$
8. (i) $\mathbf{b} - 1\frac{1}{2}\mathbf{a}$; (ii) $\frac{1}{2}\mathbf{b} - \mathbf{a}$; (iii) $\frac{1}{2}(\mathbf{a} + \mathbf{b})$; (iv) $1\frac{1}{2}\mathbf{b} - 4\mathbf{a}$.

Exercise 8.3

1. (i) $2\mathbf{a} + \mathbf{b}$; (ii) $\overrightarrow{OT} = \mathbf{a}$; (iii) $\overrightarrow{PT} = \mathbf{a} - \mathbf{b}$.
2. (i) $AC/CB = 3/4$; (ii) $AC/CB = 4/3$; (iii) $7\overrightarrow{OC}$ where $AC/CB = 5/2$.

3. (i) **a** + **b**; (ii) $\frac{1}{2}$(**a** + **b**). MN is parallel to BC and $MN = \frac{1}{2}BC$.

4. (i) **a** + **b**; (ii) $\frac{1}{2}$(**a** + **b**); (iii) **a** − **b**; (iv) $\frac{1}{2}$(**a** − **b**); (v) $\frac{1}{4}$(**a** − **b**); (vi) $\frac{3}{4}$(**b** − **a**); (vii) 3$\vec{\mathbf{b}}$ + **a**.

5. $\vec{OH} = \mathbf{O}$ the resultant is zero. The forces at A were in equilibrium. $|\vec{OK}| = 50\sqrt{2}$, $\vec{OK}$ acts at 45° to $\vec{ON}$.

6. (i) $\frac{1}{2}$(**a** + **c**); (ii) $\frac{1}{2}$(**b** + **c**); (iii) $3\vec{OG} = 2\vec{OM} + \vec{OA}$ = **b** + **c** + **a**; (iv)$3\vec{OH} = 2\vec{ON} + \vec{OB}$ = **a** + **c** + **b**. ∴ H and G are the same point; (v) $3\vec{OL} = 2\vec{OK} + \vec{OC}$ = **a** + **b** + **c**. ∴ H, G and L are the same point, i.e. all three medians intersect on a common point G called the centroid of the △ ABC.

7. (i) $\vec{HM} = \vec{HO} + \vec{OM} = \mathbf{c} - \mathbf{a}$

(ii) $\vec{LO} = -(\mathbf{a} + \mathbf{b})$

(iii) $\vec{OK} = \mathbf{b} + \mathbf{c}$, $\vec{LO} + \vec{OK} = \vec{LK} = \mathbf{c} - \mathbf{a} = \vec{HM}$

HM is equal and parallel to LK.

(iv) $\vec{HL} = \vec{HO} + \vec{OL} = -\mathbf{a} + \mathbf{a} + \mathbf{b} = \mathbf{b}$,

$\vec{MK} = \vec{MO} + \vec{OK} = -\mathbf{c} + \mathbf{b} + \mathbf{c} = \mathbf{b}$

∴ $HL$is equal and parallel to MK. ∴ $HLKM$ is a parallelogram.

Exercise 8.4

1. (i) $\sqrt{2}$, 45°; (ii) $2\sqrt{2}$, 45°; (iii) 5, 36.9° (1 d.p.); (iv) 61, 79.6° (1 d.p.).

2. $\vec{OR} = -7\hat{\imath} + 8\hat{\jmath}$; $\vec{OR}.\vec{OC} = -14 - 56 = -70$; $|\vec{OR}| = \sqrt{113}$, $|\vec{OC}| = \sqrt{53}$
$\cos \angle COR = -0.9045$; $\angle COR = 154.8°$ (1 d.p.).

3. (i) −20 − 20 = −40; (ii) 20 + 25 = 45; (iii) −25 − 20 = −45; (iv) $|\mathbf{a}| = \sqrt{41}$, $|\mathbf{b}| = \sqrt{41}$.

$$\cos\theta = \frac{\mathbf{a}.\mathbf{b}}{41} = \frac{-40}{41};\ \theta = 167.3° \text{ (1 d.p.)}$$

$\mathbf{a}(\mathbf{b}.\mathbf{c}) = \mathbf{a}(45) = 45\mathbf{a} = 225\hat{\imath} - 180\hat{\jmath}$.

4. $\vec{AB} = \vec{AO} + \vec{OB} = (-3\hat{\imath} - 9\hat{\jmath}) + (4\hat{\imath} + 17\hat{\jmath}) = \hat{\imath} + 8\hat{\jmath}$
$|\vec{AB}| = \sqrt{1 + 64}$.

∴ unit vector in direction $\vec{AB}$ is given by $\frac{1}{\sqrt{65}}(\hat{\imath} + 8\hat{\jmath}) = \frac{1\hat{\imath}}{\sqrt{65}} + \frac{8\hat{\jmath}}{\sqrt{65}}$.

5. $\vec{OA}.\vec{OB} = -6 + 30 + 9 = 33$.

Exercise 8.5

1. (i) $r^2 = 164 + 160 \cos 30°$, $r = 17.4$ (1 d.p.)
(ii) $r^2 = 164 + 160 \cos 45°$, $r = 16.6$ (1 d.p.)
(iii) $r^2 = 164 + 160 \cos 60°$, $r = 15.6$ (1 d.p.)

2. $a^2 = b^2 + c^2 - 2bc \cos A$, $b^2 = c^2 + a^2 - 2ac \cos B$,
$c^2 = a^2 + b^2 - 2ab \cos C$.

3. $AC^2 = 81 + 121 - 2 \times 9 \times 11 \times \cos 100° = 236.4$ (1 d.p.)
$AC = 15.38$ (2 d.p.)

4. At A because ABC is a right-angled triangle.

5. (i) Put H as (x, y); $\vec{AH} = x\hat{\imath} + y\hat{\jmath}$, $\vec{BC} = (-4\hat{\imath} + 8\hat{\jmath})$,
$\vec{AH}.\vec{BC} = -4x + 8y = 0$, $\vec{CH} = (x - 10)\hat{\imath} + (y - 8)\hat{\jmath}$, $\vec{AB} = 14\hat{\imath} + 0\hat{\jmath}$
$\vec{CH}.\vec{AB} = 14(x - 10) = 0$, $x = 10$, $y = 5$.
(ii) (5, 3).

Exercise 9.1

1. (a) (i) (4, −5); (ii) (−4, −5); (iii) (−4, 5); (iv) (4, 5); (v) (5, −4).
(b) (i) (−4, 5); (ii) (4, 5); (iii) (4, −5); (iv) (−4, −5); (v) (−5, 4).

2. x axis: (i) $y = -3x$; (ii) $y = -\frac{5x}{4}$; (iii) $x_1 = x_0$, $y_1 = -y_0$, $y_0 = 4x_0 + 2$ becomes $-y_1 = 4x_1 + 2$, i.e. the line $y = -4x - 2$; (iv) $y_0 = -2x_0 + 4$ becomes $-y_1 = -2x_1 + 4$, i.e. the line $y = 2x - 4$.

y axis: (i) $y = -3x$; (ii) $y = -\frac{5x}{4}$; (iii) $x_1 = -x_0$, $y_1 = y_0$, $\therefore y_0 = 4x_0 + 2$ becomes $y_1 = -4x_1 + 2$; (iv) $y_0 = -2x_0 + 4$ becomes $y_1 = 2x_1 + 4$ in Fig. 9.2, *EF* maps onto *AB*.

3. (x_0, y_0) mapped onto itself in both (i) and (ii), (iii) $(-x_0, -y_0)$ the same result as reflecting the point (x_0, y_0) in the origin (0, 0).

Exercise 9.2

1. (a) (i) $x = 5y + 3$, $(-\frac{3}{4}, -\frac{3}{4})$; (ii) $x = -3y + 5$, $(\frac{5}{4}, \frac{5}{4})$.
(b) (i) $-x = -5y + 3$, $(-\frac{1}{2}, \frac{1}{2})$; (ii) $-x = 3y + 5$, $(\frac{5}{2}, \frac{-5}{2})$.

2. $y = 3x - 6 \rightarrow x = 3y - 6 \rightarrow -y = -3x - 6$
$\therefore y = 3x - 6 \rightarrow y = 3x + 6$

3. $y = 3x - 6 \rightarrow -x = -3y - 6 \rightarrow -y = -3x - 6$
$\therefore y = 3x - 6 \rightarrow y = 3x + 6$ (same result as for Question 2). There are no invariant points because the final image is parallel to the original line $y = 3x - 6$.

Exercise 9.3

1. 2 (iv), 2 (i), 2 (ii), 2 (iii).

2. See **1**.

3. (i) $\begin{pmatrix} 6 \\ 14 \end{pmatrix}$ (ii) $\begin{pmatrix} -2 \\ -2 \end{pmatrix}$ (iii) $\begin{pmatrix} 2 \\ 2 \end{pmatrix}$ (iv) $\begin{pmatrix} -6 \\ -14 \end{pmatrix}$

(v) $\begin{pmatrix} 8 \\ -9 \end{pmatrix}$ (vi) $\begin{pmatrix} 17 \\ 13 \end{pmatrix}$ (vii) $\begin{pmatrix} 13 \\ 17 \end{pmatrix}$ (viii) $\begin{pmatrix} 8 \\ -8 \end{pmatrix}$ (ix) $\begin{pmatrix} 4 \\ -4 \end{pmatrix}$

Exercise 9.4

1. $(2, 5) \rightarrow (-2, -5) \rightarrow (2, 5)$.

2. (i) $(3, 1) \rightarrow (-1, 3) \rightarrow (1, -3)$;
(ii) $(-2, -7) \rightarrow (+7, -2) \rightarrow (-7, +2)$;
(iii) $(x_0, y_0) \rightarrow (-y_0, x_0) \rightarrow (y_0, -x_0)$.

3. (i) $(1, 2) \rightarrow (-2, 1) \rightarrow (1, -2)$;
(ii) $(x_0, y_0) \rightarrow (-y_0, x_0) \rightarrow (x_0, -y_0)$; and $f:(x, y) \mapsto (x, -y)$ is a reflection in the line $y = 0$.

4. Any original point on $y = x + 2$ is (x_0, y_0) with $y_0 = x_0 + 2$.
In R(0, 180°) $x_1 = -x_0$, $y_1 = -y_0 = -x_0 - 2 = x_1 - 2$. Image line is $y = x - 2$.

5. $x = -6 + 21 = 15$, $y = -15 - 49 = -64$; $(15, -64)$.
Any point on $y = x - 2$ is (x_0, y_0) where $y_0 = x_0 - 2$
$x_1 = 2x_0 + 3(x_0 - 2) = 5x_0 - 6$; $y_1 = 5x_0 - 7(x_0 - 2) = -2x_0 + 14$
$\therefore$ $(x_1 + 6)/5 = (y_1 - 14)/-2$, i.e. $5y_1 + 2x_1 = 58$.

Exercise 9.5

1. (i) (-1); (ii) (8); (iii) $(-1, 8)$; (iv) $(8, -1)$.

2. (i) (23); (ii) (17); (iii) $\begin{pmatrix} 23 \\ 17 \end{pmatrix}$; (iv) $\begin{pmatrix} 17 \\ 23 \end{pmatrix}$

3. (i) $\begin{pmatrix} 19 \\ 7 \end{pmatrix}$; (ii) $\begin{pmatrix} 21 \\ 6 \end{pmatrix}$; (iii) $\begin{pmatrix} 19 & 21 \\ 7 & 6 \end{pmatrix}$; (iv) $\begin{pmatrix} 21 & 19 \\ 6 & 7 \end{pmatrix}$

4. (i) (11); (ii) (11); (iii) (4); (iv) (4).

5. (i) $\begin{pmatrix} 4 & 5 \\ 8 & 10 \end{pmatrix}$; (ii) $\begin{pmatrix} 12 & 15 \\ 28 & 35 \end{pmatrix}$; (iii) does not exist.

6. (i) $\begin{pmatrix} -3 & 0 \\ 0 & -3 \end{pmatrix}$; (ii) $\begin{pmatrix} -3 & 0 \\ 0 & -3 \end{pmatrix}$

7. (i) $AB = \begin{pmatrix} 14 & 25 \\ 24 & 43 \end{pmatrix}$, $BA = \begin{pmatrix} 10 & 13 \\ 36 & 47 \end{pmatrix}$

(ii) $AC = \begin{pmatrix} -8 & 13 \\ -14 & 23 \end{pmatrix}$, $CA = \begin{pmatrix} 6 & 7 \\ 8 & 9 \end{pmatrix}$

(iii) $BC = \begin{pmatrix} -5 & 8 \\ -18 & 29 \end{pmatrix}$, $CB = \begin{pmatrix} 7 & 12 \\ 10 & 17 \end{pmatrix}$

(iv) All the same because matrix multiplication is associative.

$ABC = \begin{pmatrix} -64 & 103 \\ -110 & 177 \end{pmatrix}$

8. (i) $\begin{pmatrix} -1 & 16 & 27 \\ -1 & 28 & 47 \end{pmatrix}$; (ii) $\begin{pmatrix} -1 & 10 & 17 \\ -3 & 36 & 61 \end{pmatrix}$; (iii) $\begin{pmatrix} -3 & 6 & 11 \\ -5 & 8 & 15 \end{pmatrix}$

(iv) does not exist.

9. (i) $\begin{pmatrix} 4 & 6 \\ 8 & 10 \end{pmatrix}$; (ii) $\begin{pmatrix} 3 & 6 \\ 12 & 21 \end{pmatrix}$; (iii) $3A = \begin{pmatrix} 6 & 9 \\ 12 & 15 \end{pmatrix}$; (iv) $\begin{pmatrix} 3 & 5 \\ 8 & 12 \end{pmatrix}$

10. $X = 2A = \begin{pmatrix} 4 & 6 \\ 8 & 10 \end{pmatrix}$

11. (i) R(0, 180°); (ii) R(0, 180°); (iii) $\begin{pmatrix} -1 & 0 \\ 0 & 1 \end{pmatrix}\begin{pmatrix} 0 & -1 \\ 1 & 0 \end{pmatrix} = \begin{pmatrix} 0 & 1 \\ 1 & 0 \end{pmatrix}$, $M_{y=x}$; (iv) I

Exercise 9.6

1. (i) $AB = \begin{pmatrix} 9 & 11 \\ 17 & 18 \end{pmatrix}$, $BA = \begin{pmatrix} 29 & 11 \\ -3 & -2 \end{pmatrix}$

(ii) $(AB)C = \begin{pmatrix} 47 & 94 \\ 86 & 172 \end{pmatrix}$, $A(BC) = A\begin{pmatrix} 11 & 22 \\ 3 & 6 \end{pmatrix} = \begin{pmatrix} 47 & 94 \\ 86 & 172 \end{pmatrix}$

2. $A^{-1} = \frac{1}{5}\begin{pmatrix} 3 & -1 \\ -7 & 4 \end{pmatrix}$, $B^{-1} = \frac{-1}{5}\begin{pmatrix} -1 & -3 \\ -1 & 2 \end{pmatrix}$

det $C = 0$ so that C^{-1} does not exist.

3. $A^{-1}B = \frac{1}{5}\begin{pmatrix} 5 & 10 \\ -10 & -25 \end{pmatrix} = \begin{pmatrix} 1 & 2 \\ -2 & -5 \end{pmatrix}$, $BA^{-1} = \frac{1}{5}\begin{pmatrix} -15 & 10 \\ 10 & -5 \end{pmatrix} = \begin{pmatrix} -3 & 2 \\ 2 & -1 \end{pmatrix}$

4. Generally we know that matrices are associative with respect to multiplication, so if we put $X = A^{-1}B$ then $A(A^{-1}B) = (AA^{-1})B = IB = B$, the right-hand side.

5. We use the fact that $(D^{-1})^{-1} = D = \frac{1}{2}\begin{pmatrix} 2 & -4 \\ -1 & 3 \end{pmatrix}$

6. (i) $X = A^{-1}C = \frac{1}{5}\begin{pmatrix} 11 & 22 \\ -24 & -48 \end{pmatrix}$; (ii) $X = CA^{-1} = \frac{1}{5}\begin{pmatrix} -44 & 28 \\ -11 & 7 \end{pmatrix}$

7. (i) 0; (ii) 0; (iii) $\begin{pmatrix} 10 & 5 \\ -20 & -10 \end{pmatrix}$; (iv) 0.

Exercise 9.7

1. $x = 1, y = 3$ from $\frac{1}{3}\begin{pmatrix} 2 & -1 \\ -5 & 4 \end{pmatrix}\begin{pmatrix} 7 \\ 11 \end{pmatrix} = \frac{1}{3}\begin{pmatrix} 3 \\ 9 \end{pmatrix}$

2. $x = -2, y = 3$ from $\frac{1}{5}\begin{pmatrix} 2 & -1 \\ -3 & 4 \end{pmatrix}\begin{pmatrix} -5 \\ 0 \end{pmatrix} = \frac{1}{5}\begin{pmatrix} -10 \\ 15 \end{pmatrix}$

3. $x = 2, y = 1$ from $\frac{+1}{-11}\begin{pmatrix} -1 & -2 \\ -3 & 5 \end{pmatrix}\begin{pmatrix} 12 \\ 5 \end{pmatrix} = \frac{-1}{11}\begin{pmatrix} -22 \\ -11 \end{pmatrix} = \begin{pmatrix} 2 \\ 1 \end{pmatrix}$

4. $x = -2, y = -3$ from $\frac{1}{7}\begin{pmatrix} 1 & 5 \\ -1 & 2 \end{pmatrix}\begin{pmatrix} 11 \\ -5 \end{pmatrix} = \frac{1}{7}\begin{pmatrix} -14 \\ -21 \end{pmatrix}$

5. $A = \begin{pmatrix} 3 & 2 \\ 2 & 3 \end{pmatrix}$; $A^{-1} = \frac{1}{5}\begin{pmatrix} 3 & -2 \\ -2 & 3 \end{pmatrix}$, $\therefore$ $\quad \begin{aligned} \tfrac{3}{5}x_1 - \tfrac{2}{5}y_1 &= x_0 \\ \tfrac{-2}{5}x_1 + \tfrac{3}{5}y_1 &= y_0 \end{aligned}$

Image of (2, 5) is $\begin{pmatrix} 3 & 2 \\ 2 & 3 \end{pmatrix}\begin{pmatrix} 2 \\ 5 \end{pmatrix} = \begin{pmatrix} 16 \\ 19 \end{pmatrix}$

Image of (16, 19) is $\frac{1}{5}\begin{pmatrix} 3 & -2 \\ -2 & 3 \end{pmatrix}\begin{pmatrix} 16 \\ 19 \end{pmatrix} = \begin{pmatrix} 2 \\ 5 \end{pmatrix}$

Thus A maps (2, 5) onto (16, 19), A^{-1} maps (16, 19) back onto (2, 5).

6. $B^{-1} = \frac{1}{2}\begin{pmatrix} 2 & -4 \\ -3 & 7 \end{pmatrix}$; $B^{-1}\begin{pmatrix} 2 \\ 0 \end{pmatrix} = \frac{1}{2}\begin{pmatrix} 4 \\ -6 \end{pmatrix} = \begin{pmatrix} 2 \\ -3 \end{pmatrix}$

i.e. (x_0, y_0) is $(2, -3)$.

Exercise 9.8

1. (i)

$\times$	I	P	Q	R
I	I	P	Q	R
P	P	Q	R	I
Q	Q	R	I	P
R	R	I	P	Q

(ii) Yes; (iii) I; (iv) $I^{-1} = I, P^{-1} = R, Q^{-1} = Q, R^{-1} = P$;
(v) Yes;
(iv) Yes. $I \to 0, P \to 1, Q \to 2, R \to 3$;
(vii) $\{I, Q\}$

2. $AR = K, RK = A, RA = \begin{pmatrix} 0 & 1 \\ 1 & 0 \end{pmatrix}$ which is not in the set $\{I, A, R, K\}$ so the operation $\times$ is not closed on the set.

3. A^{-1} does not exist.

Exercise 9.9

1. (i) $\begin{pmatrix} 16 & 17 \\ 24 & 25 \end{pmatrix}$; (ii) $\begin{pmatrix} -2 & 3 \\ 18 & 5 \end{pmatrix}$

2. $X = A - B$

3. $\begin{pmatrix} 1 & -5 \\ -22 & -14 \end{pmatrix}$ for (i) and (ii).

4. $X = B + C - A$ as in **3**.

5. $X = C^{-1}(A + B) = \begin{pmatrix} -9 & 2 \\ 4 & -1 \end{pmatrix}\begin{pmatrix} 16 & 17 \\ 24 & 25 \end{pmatrix} = \begin{pmatrix} -96 & -103 \\ 40 & 43 \end{pmatrix}$

6. $X = (A + B)C^{-1} = \begin{pmatrix} 16 & 17 \\ 24 & 25 \end{pmatrix}\begin{pmatrix} -9 & 2 \\ 4 & -1 \end{pmatrix} = \begin{pmatrix} -76 & 15 \\ -116 & 23 \end{pmatrix}$

7. $X = C^{-1}(A + C) = C^{-1}A + C^{-1}C = C^{-1}A + I$

$$= \begin{pmatrix} -9 & 2 \\ 4 & -1 \end{pmatrix}\begin{pmatrix} 7 & 10 \\ 21 & 15 \end{pmatrix} + \begin{pmatrix} 1 & 0 \\ 0 & 1 \end{pmatrix} = \begin{pmatrix} -20 & -60 \\ 7 & 26 \end{pmatrix}$$

Exercise 10.1

1. (i) $2\,\overrightarrow{CA}$;
(ii) $-2\,\overrightarrow{DB}$; (iii) R(Q, 180°) followed by $-2\overrightarrow{DB} + \overrightarrow{CA}$;
(iv) R(Q, 180°) followed by $-\overrightarrow{DB} - \overrightarrow{CA}$.

2. (i) 16; (ii) 5; (iii) 3; (iv) 11.

3.

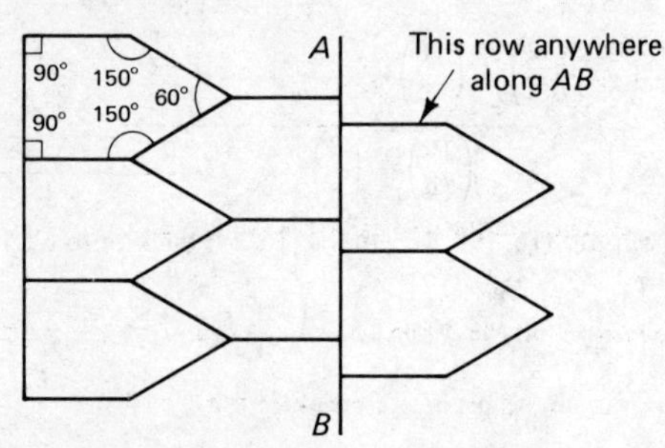

4. (ii) $ABCD$ lettered as shown with Q the midpoint of BC. Use basic vectors $\overrightarrow{CA}$ and $\overrightarrow{BD}$ together with R(Q, 180°). Thus, position 2 is gained from $\overrightarrow{CA}$. Position 1 is gained from R(Q, 180°) followed by $\overrightarrow{CA} + \overrightarrow{BD}$.

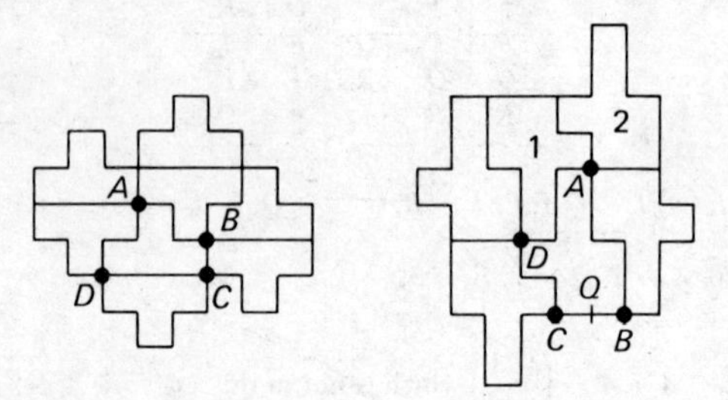

Exercise 10.2

1, 2. The diagram gives the answer to both questions.

3. Use basic vectors $\vec{CE}$, $\vec{AE}$, $\vec{BD}$ and R(S, 180°):
   (i) $\vec{BD} + \vec{AE}$; (ii) $\vec{BD} + \vec{AE} + \vec{EC}$; (iii) $\vec{CE}$; (iv) R(S, 180°) + $\vec{EC}$.
4.

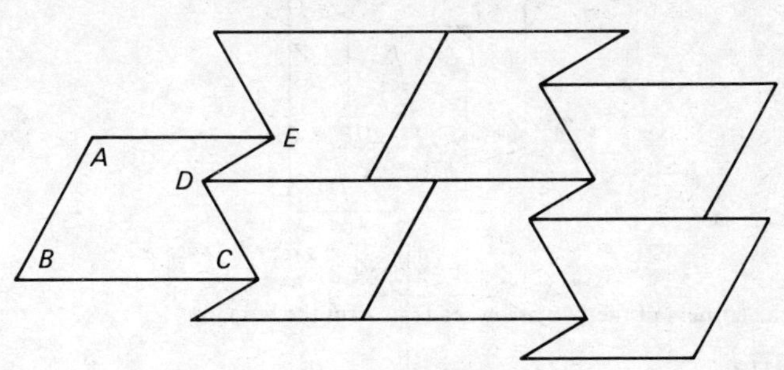

5. Each angle is 108° and no multiple of 108° makes either 180° or 360°.

Exercise 10.3

1. This title is a pentagon of the type shown in Fig. 10.9 (i), which we know will tessellate.
2. The following diagram shows a possible tessellation.

3.

4.

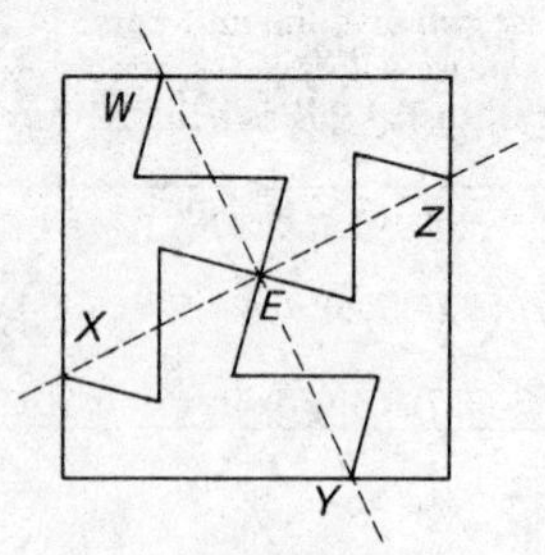

WEY ⊥ XEZ

5. Join the corners of the two arrowheads to form the hexagon.

Exercise 11.1

1. The points L, M and N will always lie on a straight line.
2. Same construction as in worked Example 3.
3. Draw any small equilateral $\triangle XYZ$ in any position having X on BA and Y on BC and the third vertex Z not on a side. Join BZ to intersect AC at Z_1. The result follows by drawing Z_1X_1 parallel to ZX, etc. There are an infinite number of possible triangles. For second part start with XY parallel to AC.
4. Let O be the centre of the chord. Construct a sample square $ABCD$ with O the mid-point of AB then use a radial enlargement from O.

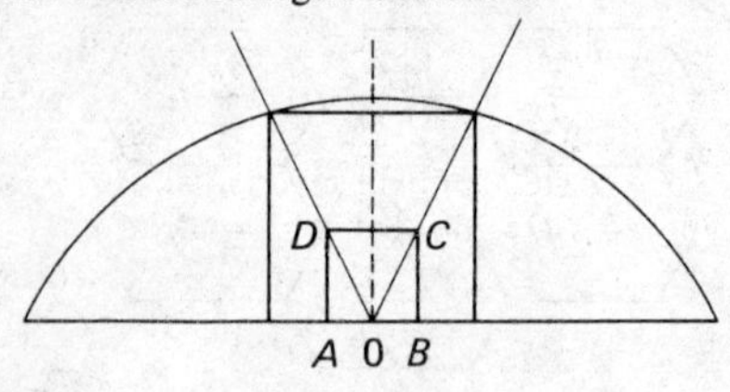

5. Draw a line through C parallel to XY and the result follows as a radial enlargement, centre E.

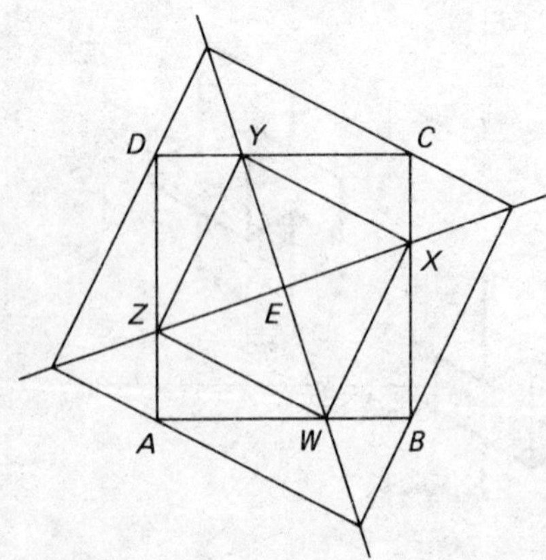

Exercise 11.2

1. (i) A B C D E K M T U V W Y; (ii) H I X; (iii) none; (iv) none.
2. (i) 1; (ii) 3; (iii) 2; (iv) 6.

3. (i) Fold A onto B; the crease will give the mid-point of AB and (ii).
(iii) Fold BA onto BC; the crease will give the bisector of the angle.

4. Fold once to produce a crease, using this as a line perform (ii) then (iii) of Question 3, bisecting an angle of 90°.

5. In $y = 0$, $A_1(3, -2)$, $B_1(7, -3)$, $C_1(5, -6)$, followed by $x = 0$; $A_2(-3, -2)$, $B_2(-7, -3)$, $C_2(-5, -6)$.
Reverse order is $A_1(-3, 2)$, $B_1(-7, 3)$, $C_1(-5, 6)$, followed by $y = 0$; $A_2(-3, -2)$, $B_2(-7, -3)$, $C_2(-5, -6)$.

6. (i) (a) In $y = x$; $A_1(2, 3)$, $B_1(3, 7)$, $C_1(6, 5)$; in $y = 0$; $A_2(2, -3)$, $B_2(3, -7)$, $C_2(6, -5)$.
(ii) In $y = 0$; $A_1(3, -2)$, $B_1(7, -3)$, $C_1(5, -6)$; in $y = x$; $A_2(-2, 3)$, $B_2(-3, 7)$, $C_2(-6, 5)$
not the same result as for (i).
(iii) In $y = x$; $A_1(2, 3)$, $B_1(3, 7)$, $C_1(6, 5)$; in $x = 0$; $A_2(-2, 3)$, $B_2(-3, 7)$, $C_2(-6, 5)$.
(iv) In $x = 0$; $A_1(-3, 2)$, $B_1(-7, 3)$, $C_1(-5, 6)$; in $y = x$; $A_2(2, -3)$, $B_2(3, -7)$, $C_2(6, -5)$.
(i) = (iv), (ii) = (iii).
(v) In $y = x$; $A_1(2, 3)$, $B_1(3, 7)$, $C_1(6, 5)$; in $y = -x$; $A_2(-3, -2)$, $B_2(-7, -3)$, $C_2(-5, -6)$.
(vi) In $y = -x$; $A_1(-2, -3)$, $B_1(-3, -7)$, $C_1(-6, -5)$; in $y = x$; $A_2(-3, -2)$, $B_2(-7, -3)$, $C_2(-5, -6)$.
(v) = (vi) as expected because the mirror lines are at right angles.

7. Using Fig. 11.7, we require BKA_1 to be a straight line.

Exercise 11.3

1. (i) $x = -5$; (ii) $x = -1$; (iii) $x = 4$; (iv) $\dfrac{x}{5\sqrt{2}} + \dfrac{y}{5\sqrt{2}} = 1$.

2. $A(0, 5)$, $C(5, 0)$; $A(5, 0)$, $O(10, 0)$, $C(10, 5)$.

3. F_1 is the reflection of F_2 in OC, and will be at opposite ends of a diameter. The bisector of the right angle F_1CO is the first mirror line.

4. Diagram needed is:

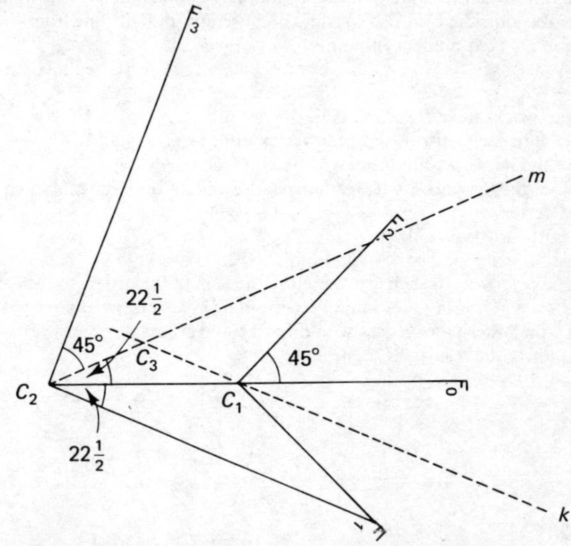

(i) F_1 is where the two arcs intersect for rotations about C_1 and C_2.
(ii) k is the bisector of $\angle F_0C_1F_1$. Reflect F_0 in k, onto F_1 then F_1 in C_2C_1, onto F_2.

(iii) m is the bisector of $\angle F_3C_2F_1$. Reflect F_0 in k, onto F_1 then F_1 in m, onto F_3. C_3 is the intersection of m and k. $R(C_3, 90°)$ maps F_0 onto F_3.

5. (i) Still on the same line but passing through to the other side of m.
   (ii) With B and C below A we have a reflection of the whole problem about AN, and the path of C is a line parallel to m_1 but a distance d below A.
6. Draw the same sample diagrams as in Fig. 11.14 but with $\angle BAC = 60°$ instead of 90°. The path of C is a straight line but inclined at 60° to m. The result can be proved in the same manner as for Example 5 by considering $R(A, -60°)$. Any isosceles triangle with $AB = AC$ will produce a straight-line path for C.
7. Yes. Draw m_2 parallel to m_1 distance d from A, to intersect n at C, and the result follows as before.
8. Join AC, draw $AC_1 \perp AC$ and $AC_1 = AC$. One side of the square lies along BC_1. Taking ABC as anticlockwise, $R(A, -90°)$ maps the side of the square through C onto the side of the square through B. $R(A, 90°)$ maps B onto B_1 (say) and the other side of the square lies on B_1C. The square can now be completed.

Exercise 12.1

1. (i) Straight on and take the fifth on the left.
   (ii) First right, and just past the fourth on the right.
   (iii) First left and just past the third on the right.
   (iv) First right and first left.
2. (a) (xiv), (xv)
   (b) (ii), (iii), (iv), (v), (vi)
   (c) (xix), (xx)
   (d) (xvi), (xvii), (xviii)
   (e) (ix), (x), (xi), (xii), (xiii)
3. (ii), (iii), (iv), (v), (vi)
4. (a) Extend the line into the inner region.
   (b) Add an outward pointing line to any point on the perimeter.
   (c) Join any two points on the perimeter with a straight line.
5. If the line intersects the curve at an odd number of points then it (the point) is inside the curve. B is outside, C is also outside, because any straight line from B and C cuts the curve in an even number of points.

Exercise 12.2

1. All eight networks have the same table entries.
   $V(4)$, $R(5)$, $A(7)$, each network has one vertex of order 2, 3, 4 and 5, $V + R - A = 2$.
   No two of the networks are topologically equivalent.
   Every one of the networks is traversable starting at an odd vertex and ending at the other odd vertex.
2. {(ii), (iv), (v)}; {(iii), (vi), (vii)}.
3. (ii) $16 \times 2 = 3 \times 8 + 2 \times 4$; (i) and (iii), $7 \times 2 = 5 \times 1 + 3 \times 3$.
4. There are two vertices to each arc; therefore the sum of the orders of the vertices must always be even. (Hint in Question 3.) Hence in a three-vertex network with only one odd vertex the other two vertices would have to be even and even + even + odd = odd contradicts the result of Question 3.
5.

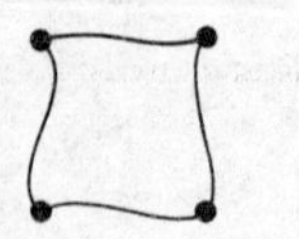
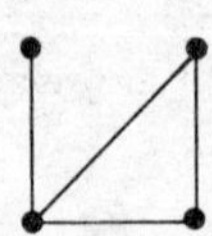

Exercise 12.3

1. (i) Traversable, two odd vertices; (ii) traversable, two odd vertices; (iii) unicursal, all vertices even.

2. Add two arcs to join the odd vertices in pairs and so make all vertices even.
3. (i) Yes; (ii) No.
4. (i) Vertices of order $4 \times 6 + 6 \times 2$; unicursal.
(ii) Vertices of order $4 \times 4 + 3 \times 2 + 6 \times 1$; traversable.
5. Start at A and finish at D or vice versa. CA, CD, AD, AB, BD.
6. (i) Network has vertices of order 5, 7, 6, 6; traversable starting at the upper bank and finishing at the left island or vice versa.
(ii) Remove a bridge connecting left island to upper bank to make all vertices of the network even.
7. (i) Two odd vertices. Threading is possible.
(ii) All even vertices, start and end at the same place anywhere in the diagram.

Exercise 12.4

1.

(i)	*a*	*b*	*c*	*d*	
a	0	1	0	2	
b	1	0	1	2	
c	0	1	0	2	
d	2	2	2	0	
Total	3	4	3	6	= 16

(ii)	*a*	*b*	*c*	*d*	*e*	
a	0	0	1	0	0	
b	0	0	1	0	0	
c	1	1	0	2	2	
d	0	0	2	0	2	
e	0	0	2	2	0	
Total	1	1	6	4	4	= 16

(iii)	*a*	*b*	*c*	*d*	*e*	*f*	*g*	
a	0	1	0	0	0	1	2	
b	1	0	2	0	0	0	1	
c	0	2	0	2	0	0	0	
d	0	0	2	0	1	0	1	
e	0	0	0	1	0	2	1	
f	1	0	0	0	2	0	1	
g	2	1	0	1	1	1	0	
Total	4	4	4	4	4	4	6	= 30

2. $\begin{pmatrix} 2 & 2 & 1 \\ 1 & 1 & 2 \end{pmatrix} \begin{pmatrix} 2 & 2 \\ 3 & 1 \\ 1 & 0 \end{pmatrix} = \begin{matrix} \\ x \\ y \end{matrix} \overset{\begin{matrix} a & b \end{matrix}}{\begin{pmatrix} 11 & 6 \\ 7 & 3 \end{pmatrix}}$

3. (i) $\begin{matrix} \\ a \\ b \end{matrix} \overset{\begin{matrix} x & y \end{matrix}}{\begin{pmatrix} 10 & 10 \\ 5 & 5 \end{pmatrix}}$; (ii) $\begin{matrix} \\ x \\ y \end{matrix} \overset{\begin{matrix} p & q \end{matrix}}{\begin{pmatrix} 1 & 0 \\ 1 & 0 \end{pmatrix}} \begin{matrix} \\ p \\ q \end{matrix} \overset{\begin{matrix} a & b \end{matrix}}{\begin{pmatrix} 2 & 1 \\ 3 & 0 \end{pmatrix}} = \begin{matrix} \\ x \\ y \end{matrix} \overset{\begin{matrix} a & b \end{matrix}}{\begin{pmatrix} 2 & 1 \\ 2 & 1 \end{pmatrix}}$

4. (i) R and A decreased by 1.
(ii) V and A increased by 2.
(iii) A increased by 3, V increased by 2, R increased by 1. $V + R - A = 2$ remains true in each case.

Exercise 12.5

1. In order the results are: 2 loops interlocked; 2 loops interlocked; 2 loops interlocked; 1 loop, 1 knot; 2 loops interlocked, 1 knot; 2 loops interlocked; 2 loops interlocked.
2. The figure forms two interlocked holes.
3. The figure separates into one strip with two holes.

Appendix

SOME ABBREVIATIONS AND DEFINITIONS

$\approx$ may be read 'is approximately'; e.g. $3.98 \approx 4$ (see page 248).

$//$ may be read 'is parallel to' (see page 208).

$\mathbb{R}$ represents the infinite set of real numbers. It is beyond the scope of this book to offer a definition of a real number and so at this level we must rely on offering some examples like 0, $+2.4$, $-1.\bar{9}$, $\sqrt{2}$, $4.7/5.9$ -2.1^3 to suggest what we mean (see page 98).

$\mathbb{R}^+$ represents the infinite set of positive real numbers (see page 126).

The infinite set of integers $= \{\ldots -3, -2, -1, 0, 1, 2, 3, 4, \ldots\}$ (see page 118).

A **quadrilateral** is a plane figure bounded by four straight lines. By plane we mean that the whole figure lies in one plane such as this page laid flat on a table.

A **parallelogram** is a quadrilateral whose opposite sides are parallel. It is possible to prove the opposite sides are also equal in length see (page 161).

A **rhombus** is a parallelogram with all four sides equal in length and none of its angles a right angle (90°) (see page 209). Thus all rhombuses are parallelograms but not all parallelograms are rhombuses.

A **rectangle** is a parallelogram with each of its angles a right angle (see page 10).

A **square** is a rectangle with all four sides equal in length. Thus all squares are rectangles but not all rectangles are squares (see page 10).

A regular figure has all sides equal in length and all angles equal in magnitude. Thus the hexagon on page 200 is not regular but the one on page 159 is regular. (A square is a regular figure.)

Index

Index